Modern Techniques in Electroanalysis

CHEMICAL ANALYSIS

A SERIES OF MONOGRAPHS ON
ANALYTICAL CHEMISTRY AND ITS APPLICATIONS

Editor

J. D. WINEFORDNER

VOLUME 139

A WILEY-INTERSCIENCE PUBLICATION

JOHN WILEY & SONS, INC.

New York / Chichester / Brisbane / Toronto / Singapore

Modern Techniques in Electroanalysis

Edited by

PETR VANÝSEK

Department of Chemistry
Northern Illinois University
DeKalb, Illinois

A WILEY-INTERSCIENCE PUBLICATION

JOHN WILEY & SONS, INC.

New York / Chichester / Brisbane / Toronto / Singapore

Library of Congress Cataloging in Publication Data:
Modern techniques in electroanalysis / edited by Petr Vanýsek.
     p. cm.—(Chemical analysis ; v. 139)
    "A Wiley-Interscience publication."
    Includes index.
    ISBN 0-471-55514-2 (alk. paper)
    1. Electrochemical analysis.   I. Vanýsek, P., 1952–
II. Series.
QD115.M55   1996
543′.0871 – dc20   96–8027

Printed in the United States of America

10 9 8 7 6 5 4 3 2 1

CONTRIBUTORS

Howard D. Dewald, Department of Chemistry, Clippinger Laboratories, Ohio University, Athens, Ohio 45701

Sean D. Garvey, Department of Chemistry, University of Arizona, Tucson, Arizona 85721

David K. Gosser, Jr., Department of Chemistry, City College of the City University of New York, New York 10031

Jeanne E. Pemberton, Department of Chemistry, University of Arizona, Tucson, Arizona 85721

Manuel P. Soriaga, Department of Chemistry, Texas A&M University, College Station, Texas 77843

John L. Stickney, Department of Chemistry, University of Georgia, Athens, Georgia 30602

Marek Trojanowicz, Laboratory for Flow Analysis and Chromatography, Department of Chemistry, University of Warsaw, 02-093 Warsaw, Poland

Petr Vanýsek, Department of Chemistry, Northern Illinois University, DeKalb, Illinois 60115

Michael D. Ward, Department of Chemical Engineering and Materials Science, University of Minnesota, Minneapolis, Minnesota 55455

Henry S. White, Department of Chemistry, University of Utah, Salt Lake City, Utah 84112

Cynthia, G. Zoski, Department of Chemistry, University of Rhode Island, Rhode Island 02881

CONTENTS

PREFACE

Electrochemistry refers to that field of science in which during the course of our studies we intervene with a reaction by applying an outside electric circuit, probe the system with attached electrodes, or, at least, interpret the reactions by concepts of charged species. Mostly it deals with processes on interfaces.

Hence, homogeneous oxidations in organic chemistry will seldom be treated by electrochemists although they are redox processes. They do not require the outside electric current and they do not deal with surfaces. On the other hand, stepwise redox processes in biochemistry energy cycles are of interest to an electrochemist. First, the biochemical process is attached to a structure, an interface, that is attractive to an electrochemist; second, energy exchange stimulates the ideas of the large group of electrochemists interested in energy conversion and storage. Electrochemistry is rather wide field by today's standards, and each specialty narrows down the object and the mode of its study.

Electroanalytical chemistry is such a case, dealing with electrochemistry applied to analytical chemistry. However, just as analytical chemistry changes and diversifies, electroanalysis changes with it. In today's field of electro-analytical chemistry, three apparent subjects can be found; detection and determination of various compounds, studies of interfaces and surfaces (structure and composition), and electrochemical instrumentation.

This book is a collection of several topics that are part of the modern aspects of electrochemistry. No doubt, in the years to come this novelty will vanish. The topics, though not "modern" anymore, will still be useful to the reader because the material described in these chapters is not usually covered in basic textbooks or if it is, then in much lesser detail.

The chosen topics are intended to fill gaps in current available literature and provide in one volume a collection of resources. The initial few pages of each chapter begin with the rudiments and only later build up into complexity. There are some contemporary topics missing from the book. In today's busy world the authors for chapters are not easy to find and my thanks go to all the contributors who found the time in their schedules to take part in this project.

The very basic introduction to electrochemistry of metal surfaces can be made through the description of two fundamental equations: (1) the Nernst equation, dealing with equilibrium, and (2) the Butler–Volmer equation, dealing with chemical changes, dynamics of a process, its kinetics.

The Nernst equation in one of its renditions describes the equilibrium potential of an inert electrode, such as a piece of platinum wire, immersed in a solution of a mixture containing both forms of a redox couple. So, for example, a Pt wire immersed in a solution of mixture of potassium ferrocyanide and potassium ferricyanide (mixture of complexed iron(II)/iron(III) ions) will attain potential

$$E = E^0_{Fe^{3+}/Fe^{2+}} - (RT/nF) \ln \frac{a_{Fe^{2+}}}{a_{Fe^{3+}}} \tag{1}$$

where $E^0_{Fe^{3+}/Fe^{2+}}$ is the standard reduction potential for the iron(III)/iron(II) hexacyanocomplex couple, which equals 0.358 V, n is the number of electrons needed for the reduction of the oxidation state of iron(III) to iron (II) $[n = 1]$, and $a_{Fe^{2+}}$ and $a_{Fe^{3+}}$ are the activities of the reduced (Fe(II)) and oxidized (Fe(III)) forms of the redox system. These activities are, strictly speaking, the activities of these compounds on the surface of the electrode. However, because this is an equilibrium state, the activities are the same in the bulk as they are on the surface of the electrode. Very often the equation is written with concentrations, instead of activities. The error of this assumption is negligible at low concentrations (< 0.001 mol/liter). Although at higher concentrations the error increases, the predicted trends still hold.

This equation exists in a number of variations. For equilibrium between a less noble metal electrode in contact with a solution of its salt, the equation attains a simpler form (e.g., zinc in zinc sulfate):

$$E_{Zn^{2+}} = E^0_{Zn^{2+}} + (RT/2F) \log a_{Zn^{2+}} \tag{2}$$

which is just a rewritten version of Eq. (1), where the activity of the reduced form (which is the solid metal, Zn(0)) is simply defined as unity.

Equation (1) can be extended to potential difference experienced on a membrane of a membrane-selective electrode or the interface between two liquids (see the chapter 8 on liquid–liquid electrochemistry). It is a general relationship that can be used in equilibrium situations, such as those found in stationary measurements with ion-selective electrodes or during potentiometric titrations.

One should be aware that this simplification does not give the full truth about the nature of the things at equilibrium. However, it does not hide anything crucial, and with an open mind it should not lead to misconceptions. Some of the curious deviations from the simplified reality can be mentioned here. For example, the "inert" platinum electrode is never quite inert. It is almost always covered by a layer of platinum oxide, and more of this oxide is formed during anodic polarization of a Pt electrode. Since the oxide formation is not very deep (only several monolayers), the electrode will not dissolve or corrode, which earns it the label "inert." However, the oxide presence has a profound effect on some electrochemical reactions or on its potential measured in a solution that contains dissolved oxygen, because in that case it can be the platinum oxide/oxide redox couple that determines the equilibrium.

Mathematical manipulation of Eq. (1) will reveal that if only one form of the redox pair species is present (the other would be zero), the logarithm will have a singularity and the potential, at least mathematically, would be infinity. Since an electrochemical potential of any known system cannot attain a value larger than about ± 3 V, there is clearly something wrong with this calculation. Here are several arguments that can be used to set this problem straight: (1) If only one species is to be considered, then the condition of equilibrium between two forms cannot be used; that is, the single oxidation form does not play any role in the equilibrium potential and cannot be inserted into the Nernst equation. In such a case, the potential will most likely be determined by equilibrium of surface oxides and traces of dissolved oxygen. (2) In the case where only one oxidation form is initially present, but a second oxidation form can exist, a small amount of the other form will be generated. For example, a platinum electrode immersed in Fe(II) solution would in theory attain a large positive value that would cause trace oxidation of Fe(II) to Fe(III), thus enabling the Nernst equation to be used.

When an outside potential is applied and electric current flows through the system, the situation changes, because either oxidation or reduction on the electrode is in progress and the amounts of the oxidized and reduced forms of various species in the vicinity of the electrode change. It is still possible to use the Nernst equation to calculate the instantaneous surface concentrations, but for most applications this is not sufficient.

The best description of the dynamic situation on an electrode with current flow comes from the Butler–Volmer approach, which describes the current–potential characteristics. The current I as a function of time t depends of the applied interfacial potential E as follows:

$$I = nFAk^0[c_{ox}(0,t)e^{-\alpha nF(E-E^0)/RT} - c_{red}(0,t)e^{(1-\alpha)nF(E-E^0)/RT}] \qquad (3)$$

The parameters have the following meaning: A is the electrode surface, n is the number of electrons involved in the redox process, k^0 is standard heterogeneous rate constant of the electrode process, $c_{ox}(0,t)$ and $c_{red}(0,t)$ are concentrations of the oxidized and the reduced species at the surface of the electrode ($x = 0$) at a given time t, E^0 is the equilibrium potential of the system,

E is the applied potential, and α is a transfer coefficient, often close to 0.5, signifying symmetry between oxidation and reduction.

Although electrochemistry deals in a classical sense with metal–solution processes, in modern terms it embraces many aspects of solid-state chemistry and physics, such as micromachining, etching for solid-state electronics, high-vacuum studies of electrode surfaces, and so on.

In this book we are presenting eight chapters giving a cross section of new electroanalytical subjects. The solid-state approach is represented in the discussion of the vacuum surface techniques, as well as the spectroscopic methods and the scanning tunneling and atomic force microscopy. The wet analysis is represented by a very sensitive technique of stripping analysis. Detection in analytical chemistry is often automated, which is described in the chapter on electrochemical detectors. Practice also needs a good underlying theory. The chapter on steady-state voltammetry chooses a modern tool of electroanalysis, the microelectrode, to deal with theoretical aspects of various conditions on the microelectrode that could arise during electroanalytical experiments. Only a step away from theory, almost a bridge between theory and experiment, are numerical simulations. A chapter in this book deals with such simulations on electrodes on which processes occur due to various electrochemical mechanisms. Electrochemistry is often associated with metal–solution processes. However, other possibilities have been explored, as is shown in the chapter on liquid–liquid electrochemistry.

DeKalb, Illinois   PETR VANÝSEK

CHEMICAL ANALYSIS

A SERIES OF MONOGRAPHS ON
ANALYTICAL CHEMISTRY AND ITS APPLICATIONS

J. D. Winefordner, *Series Editor*

xiii

Modern Techniques in Electroanalysis

CHAPTER

1

VACUUM SURFACE TECHNIQUES IN ELECTROANALYTICAL CHEMISTRY

MANUEL P. SORIAGA

Department of Chemistry
Texas A & M University
College Station, Texas 77843

JOHN L. STICKNEY

Department of Chemistry
University of Georgia
Athens, Georgia 30602

Modern Techniques in Electroanalysis, Edited by Petr Vanýsek, Chemical Analysis Series, Vol. 139.
ISBN 0-471-55514-2   © 1996 John Wiley & Sons, Inc.

I. INTRODUCTION

A comprehensive understanding of an electrochemical process needs to take into account the various physical and chemical interactions that arise between an electrified surface and its environment.[1-9] These interactions will be influenced by the solvent, supporting electrolyte, electrode potential, reactant concentration, crystallographic orientation, and electrode material. The traditional approach to the study of the electrode–solution interface is based upon the thermodynamic characterization of the interface and its response to perturbations in terms of current–charge–potential measurements.[10-13] Analysis of results from such measurements have relied on phenomenological models which incorporate, to varying levels of approximation, the interfacial parameters; the information thus derived represents only the macroscopic properties, without atomic-level specificity, of the electrode–electrolyte interface. The need for an atomic-level view of electrode reactions is now well established.[5-9]

Although much is presently known about heterogeneous processes at gas–solid interfaces, it is remarkable to note that barely two decades ago, these fields were beset by problems similar to what handicaps interfacial electrochemistry. For example, vacuum–solid-surface science traditionally employed thermal desorption spectroscopy and work function measurements which, respectively analogous to cyclic voltammetry and capacitance measurements, do not yield atomic-level information. Basic questions, such as the chemical nature of the adsorbed species or the atomic geometry of the adsorbate–substrate interface, could thus not be addressed. It was not until the development of surface-specific experimental[14-19] and theoretical[20-27] tools that tremendous advances in the study of the gas–solid interface were achieved. This overwhelming success motivated the adaptation of such surface-sensitive probes to the study of the electrode–solution interface.[28-35]

Most of these surface analytical methods are based upon the mass selection of molecules or the energy discrimination of electrons, ions, or atoms scattered from solid surfaces. These particles have shallow escape depths; hence the information they bear is characteristic of the near-surface layers. Their short mean free paths however, necessitate a high-vacuum environment ($\leqslant 10^{-6}$ torr).* The application of these surface analytical techniques to electrochemistry requires that the characterization be performed outside the electrochemical cell. The possibility of structural and compositional changes that accompany the removal of the electrode from solution is the major limitation in coupled ultrahigh-vacuum–electrochemical (UHV-EC) studies. Nevertheless, the results obtained have been so dramatic in atomic-level detail that experimental strategies based upon the use of vacuum surface analytical techniques are no longer uncommon.[28–35] An extensive monograph of the UHV-EC approach has recently been published[36]; less comprehensive reviews have also been written.[37–39]

II. EXPERIMENTAL PROTOCOLS

A. Electrode-Surface Preparation

While many vacuum-based surface analytical methods do not require the use of single-crystal surfaces,[37–39] for fundamental work such as those that involve surface crystallography, the use of uniform (monocrystalline) surfaces is a desirable component. In such investigations, the preparation and verification of clean and well-ordered electrode surfaces constitute critical initial steps. The low-index surface crystallographic faces, such as the (111), (110), and (100) planes of face-centered cubic crystals shown in Figure 1,† have been widely used because of their low surface free energies, high symmetries, and relative stabilities. In addition, it has been suggested that the resultant macroscopic behavior of smooth polycrystalline electrodes can be constructed in terms of the individual properties of these three surfaces planes.[40–46]

Three procedures have been used to prepare oriented monocrystalline electrode surfaces. In one method, single-crystal rods or boules, usually grown by zone-refining, are oriented by the L`aué back-reflection technique[48] and then cut along the desired crystal face. A second procedure is based on the fact

---

* Ultrahigh vacuum, typically at pressures well below 10^{-9} torr, is not required for most electron-based spectroscopies; it is necessary, however, to minimize, if not totally prevent, surface contamination.

† The numbers in parentheses—(111), (110), and (100)—represent a particular (hkl) plane, where (hkl) is referred to as a Miller index.[47] The reciprocal of a Miller index $(1/h, 1/k, 1/l)$, are the intercepts on the axes of a unit cell in terms of the lattice constants x, y, and z. As can be seen by the shaded planes in Figure 1, the (111) plane intersects all three axes, the (110) intersects two axes and is parallel to the third, and the (100) plane intersects only one axis and is parallel to the two other axes.

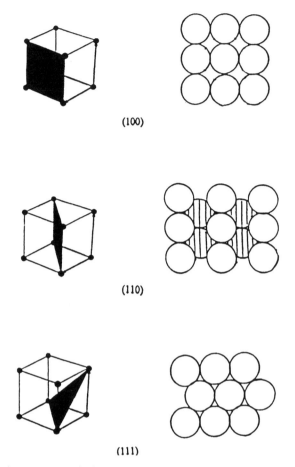

Figure 1. Atomic arrangements in the (100), (110), and (111) planes of a face-centered cubic metal. The vertical markings indicate the second-row atom layer. (Reproduced from reference 18, with permission.)

that single-crystal facets are produced when a polycrystalline wire is melted into a bead.[42,49] Metallographic polishing is subsequently performed on the oriented crystals to obtain a larger surface. The oriented and polished mono-crystalline electrodes are reannealed at near-melting temperatures to repair the damage selvedge. A third scheme, limited to (111) faces, involves epitaxial growth by vapor deposition onto a hot (400°C) mica substrate. Atomically smooth and well-ordered Au(111) and Ag(111) single-crystal thin-film electrodes have been prepared in this manner.[50-52]

If the entire single-crystal electrode is to be immersed in solution for electrochemical experiments, it is necessary to prepare a *parallelipiped* crystal in which all six faces are oriented identically. An alternative is to fabricate an electrochemical cell such that only one oriented face will be in contact with electrolyte. This configuration would permit the use of a disc electrode (in which only one face is oriented and polished) or a multifaceted crystal (in which each face represents a distinct crystallographic orientation).

The oriented single-crystal electrodes require further pretreatment before reliably clean and ordered surfaces are obtained. There are two general schemes: One employs high temperatures (thermal annealing), whereas the other uses electrode potentials (electrochemical annealing). Thermal annealing in UHV serves a dual purpose: to segregate contaminants from the bulk onto the surface (where they can be oxidatively desorbed or sputtered away) and to attain atomic smoothness. Surface analysis is subsequently performed to ascertain interfacial structure and composition; surfaces subjected to such analysis are considered truly well-defined. Interfacial characterization at a qualitative level can be done by voltammetric methods if reference data for well-characterized electrode surfaces are available.[42,53–55] Electrochemcal annealing is based on the possibility that, at appropriate potentials, disordered interfacial atoms can either be activated to diffuse to stable (ordered) sites or, as in electropolishing, be dissolved to expose ordered layers. Electrode-potential-induced surface reconstruction[56] may occur unassisted or assisted by electrolyte. The electrochemical ordering of Au(111) electrodes by sequential voltammetric scans between the oxygen and hydrogen evolution regions[57,58] is an example of the former case. The ordering of Pd(111) surfaces by potentiodynamic scans in the region where the iodide electrolyte undergoes reversible oxidative adsorption–reductive desorption is electrolyte-assisted since it is the strong chemisorption of iodine which provides the driving force in the disorder-to-order transition; subsequent cathodic desorption of the absorbed iodine yields a clean and well-ordered Pd(111) surface.[59–61] In the electrochemical annealing of Au(111) and Pd(111), no electrode dissolution occurs, as opposed to anodic dissolution exhibited by reactive materials such as Ag (in $NaCN/H_2O_2$ solution[62]) or Cu in acidic media[63] in which the damaged surface layers are oxidatively etched away.

The preservation of the single crystallinity of electrode surfaces is also an important consideration. In most cases, conditions are known at which the surface single crystallinity can be maintained. For example, the single-crystal structure remains unchanged unless excursions are made to potentials which lead to extensive surface-oxide formation. However, even if surface-oxidation potentials are averted, prolonged exposure to electrolytic solutions invariably results in the accumulation of surface impurities. In UHV-EC experiments, the regeneration of clean and ordered single-crystal surfaces from spent electrodes consists of high-temperature oxygenation or Ar^+ ion sputtering followed by

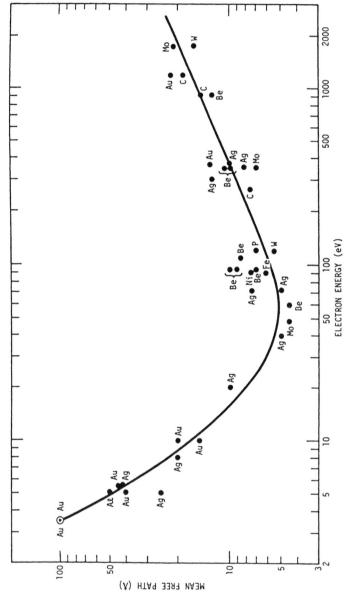

Figure 2. "Universal curve." Electron mean free path as a function of electron kinetic energy. (Reproduced from reference 14, with permission.)

6

a thermal treatment to restore atomic smoothness. *In situ* reordering options exist for other materials: (a) Cu(111) by microscopic electropolishing[62,63] and (b) Au(111) and Pd(111) by electrochemical annealing.[57-61]

B.   Interfacial Characterization Techniques

The cleanliness and single crystallinity of electrode surfaces cannot simply be assumed even if the preparative steps outlined above are followed. The verification or identification of initial, intermediate, and final interfacial structures and compositions is an essential ingredient in electrochemical surface science. Interfacial characterization in UHV-EC studies rely largely on *ex situ* surface-sensitive methods. Although a myriad of such analytical techniques is currently available,[14-19] those actually employed in UHV-EC experiments have been limited to low-energy electron diffraction, Auger electron spectroscopy, x-ray photoelectron spectroscopy, high-resolution electron energy loss spectroscopy, reflection high-energy electron diffraction, work-function changes, and thermal desorption mass spectrometry.

1.   Surface Spectroscopy with Low-Energy Electrons

The main difficulty in the surface characterization of single-crystal surfaces lies in the exceedingly low population of surface atoms (10^{15} atoms/cm^2) relative to that of bulk species (10^{23} atoms/cm^3). Experiments intended to examine the physical and chemical properties of surfaces must employ methods that interact only with the interfacial layers. The majority of interfacial characterization techniques takes advantage of the unique surface sensitivity of low-energy electrons. This surface influence arises because the mean free path of an electron through a solid is dependent upon its kinetic energy. As shown in the so-called "universal curve" reproduced in Figure 2, the electron mean free path falls to a minimum (4–20 Å) when the kinetic energy is between 10 and 500 eV. This signifies that all experimental techniques based upon the low-energy electron incidence onto and/or emergence from surfaces will bear information on the topmost surface layers.

A solid surface subjected to a beam of electrons of incident or primary energy E_p gives rise to the appearance of backscattered (primary) and emitted (secondary) electrons; the energy distribution [a plot of the number of electrons $N(E)$ as a function of energy E] of these electrons is shown in Figure 3. This spectrum can be divided into four regions according to the origin of the scattered electrons: (i) true *secondary* electrons, created as a result of multiple inelastic interactions between the incident and bound electrons; these electrons give rise to the prominent broad band at the lower end of the spectrum; (ii) Auger electrons emitted and primary electrons elastically scat-

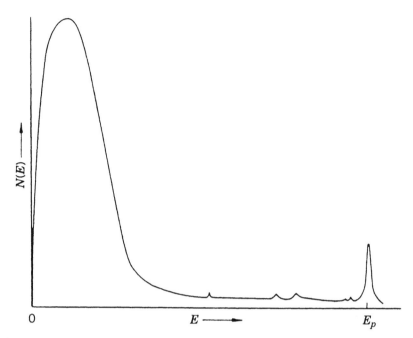

Figure 3. Experimental number [$N(E)$] of scattered electrons of energy E versus electron energy. (Reproduced from reference 14, with permission.)

tered due to interactions with electronics states in the solid; the small peaks in the medium-energy range of the spectrum are attributed to these electrons; (iii) primary electrons enelastically scattered upon interactions with the vibrational states of the surface; peaks resulting from these electrons reside close to the elastic peak since their energy losses are comparatively minute; (iv) primary electrons scattered elastically; these electrons, which comprise only a few percent of the total incident electrons, give rise to the elastic peak at E_p. Regions ii to iv of the energy distribution spectrum have been exploited in modern surface analysis. The elastic peak, for example, is used in diffraction experiments; the other peaks provide information on electronic and vibrational structures.

2. Low-Energy Electron Diffraction (LEED)

In this method[14–19,64,65] the surface is irradiated with a monoenergetic beam of electrons and the elastically backscattered electrons are collected onto a phosphor screen. The virtue of LEED as a surface structural technique is a result of the low kinetic energies used (50–500 eV) because (i) the electron mean free path is at a minimum which affords LEED its surface sensitivity; (ii) the de Broglie wavelengths, $\lambda_e \approx (150/E_e)^{1/2}$ (where E_e is in electron volts and λ_e is in angstroms), correspond to crystal lattice dimensions which render the low-energy electrons suitable for diffraction studies, and (iii) electron back-scattering is strong, which minimizes incident electron fluxes at, and subsequent scattering from, nonsurface layers. In LEED, therefore, the presence (or absence) of diffraction patterns on the fluorescent screen is a consequence of the order (or disorder) of the atomic arrangements near the surface.

The locations of the diffracted beams define the reciprocal lattice of the real surface. The real-space surface structure itself can be reconstructed from the real-space unit cell vectors generated from the reciprocal lattice vectors according to the following relationships:

$$\mathbf{a}^* = \frac{\mathbf{b} \times \mathbf{z}}{\mathbf{a} \cdot \mathbf{b} \times \mathbf{z}} \tag{1}$$

$$\mathbf{b}^* = \frac{\mathbf{z} \times \mathbf{a}}{\mathbf{a} \cdot \mathbf{b} \times \mathbf{z}} \tag{2}$$

where $\mathbf{a}$ and $\mathbf{b}$ are the real-space lattice vectors, $\mathbf{a}^*$ and $\mathbf{b}^*$ are the reciprocal unit mesh vectors, and $\mathbf{z}$ is the surface normal. The coherence width of electron beam sources in LEED is typically 100 Å. That is, sharp diffraction features appear only if well-ordered domains are at least $(100\,\text{Å})^2$ in size; diffraction from smaller domains leads to beam broadening.

The analysis of LEED data based solely upon the geometry of the diffraction spots provides information on the periodicity of the electron scatterers on the surface. In some favorable instances, other information such as adsorbate coverages or point group symmetries can also be inferred. However, the actual location of the atoms within the surface lattice cannot be determined without an analysis of the intensities of the diffracted beams. Surface crystallography by LEED can only rely upon a comparison of the measured diffraction intensities with those calculated for model structures. These LEED simulations are extremely difficult because of complications brought about by the possibility of multiple electron scattering.[65]

For structures formed under electrochemical conditions, only single-scattering (kinematic) LEED simulations for simple atomic adsorbates have been carried out; the primary intent has simply been the qualitative verification of proposed structures. The calculations are based on the following equation[47,66]:

$$I_s = \left\{ \frac{1}{J} \sum_{i=1}^{J} a_i \exp[2\pi i (X_i X_s + Y_i Y_s + Z_i Z_s)/\lambda_e] \right\}^2 \tag{3}$$

where I_s is the intensity of each beam s calculated for selected kinetic energies, J is the number of atoms in the unit mesh, and a_i is the scattering factor of the ith atom. X_s, Y_s, and Z_s are the Cartesian coordinates of the scattered beam where $Z_s = 1 + \cos \theta_d$, with θ_d being the angle between the incident and diffracted beams.

There are two schemes for the notation of interfacial adlattice structures. The matrix notation, which is applicable to any system, is based upon the relationship between the real-space lattice vectors of the *adsorbate* mesh (two-dimensional lattice) and the *substrate* mesh. For example, if the adsorbate unit cell vectors $\mathbf{a}'$ and $\mathbf{b}'$ are related to those of the substrate mesh according to

$$\mathbf{a}' = m_{11}\mathbf{a} + m_{12}\mathbf{b} \tag{4}$$

$$\mathbf{b}' = m_{21}\mathbf{a} + m_{22}\mathbf{b} \tag{5}$$

then the matrix M defined by the coefficients m_{ij},

$$M = \begin{pmatrix} m_{11} & m_{12} \\ m_{21} & m_{22} \end{pmatrix} \tag{6}$$

denotes the real-space surface structure. The other method, known as the *Wood notation*,[48] is more widely used but is applicable only if the angle between $\mathbf{a}'$ and $\mathbf{b}'$ is the same as that between $\mathbf{a}$ and $\mathbf{b}$. The surface structure is labeled using the general form $(n \times m)R\phi°$ or $c(n \times m)R\phi°$, where c designates a centered unit cell, $R\phi°$ is the angle of rotation of the adsorbate unit cell relative to the substrate unit mesh, and n and m are scale factors relating the adsorbate and substrate unit cell vectors:

$$|\mathbf{a}'| = n|\mathbf{a}| \tag{7}$$

$$|\mathbf{b}'| = m|\mathbf{b}| \tag{8}$$

A schematic diagram of a typical LEED instrument is shown in Figure 4.[67] The LEED "optics" consists of a phosphor-coated hemispherical screen, at the center of which is a normal incidence, electrostatically focused electron gun. In front are three concentric grids; the outer grid is held at ground potential, while the inner two are maintained at a voltage just below that of the electron gun in order to reject inelastically backscattered electrons. The elastically diffracted electrons which pass through the suppressor grids are accelerated onto the fluorescent screen by a 5-kV potential applied to the screen. For quantitative

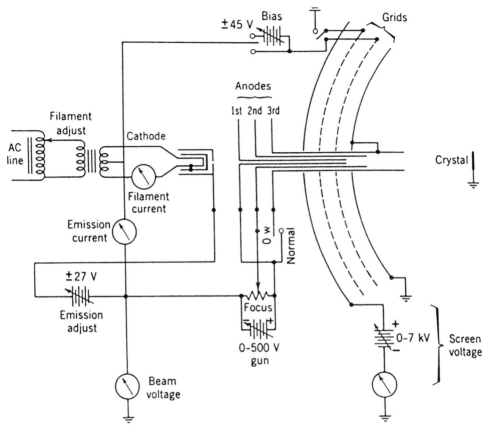

Figure 4. Schematic diagram of a LEED apparatus. (Reproduced from reference 67, with permission.)

LEED intensity measurements, additional provisions are required such as the use of a movable Faraday cup, a spot photometer, or a computer-interfaced video camera; all these allow for quantification of the LEED spot intensities and profiles.[65]

The LEED pattern for an iodine-coated Pd(111) electrode surface is provided as an example in Figure 5; included in this figure is the suggested real-space surface structure of the Pd(111) $(\sqrt{3} \times \sqrt{3})R30°$–I adlattice.[68] It can be seen here that the iodine adlattice has the same hexagonal symmetry as the Pd(111) substrate except that (i) it is rotated 30° and (ii) it has an edge that is $\sqrt{3}$ times longer.

Pd(111)-(√3x√3)R30°-I

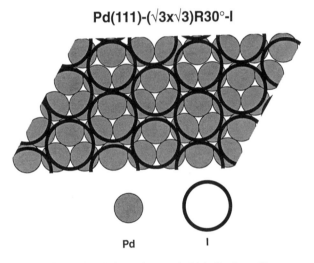

Pd   I

Figure 5. LEED pattern for a Pd(111) electrode coated with iodine from dilute aqueous NaI; also shown is the postulated real-surface structure. (Reproduced from reference 68.)

3. Reflection High-Energy Electron Diffraction (RHEED)

As a method for the determination of near-surface structural order, RHEED[14,17,19,69] represents an alternative to LEED. The principal difference between the two structural techniques is that, while low-energy electrons are utilized in LEED, RHEED employs high-energy (30–100 keV) electrons. At such energies, the inelastic mean free paths of the incident electrons are long

(100–1000 Å) and only a very small fraction of electrons is backscattered; that is, elastic scattering is predominantly in the forward direction. Hence, to afford the required surface sensitivity, RHEED experiments are performed with very small angles ($< 5°$) of incidence and diffraction. The requirements for energy filtering are far less stringent in RHEED than in LEED because of the large energy difference between the elastically and inelastically scattered electrons; post-acceleration is likewise unnecessary in RHEED because the primary electrons are sufficiently energetic to produce fluorescence on the phosphor screen.

Figure 6 shows LEED and RHEED patterns of gold films evaporated on glass and on mica[52]; the film deposited on glass at room temperature is rough, but that on mica at elevated temperatures is smooth and well-ordered in the (111) plane. It can be seen in this figure that ordered-surface diffraction is manifested in LEED by distinct spots and in RHEED by sharp streaks. This difference can be understood by recalling that the diffraction pattern actually corresponds to a section through the reciprocal lattice. From an Ewald sphere* construction,[47,69] it can be noted that the reciprocal net points of an ordered surface layer are elongated in the direction perpendicular to the surface plane; that is, in reciprocal space, the surface layer is represented by perpendicularly oriented rods that pass through the points in the reciprocal net. In normal-incidence diffraction as in LEED, only a section *perpendicular* through these reciprocal lattice rods is displayed, leading to the spot pattern. In grazing-incidence diffraction as in RHEED, a section *parallel* through the rods is displayed, resulting in a streak pattern. For the same reason, RHEED is unable to detect changes in periodicity along the plane of incidence. Hence, if the full two-dimensional periodicity is to be established, it will be necessary to rotate the sample about the surface normal.

RHEED is most useful in studies related to the structure and morphology of thin films and surface coatings. It is possible to continuously monitor film formation under deposition conditions since the front of the sample is unimpeded by either the electron source or analyzer. In view of its low-scattering-angle geometry, RHEED is quite sensitive to surface asperities. For example, surface roughness or ilsand formation will obscure the flat-surface

---

* As the lattice vector in any direction is increased, the corresponding reciprocal lattice vector will decrease; if it is increased towards infinity, the reciprocal lattice points will converge towards zero to form a continuous line. If the separation between planes is increased towards infinity, the two-dimensional structure is approached. The reciprocal lattice is given by the discrete two-dimensional mesh plus continuous lines or *rods* in the third dimension, aligned in a direction normal to the real-space plane. It is then possible to view a geometrical construction in reciprocal space which would simulate the diffraction process. Such construction is known as "Ewald construction." The sphere which defines the locus of beam vectors which have the same energy as the incident beam is known as the *Ewald sphere*.[47,69]

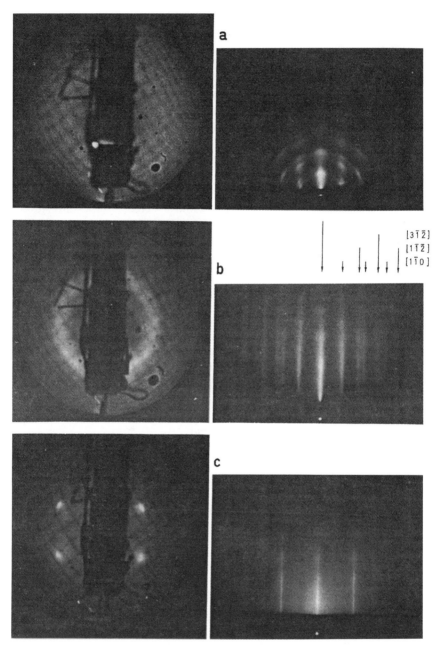

Figure 6. LEED and RHEED patterns for Au vapor deposited onto (a) glass at room temperature, (b) glass at 400°C, and (c) mica at 360°C. (Reproduced from reference 52, with permission.)

RHEED streak patterns; a pattern of diffraction spots will be generated due to transmission electron diffraction through the tips of the asperities. Dynamical theories for RHEED[14,17,19] have been developed, but accurate experimental data are not available for analysis.

4. Auger Electron Spectroscopy (AES)

AES is one of the more widely used techniques for surface elemental analysis.[14-19,70-82] In the Auger process, illustrated schematically in Figure 7, a core (K) level electron is emitted when a beam of electrons, typically with energies between 2 and 10 keV, is impinged onto the sample surface. In the decay process, an electron in an upper (L_I) level falls into the vacant core level and another electron in a different upper (L_{III}) level is ejected; the second, emitted electron is the Auger electron, and this particular process is labeled as a KL_IL_{III} in order to specify which energy levels are involved. The kinetic energy of the Auger electron is dependent upon the binding energies of the K, L_I, and L_{III} electrons but not upon the energy of the incident or primary electrons. The appropriate relationship is given by

$$E_{KL_IL_{III}} = E_K - E_{L_I} - E_{L_{III}} - e\varphi_{sp} \qquad (9)$$

where e is the electronic charge and φ_{sp} is the spectrometer work function. The exact application of Eq. (3.9) must realize that the energy difference is actually between singly ionized (one-hole) and doubly ionized (two-hole) binding

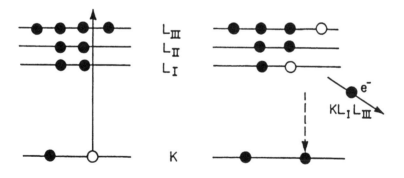

(a) EXCITATION   (b) ELECTRON EMISSION

Figure 7. Schematic diagram of the Auger emission process. The core (K) level electron is that which is ejected in x-ray photoelectron spectroscopy, while the Auger electron is the ejected electron designated as KL_IL_{III}. (Reproduced from reference 14, with permission.)

energy states. Nevertheless, $E_{KL_IL_{III}}$, as obtained from empirical spectra, is characteristic of a given atom which affords AES its element-specificity. Since the overall Auger process involves three electrons, AES is clearly not applicable for the analysis of H and He. It should also be noted that, although the incident electrons are of high energies, AES is still a surface-sensitive method because the emitted Auger electrons are generally of much lower energies and correspond to the minimum in the universal curve (Figure 2).

An inherent difficulty in AES arises from the fact that the Auger emission peaks are actually of very low intensities superimposed on a large secondary emission background (Figure 3). The usual approach to circumvent this problem is the combination of electron-energy analysis with suitable modulation techniques; in this manner, spectra exclusive of the original background can be obtained. The two more common energy analyzers used in AES are the retarding field analyzer (RFA) and the cylindrical mirror analyzer (CMA). In a multitechnique instrumentation that includes LEED, and RFA is most affordable since it makes use of the LEED optics (Figure 4). Energy analysis with an RFA involves the application of a voltage ramp to the suppressor grids such that only electrons of energies higher than the applied potential are transmitted through and accelerated onto the screen which is kept at a high positive (1 kV) voltage. In modulated RFA, the retarding voltage is modulated, typically at 1 kHz, with a 5-V (peak-to-peak) signal. The modulated component of the signal arriving at the screen is then passed to a lock-in amplifier tuned to a frequency twice that of the suppressor grid modulation (second harmonic). The end result is a derivative signal, $dN(E)/dE$, devoid of the original large background.

Because of important advantages over an RFA, such as higher sensitivity and resolution, AES with a CMA is now more widely used. A schematic diagram of a CMA is given in Figure 8. Energy analysis with a CMA is achieved by a negative ramp voltage applied to the outer cylinder while the inner cylinder is held at ground. Only electrons of the appropriate energy can pass unhindered through the CMA and into the detector which is usually a channel electron multiplier. As with modulated RFA, the pass energy of the CMA is modulated and then synchronously demodulated with a lock-in amplifier. The resultant spectrum is also a derivative spectrum. It should be noted that modern instruments now employ software-based modulation and filtering. The resolution of a CMA is dependent upon its entrance and exit slits; improved resolution can be achieved by a double-pass CMA. Improved resolution is also afforded by the use of a cylindrical hemisphere analyzer (CHA) which, as described below, is inherently a double-focusing analyzer.

A surface of a given composition possesses a signature Auger $dN(E)/dE$ spectrum; this renders AES a powerful technique for qualitative surface

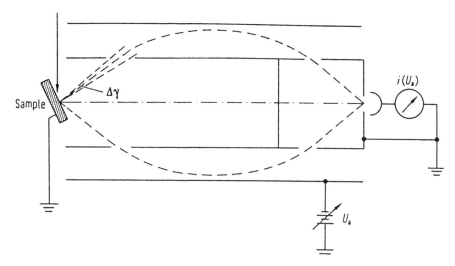

Figure 8. Schematic cross section of a cylindrical mirror analyzer. U_a is the potential applied between the two coaxial cylindrical electrodes, $\Delta\gamma$ is the angular spread of the electrons at the entrance slit, and $i(U_a)$ is the current at the exit aperture. (Reproduced from reference 16, with permission.)

elemental analysis. An example is given in Figure 9 for a Au(100) surface electrodeposited with a compound semiconductor CdTe.[83] For quantitative and/or molecular compositional analysis[70-73] the derivative spectrum is difficult to process. An approach to the collection of standard (nonderivative) Auger spectra involves pulse counting electronics or direct current measurements; spectra generated in this manner have been deconvoluted by a fast Fourier transform algorithm[74,75] to obtain information on chemical shifts and lineshapes. Changes in Auger lineshapes reflect modifications in the valence band density of states.

The use of derivative Auger spectra for the determination of adsorbate surface coverages has been the subject of numerous studies.[76-79] One method, specifically used in surface electrochemical studies, makes use of the following equation[80-82]:

$$\Gamma_a = \frac{I_a}{I_p\phi_c G_a} \tag{10}$$

where Γ_a is the absolute packing density of the adsorbate (mol/cm^2), I_a is the Auger current for the adsorbate, I_p is the primary beam current, ϕ_c is the measured collection efficiency of the Auger spectrometer, and G_a is the

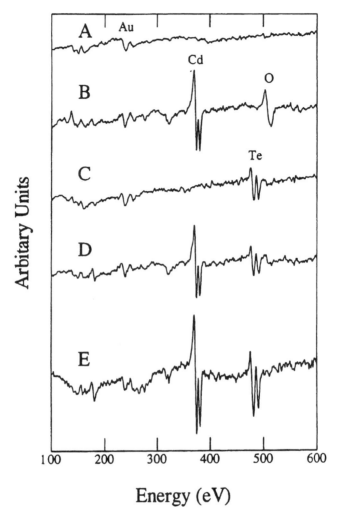

Figure 9. Auger electron spectra for Au(100): (a) after ion bombardment and annealing; (b) after Cd UPD; (c) after first Te UPD; (d) after Cd UPD on first Te UPD layer; (e) after Cd UPD on second Te UPD layer.

calculated Auger electron yield factor.[84] I_a is obtained by double integration of the adsorbate second-harmonic amplitude A_2 corrected for the clean-surface signal A_{2c}[78,79]:

$$I_a = \frac{4}{k^2} \int_0^{E_p} \int_0^E (A_2 - \phi_b A_{2c}) dE' \, dE \qquad (11)$$

where ϕ_b is the observed attenuation of the substrate signal by the adsorbed species, and k is the modulation amplitude. For simple adsorbates for which well-characterized surface layers are available for calibration purposes, Eq. (10) can be expressed in purely empirical terms[81,82]:

$$\Gamma_a = \left(\frac{I_a}{I_M^0}\right)\frac{1}{B_a} \tag{12}$$

where I_M^0 is the Auger signal for the clean substrate, and B_a is a calibration factor.

5. X-Ray Photoelectron Spectroscopy (XPS)

This technique,[14-19,85,86] originally referred to as electron spectroscopy for chemical analysis (ESCA),[87] is the other widely used method for surface compositional analysis. In XPS, which is based upon the photoelectric effect, the solid surface is irradiated with x-rays which results in the ejection of a core-level electron. The kinetic energy E_{kin} of the emitted photoelectron is given by

$$E_{kin} = h\nu - E_B - e\varphi_{sp} \tag{13}$$

where $h\nu$ is the energy of the incident x-ray photon and E_B, within the framework of Koopman's theorem, is the binding energy of the core-level electron. If E_B is to be referred to the vacuum level, the spectrometer work function φ_{sp} must be known. For studies with metals, it is more convenient to reference E_B with respect to the Fermi level; the latter is readily determined since it is marked by the onset of electron emission at the highest kinetic energy.

The XPS source consists of an anode material which, upon bombardment by high-energy electrons, emits x-rays. The emitted radiation can be rendered monochromatic either by Bragg diffraction or by the use of the characteristic emission lines of the anode; for Mg and Al, commonly used as anodes, these lines are 1253.6 eV (Mg-$K\alpha$) and 1486.6 eV (Al-$K\alpha$), respectively. With the availability of synchrotron radiation, continuous, high-flux x-rays can be obtained. It is important to note that, for $E_B < 700$ eV and using either an Mg or an Al source, the E_{kin} of the ejected photoelectron will not fall within the minimum of the universal curve. In such cases the surface sensitivity of XPS becomes minimal. This can be remedied either by the use of near-grazing incidence or by the detection of electrons emitted at small angles with respect to the surface plane. Such techniques result in a greater effective

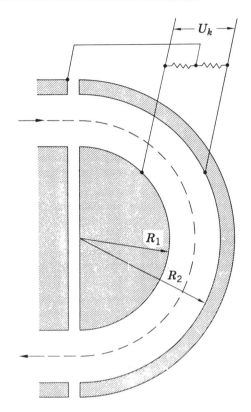

Figure 10. Schematic cross section of the concentric hemisphere analyzer. (Reproduced from reference 16, with permission.)

thickness of the interfacial layer that the incoming or outgoing electron encounters.

To afford the high resolution required for meaningful XPS studies, energy analysis is usually based upon a CHA, a diagram of which is presented in Figure 10. A potential difference U_k is applied across the inner and outer hemispheres of radii R_1 and R_2, respectively. Electrons of energy eV_e are focused at the exit slit only if the following equation is satisfied:

$$U_k = V_e \left(\frac{R_2}{R_1} - \frac{R_1}{R_2} \right) \tag{14}$$

The CHA is double-focusing since it focuses in two planes. The resolution of a CHA can be improved significantly by electron pre-retardation with an RFA or a retarding lens system. XPS has also been performed with a double-pass

CMA. Detection is typically done with a channel electron multiplier. Due to inherently weak intensities, signal averaging and other data processing routines are always employed.

Qualitative elemental analysis of sample surfaces relies upon the comparison of measured E_B values with those for reference materials. Quantitative analysis is based on the fact that the ionization cross section of a core electron is essentially independent of the valence state of the element. Hence, the intensity will always be proportional to the number of atoms within the detected volume. For quantitative purposes, the area under the background-corrected peak is taken as the intensity. The intensity is a complicated function of several parameters, some of which can be eliminated by the use of a reference state analyzed under identical conditions as the sample; however, such favorable cases are infrequent. If the spectrometer has a small aperture and the surface is uniformly irradiated, the equation for the intensity can be simplified to

$$I_A = \sigma_A D L_A J_0 N_A \lambda_M G_1 \cos \theta_1 \qquad (15)$$

where I_A is the integrated peak intensity for an element A, σ_A is the photoionization cross section, D is the spectrometer detection efficiency, L_A is angular asymmetry of the emitted intensity with respect to the angle between the incidence and detection directions, J_0 is the flux of primary photons, N_A is the density of atoms A, λ_M is the escape depth of the photoelectron, G_1 is the spectrometer transmission, and θ_1 is the angle between the surface normal and the detection direction. The application of Eq. (15) for various adsorbate–substrate configurations has been the subject of extensive discussion.[14-19,88]

As a surface elemental analysis tool, XPS is complementary to AES. Since the ionization cross secton for an Auger process decreases with E_B, which, in turn, increases with atomic number Z, AES is most sensitive to $Z < 45$ elements; for heavier elements, XPS provides higher sensitivity. The one distinct advantage that XPS offers is in the determination of oxidation states of the elements under examination. This is easily done in XPS because the binding energies of the core-level electrons are influenced by changes in chemical environment. In principle, identical information can be obtained from Auger peak energy shifts and lineshapes; in practice, however, the extraction of such information from raw derivative or nonderivative AES spectra is not a trivial task.[70-75]

Examples of XPS spectra, those of a smooth polycrystalline Ir foil electrode surface before and after pretreatment with iodine, are shown in Figure 11; the peaks at 62.1 and 65.2 eV represent surface iridium oxide.[89]

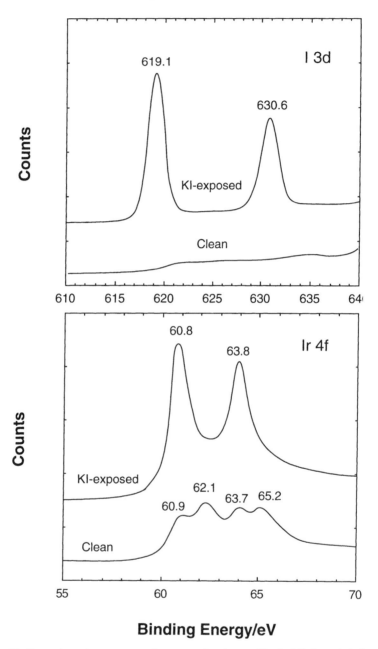

Figure 11. X-ray photoelectron spectra for a smooth polycrystalline Ir foil electrode before and after pretreatment with iodine.

6. High-Resolution Electron Energy Loss Spectroscopy (HREELS)

Almost all of the incident electron impinged at a solid surface undergo inelastic events that cause them to be backscattered at energies lower than the primary energy E_p. If E_l is the energy lost to the surface, peaks would appear in the energy distribution spectrum (Figure 3) at energies $\Delta E = E_p - E_l$. Such peaks, commonly referred to as *electron energy loss peaks*, are of several types according to the origin of the energy loss; these types include core-level ionization, valance-level excitations, plasmon losses, and vibrational excitations. For the latter, the energy losses are small since $E_{vib} < 4000 \, cm^{-1} < 0.5 \, eV$. Hence, the loss peaks due to vibrational interactions at $\Delta E = E_{vib}$ lie close to the elastic peak and can be observed only if electron energy loss measurements are done at high resolution.

At solid surfaces, there are two mechanisms that give rise to vibrational HREELS spectra[14-19,90,91]: dipole scattering and impact scattering. In dipole scattering, the incident electron interacts with the oscillating electric dipole moment induced by the vibration of species at the surface. Such interactions occur at long range and can be described either classically or quantum mechanically. Two important selection rules apply for surface dipole scattering: (i) Only vibrations whose dynamic dipole moments perpendicular to the surface are nonzero contribute to HREELS spectra. This selection rule is the same as that for surface infrared reflection–absorption spectroscopy (IRAS).[90-99] (ii) The intensity distribution with respect to scattering angle is sharply peaked in the specular direction; that is, loss peaks due to dipole scattering disappear when the backscattered electrons are collected at an angle different from that of the specularly reflected beam.

The mechanism for impact scattering at solids, which can only be treated quantum mechanically, involves exceedingly short-range interactions between the incident electron and the oscillator at the surface. The surface dipole selection rules do not apply to impact scattering. Theoretical considerations have predicted, and experimental studies have confirmed, the following properties of this type of scattering mechanism[90,91]: (i) Impact scattering vanishes in the specular direction; that is, loss peaks due to impact scattering can be observed only if the scattered electrons are detected at angles removed from the specular direction. The dependence of dipole and impact scattering on off-specular scattering angle (φ) is demonstrated by the data in Figure 12 [110]. (ii) Impact scattering is more likely to prevail at higher energies. (iii) Strong dipole scatters are weak impact scatterers; conversely, weak dipole scatterers are strong impact scatterers.

It is clear that the combination of specular and off-specular HREELS could provide a means for the complete identification of the normal modes of an adsorbed molecular species; point and space group theoretical considerations

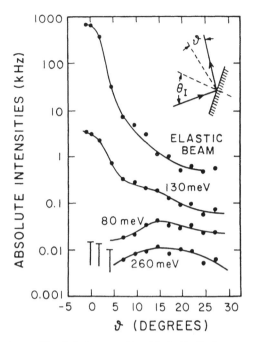

Figure 12. The angular profile of electrons scattered inelastically from an ordered monolayer of H on W(100). The 80-meV mode is parallel whereas the 130-meV vibration is perpendicular to the surface. The 260-meV mode is the first overtone. (Reproduced from reference 90, with permission.)

would of course be required. HREELS is an extremely sensitive technique. The limit of detection for strong dipole scatterers such as CO can be as low as 0.0001 monolayer; for weak scatterers such as hydrogen, the limit is 0.01 monolayer. In comparison, IRAS for chemisorbed CO, a strong infrared absorber, is restricted to coverages above 0.1 monolayer. HREELS studies of non-CO organic molecules adsorbed at atomically smooth electrode surfaces are abundant; similar experiments using IRAS are meager. The energy accessible to HREELS ranges from $100\,cm^{-1}$ to $4000\,cm^{-1}$ ($1\,eV = 8066\,cm^{-1}$); present IRAS detectors are not useful below $600\,cm^{-1}$. On the other hand, IRAS has higher resolution (nominally $4\,cm^{-1}$) than HREELS (at best $30\,cm^{-1}$) and can be utilized for experiments under electrochemical conditions. [95,99,101,102] Figure 13 is a schematic diagram of an HREELS spectrometer.[103] The energy of incident electrons can be varied from 1 to 10 eV. To afford high resolution, energy monochromation and analysis are done with either a CMA, a cylindrical deflector analyzer, or a spherical deflector analyzer in combination with retarding field optics. Off-specular collection of the

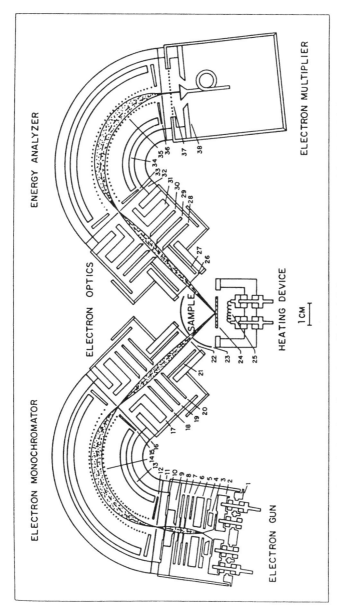

Figure 13. Schematic drawing of a tandem cylindrical deflector spectrometer used for HREELS. (Reproduced from reference 103, with permission.)

backscattered electrons is afforded by rotation of either the sample or the analyzer. Due to extremely low signals (10^{-10} A), continuous dynode electron multiplier detectors have been recommended.

7. Work Function Measurements

The work function of a uniform crystal surface is the work to remove a Fermi-level electron from the bulk to the vacuum just outside the surface. This quantity is really a difference in electrochemical potentials of the electron in two places. In the bulk metal the electron has an electrochemical potential $\bar{\mu}_e$ which is equal to the Fermi energy.[104,105] Once the electron is removed to outside the crystal and is at rest, its electrochemical potential has no entropy component and is just the electrostatic potential energy $-e\Phi_0$. Here Φ_0 is the electrostatic potential just outside the surface—that is, far enough away from the surface that the electron does not feel its image charge. Thus the work function is defined by[106]:

$$\phi = -e\Phi_0 - \bar{\mu}_e = -e(\Phi_0 - \Phi_i) - \mu_e \tag{16}$$

where the second form follows from writing the electrochemical potential as a sum of its chemical (μ_e) and electrical ($-e\Phi_i$) parts. In the second form the bulk contribution μ_e is clearly separated from the surface contribution $e(\Phi - \Phi_i)$. Note that the Fermi energy (or equivalently the electrochemical potential) is not a property only of the bulk, because it contains the electrostatic potential Φ_i inside the metal, which is determined by the surface dipole layers.

In surface adsorption studies, only the surface dipole part $e(\Phi_0 - \Phi_i)$ changes, and consequently the change in work function is equal to the change in this quantity; absolute work functions are less important. For a clean metal surface, the exponential decay of the wave function into the vacuum (electron "overspill") creates the surface double layer. The dipole has the negative end outward and is dependent on the surface crystallography.

The adsorption of an electronegative atom such as oxygen changes the surface dipole, making the outside of the crystal more negative and thus increasing the work function. Conversely, and electropositive adsorbate such as Cs increases the work function. For atomic adsorbates, the sign of $\Delta\phi$ is therefore correlated with the direction of charge transfer: Positive $\Delta\phi$ is associated with adsorbate to substrate charge transfer and negative $\Delta\phi$ with substrate to adsorbate charge transfer. This reasoning regarding charge transfer has been extended to the case of molecular adsorbates. It should be used with caution, because the permanent dipole moment of the molecule is typically a more important contributor to $\Delta\phi$ than the surface chemical bond.

The work function also contains potential information about the orientation of the adsorbed molecule, since it probes only the component of the dipole moment normal to the surface.

Drawing quantitative conclusions about adsorbate structure and bonding from work function measurements is difficult, for several reasons. The work function change is often not proportional to the number of adsorbed molecules, expect at low coverages. Accordingly, it reflects not only the properties of individual molecules, but also their interactions; the surface dipoles mutually depolarize one another. Even at low converges, the dipole moments of adsorbed molecules estimated from $\Delta\phi$ are often much smaller than gas-phase dipole moments. Subtle effects of the energy levels involved in the bonding may also be significant, as evidenced by some systems in which even the sign of $\Delta\phi$ is dependent on the type of adsorption site.

Nonetheless, work function measurements lead to useful qualitative conclusions, and empirical correlations between $\Delta\phi$ and the nature of adsorption have been established. In studies of the structure of the electrode–electrolyte interface, work function change measurements are invaluable since $\Delta\phi$ is very sensitive to both the charge at the electrode and the geometry of the electrochemical double-layer.[104,105] At a simpler level, work function changes are often used to monitor adsorption. After the $\Delta\phi$-versus-coverage relationship is established in a calibration experiment, $\Delta\phi$ may then be used as a sensitive measure of surface coverage. Abrupt changes in the slope of the work-function-versus-coverage relationship are diagnostically useful for changes in the type of adsorption—for example, completion of an ordered quarter monolayer structure.

Experimental measurements of ϕ and $\Delta\phi$ have been based upon the diode method, field emission, contact potential difference, and the photoelectric effect; the latter two are more commonly utilized. The photoelectric method, whcih measures absolute values of the work function, is based upon the determination of the threshold energy hv_0 for photoelectron ejection; the work function is then calculated from the equation $hv_0 = e\phi$. The contact potential difference method, which monitors changes in work function, depends on the measurement of the potential difference between two plates in electrical contact. If one of the plates is used as a reference of constant work function, $\Delta\phi$ at the other plate is manifested as a change in the contact potential. The most common way of measuring the change in contact potential uses a vibrating tip close to the surface as the reference plate (the Kelvin probe method). In its balanced condition there is an electric-field-free region between the sample and the tip, and there is no induced alternating current flows in the circuit. Adsorption leads to a momentary field and an alternating current. Electronic feedback is used to adjust the potential on the sample until the field-free condition again applies. The potential adjustment required is equal to the change in work function. Feedback time constants of less than a second are typical.

8.   Temperature-Programmed Desorption (TPD) or Thermal Desorption Mass Spectrometry (TDMS)

Thermal desorption techniques[14-19] exploit the fact that species adsorbed on a surface will desorb at a rate which increases with temperature. A study of the temperature-dependence of the desorption rate yields data on desorption energies which, for most cases, lead to information on adsorption binding energy states. Thermal desorption can also be used to obtain surface coverages and, in combination with mass spectrometry, determine the composition of species desorbed from the surface. Thermal desorption is a destructive surface analytical tool, although it does not always provide a clean surface after one heating cycle in ultrahigh vacuum; at catalytic surfaces, complex organic compounds may decompose to form stable graphitic or carbidic layer upon anaerobic heating.

Temperature-induced desorption methods can be classified according to whether the rise in temperature is fast (flash desorption) or is gradual (TPD). In flash desorption, the desorption rate is much greater than the rate at which desorbed gas is pumped out of the system; in this case, the desorption of a given binding state is marked by a plateau in the pressure-temperature curve. In TPD, the slow heating (desorption) rate allows the evolved gases to be pumped out; as a result, the desorption of a particular binding state appears as a peak instead of a plateau in the desorption curve.

In the simplest method, the extraction of the activation energy for desorption E_d from the peak temperature T_p in the TPD curves involves the use of either of the following equations[14,17,106]:

Zeroth-order desorption:
$$\frac{E_{d,0}}{R} = \frac{v_0}{\Gamma\alpha}\exp\left(\frac{-E_{d,0}}{RT_p}\right) \tag{17}$$

First-order desorption:
$$\frac{E_{d,1}}{RT_p^2} = \frac{v_1}{\alpha}\exp\left(\frac{-E_{d,1}}{RT_p}\right) \tag{18}$$

Second-order desorption:
$$\frac{E_{d,2}}{RT_p^2} = \frac{v_2\Gamma}{\alpha}\exp\left(\frac{-E_{d,2}}{RT_p}\right) \tag{19}$$

where v is the preexponential factor, Γ is the initial adsorbate concentration, α is the heating rate for the linear temperature ramp, and R is the gas constant. The desorption activation energy is related to the heat of adsorption ΔH_{ads} by

$$E_d = E_a + \Delta H_{ads} \tag{20}$$

where E_a is the activation energy for adsorption. Several instances can be found in which the adsorption is nonactivated; for such cases, E_d can be equated to ΔH_{ads}.

Thermal desorption experiments based upon measurement of gas pressures are conceptually simple but the experiments are beset with difficulties, such as selective pumping of gases at rates which vary during desorption as well as outgassing from the various elements of the vacuum system due to radiation from the hot sample, which can invalidate quantitative conclusions. Most of these problems can be circumvented by the use of a mass spectrometer to monitor species emitted from the surface. A comparatively inexpensive means for mass spectrometric detection is afforded by a quadrupole mass analyzer, a schematic diagram of which is shown in Figure 14.[107] This mass filter consists of four parallel rod-shaped electrodes arranged at the apices of a diamond. The existence of an appropriately varying electrical field between the pairs of opposite electrodes will cause all ions to impact on the rods during transit except those of a particular mass-to-charge ratio m/z. Time-of-flight mass spectrometers are more widely used in laser-induced desorption studies.

In TDMS, the substrate is positioned as close as possible to the mass spectrometer; provisions must be made to minimize temperature gradients at the sample surface. The temperature is monitored by a thermocouple wire placed in direct contact with the crystal. Problems associated with degassing from parts of the sample manipulator close to the crystal can be solved by masking the mass spectrometer with a small aperture such that only line-of-sight detection is possible. For programmed TDMS, microprocessor control

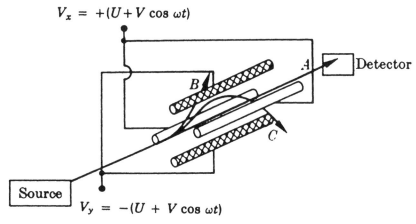

Figure 14. Schematic diagram of a quadrupole mass filter. Ions B and C have incorrect m/z ratio and are thrown against the rods; ion A has the proper m/z ratio and is transmitted through the rods onto the detector. (Reproduced from reference 107, with permission.)

allows multiplexed or simultaneous data acquisition of several preselected masses.

C. Instrumentation Designs

The most critical step in UHV-EC experiments is the transfer of the electrode between the electrochemical cell (at ambient pressures) and the surface analysis chamber (in ultrahigh vacuum). Ideally, the transfer is not accompanied by changes in surface structure and/or composition. The simplest approach, which would not require a dedicated surface analysis instrument, is the transfer of the sample electrode through air. Clearly this procedure is applicable only if the surface is inert (such as an oxide film) or is covered with a protective film (of solvent or electrolyte) which can be removed by evacuation inside the UHV chamber. An alternative procedure that would allow electrode transfer in a controlled environment makes use of an inert-atmosphere glovebox attached to the surface-analysis instrument; electrochemical experiments are perfomed inside the box. However, regardless of the purity of the inert gas, the environment is not completely free of contaminants since the walls of the glovebox contains impurities that are slowly desorbed. The transfer-through-air and drybox methods have been used extensively in XPS and AES studies with oxided or chemical modified polycrystalline surfaces; they have not been adopted for single-crystal work.

The approach employed most successfully in UHV-EC studies of single-crystal electrode surfaces involves the fabrication of a multitechnique surface-analysis apparatus to which an electrochemistry chamber is physically appended. The entire assembly is constructed to stainless steel and can be baked to about 200°C in UHV to attain ultraclean conditions. Ultrahigh vacuum is maintained by a combination of a titanium sublimation pump and either an ion or turbomolecular pump. Transfer of the electrode between the analysis and electrochemistry chambers is accomplished by a sample manipulator–translator. In some instruments, the crystal remains attached to the same sample holder as it is moved between the two compartments; in other systems, the crystal has to be transferred between two different manipulators. A gate valve isolates the electrochemistry compartment, whenever necessary, from the rest of the system. It is preferable to keep the electrochemistry chamber under UHV when not in use in order to preserve its cleanliness. The electrochemical cell itself is located inside a bellows-enclosed compartment separated from the electrochemistry chamber by another gate valve; the cell is inserted only after the electrochemistry chamber is brought to ambient pressures with ultrahigh-purity inert gas. Based upon these considerations, various types of UHV-EC instruments have been constructed[28–35,109–113]; three of these are shown for illustrative purposes in Figures 15–17.

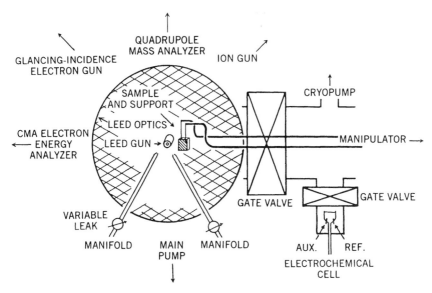

Figure 15. Schematic drawing of an experimental arrangement for UHV-EC studies. In this design, the crystal remains attached to only one sample manipulator. (Reproduced from reference 28b, with permission.)

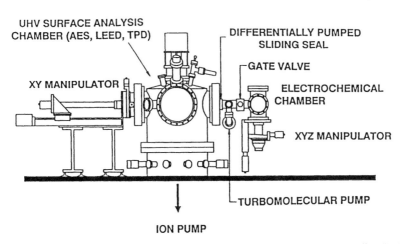

Figure 16. Schematic drawing of an experimental arrangement for UHV-EC studies. In this design, the electrode remains attached to only one sample manipulator. (Reproduced from reference 113, with permission.)

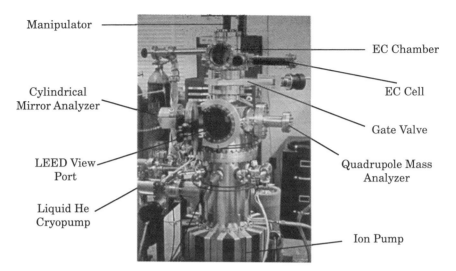

Manipulator

EC Chamber

EC Cell

Cylindrical
Mirror Analyzer

Gate Valve

LEED View
Port

Quadrupole Mass
Analyzer

Liquid He
Cryopump

Ion Pump

Figure 17. Photograph of a UHV-EC instrument.

A typical UHV-EC experiment following instrument bakeout (at which point the base pressure should be less than 5×10^{-10} torr) would include the following steps. After electrode preparation and initial surface characterization, the electrode is transferred into the electrochemistry chamber which is then isolated from the UHV system, by closure of the appropriate gate valve, and backfilled with high-purity inert gas. The external gate valve is opened and the electrochemical cell is inserted into the chamber. After completion of the electrochemical experiments, the cell is retracted, the external gate valve is closed, and the chamber is pumped down by a turbomolecular or liquid–helium cryogenic pump to less than 10^{-6} torr. At this point, the main gate valve can be opened to complete the evacuation of the electrochemical chamber and to transfer the electrode into the surface analysis compartment. Pumpdown from ambient pressure to 10^{-8} torr vacuum can be achieved in less than 15 minutes.

In the UHV-EC system shown in Figure 16, an isolable, differentially pumped antechamber is situated between the UHV and electrochemistry compartments. The main function of this antechamber is to minimize the influx of solvent and/or electrolyte vapor into the surface analysis compartment. In this context, it is important to mention that the pressure in the EC chamber is usually an order of magnitude higher than that in the UHV chamber; mass spectrometric analysis of the residual gas has revealed that the pressure difference arises primarily from higher amounts of water in the electrochemistry compartment.[31] Because water is only weakly surface-active, it is generally not of major concern in UHV-EC studies. However, in the presence of comparatively high quantities of water, impurity species may be dislodged from the walls of the chamber and onto the sample surface. Similar "knockoff" effects can arise when the chamber is backfilled with high-purity inert gas. Hence, it is critical to maximize the cleanliness of the electrochemistry chamber and its associated manifold; this can be accomplished by frequent bakeout and continuous evacuation of the electrochemistry chamber when it is not in use.

It is also important to ensure that the backfill gas is of the highest purity to minimize surface contamination by trace-level impurities; argon of at least 99.99% purity is usually employed. Background contamination is metal specific; for example, Cu is more sensitive to residual O_2 while Pt is more susceptible to carbonaceous impurities. Hence, depending upon the nature of the investigation, it may be necessary to pass the high-purity inert gas through molecular scavengers such as a Ti sponge heated to 900°C for still further purification.[31] While efforts to ensure the cleanliness of the UHV-EC system cannot be overemphasized, it must also be realized that electrode-surface

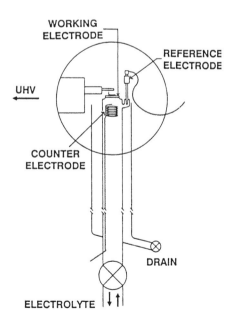

Figure 18. Schematic view of an electrochemical cell for use with a single-crystal disc electrode. (Reproduced from reference 113, with permission.)

contamination can also result from trace-level impurities in the electrolyte solution. Such impurities can originate from the solvent, electrolyte, glassware, and/or the inert gas employed for solution deaeration. The level of solution-based impurities can be minimized by the use of highly purified chemical reagents and gases; in the case of aqueous solutions, the utilization of pyrolytically triply distilled water is recommended,[114] although the use of Millipore Milli-Q water is now an acceptable alternative.

Electrochemical experiments have been performed with cells in either the standard arrangement or the thin-layer arrangement. The latter significantly reduces the level of surface contamination from solution-borne impurities. If the entire electrode is to be immersed in solution, all faces of the single-crystal should be oriented identically in order to obtain characteristic voltammetry. As an alternative, the electrode can be positioned on top of the electrochemical cell in such a way that only one crystal face is exposed to solution, as illustrated in Figure 18. Such a configuration, however, often results in the adherence of a droplet of electrolyte on the crystal surface when the electrode is withdrawn from the solution; this problem, on the other hand, does not arise when the electrode is withdrawn slowly (1 mm/sec) in the vertical position since the droplet would form at the bottom, not on the surface, of the crystal.[34]

III. FUNDAMENTAL ASPECTS

A.   The Emersion Process

It is important to determine the changes in the interfacial properties when the electrode is removed, at a given potential, from the electrolyte solution. The electrode-withdrawal process under potential control has been termed *emersion*. In the ideal process, the emersed electrode retains an interfacial layer identical in composition and structure to that when the electrode was still in solution. Under electrochemical conditions, the electrode–solution interface is a structured assembly of solvent, electrolyte, and reactant. In the traditional view, this ensemble, commonly referred to as the *electrochemical double layer* and nominally 10 Å in thickness, is subdivided into an inner (compact) layer consisting of (a) field-oriented adsorbed solvent molecules and specifically adsorbed anions and (b) an outer layer composed of solvated cations. The locus of the centers of the adsorbed anions delineates the so-called inner Helmholtz plane (IHP), whereas the line of centers of the nearest solvated cations defines the outer Helmholtz plane (OHP). Charge transfer reactions of electroactive species are thought to occur at this outer (reaction) plane. The solvated ions interact with the charge metal only through long-range electrostatic forces and, because of thermal agitation in the solution, are distributed in a three-dimensional region that extends from the OHP into the bulk of the solution. This region is identified as the *diffuse layer*, and its thickness is a function of electrolyte concentration; it is less than 300 Å for concentrations greater than 10^{-2} M. Clearly, the electrode-withdrawal process involves a delicate balance with respect to the thickness of the emersion layer: it must be sufficiently thick to incorporate the intact electrochemical double layer, but it should also be thin enough to exclude residual (bulk) electrolyte. Numerous studies (*vide infra*) have helped established the fact that the electrochemical double layer can, under appropriate electrolyte concentrations, be retained intact when the electrode is withdrawn from solution under potential control. The optimum concentration depends upon whether emersion is hydrophobic or hydrophilic.[104,105,115-120] For the latter type, the concentration must not be much higher than 10^{-3} M if inclusion of bulk electrolyte is to be circumvented. For hydrophobic emersion, the concentration must not be much lower than 10^{-3} M if double-layer discharge is to be avoided. In cases where the mode of emersion is not known, an electrolyte concentration of 10^{-3} M appears to be a logical choice.

Investigations of hydrophobic emersion based upon electrode resistance measurements,[118] electroreflectance spectroscopy[115,116,119] XPS[120] and work function change determinations[104,105,117] have been able to (i) demonstrate the existence of an emersed double layer, (ii) determine its stability, and (iii) monitor changes in its structure and composition brought about by the emersion process. The evidence has been compelling that the structure and composition of the double layer in the emersed phase are very similar, if not identical, to those in the solution state; that is, only little or no double-layer discharge occurs upon emersion. More recent studies have focused on the effect of the emersion process on the structure of adsorbed molecular species. Experiments based upon use *in situ* IRAS[121,122] and surface-enhanced Raman spectroscopy (SERS)[123,124] have provided data which demonstrate that the structure and orientation of molecular adsorbates at electrode surfaces are essentially unperturbed by the emersion process.

Upon emersion, the intact double loses electrical contact with the bulk electrolyte but not with the electrode. Hence, the overall charge within the interface must remain neutral. This requirement for neutrality, however, does not disallow the occurrence of spontaneous faradaic reactions within the emersed layer. Such reactions can take place spontaneously provided that they do not result in charge imbalance within the layer, even if they are accompanied by loss of material. One example is the spontaneous oxidation of electrodeposited Cd in aqueous media:

$$\mathrm{Cd}_{(s)} + 2\mathrm{H}_2\mathrm{O} \rightarrow \mathrm{Cd(OH)}_{2(s)} + \mathrm{H}_{2(g)} \tag{21}$$

In this reaction, the water is from either the diffuse layer or the residual gas in the UHV chamber.

B.   Perturbations Caused by Evacuation and Surface Analysis

Another critical issue in coupled UHV-EC experiments pertains to perturbations of the emersed double layer caused by the evacuation and surface analytical processes. Alterations in the surface electronic structure can be studied by work function change measurements; representative results are shown in Figure 19, in which a plot of the work function of a polycrystalline Au emersed from 0.1 M HClO$_4$ into UHV as a function of the emersion

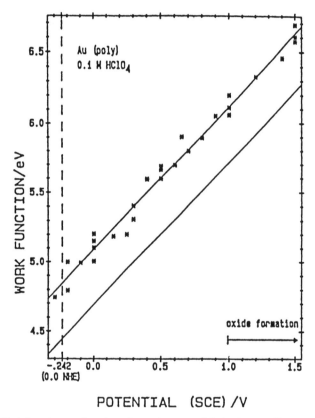

Figure 19. Work function as a function of emersion potential of polycrystalline Au emersed from 0.1 M HClO$_4$. The work function of the clean metal was 5.2 eV. The lower and upper lines, respectively, represent the solution inner potential if the absolute NHE half-cell potential is 4.45 or 4.85 V. (Reproduced from references 105 and 125, with permission.)

potential in presented.[105,125] It can be seen here that the work function tracks the applied potential over a wide range, even into the oxide formation region. This and other sets of data demonstrate that the electronic properties of the electrochemical double layer are unaffected by emersion either into the ambient or into UHV.

An expected effect of evacuation is the change in the composition within the electrochemical double layer by UHV-induced desorption; the extent of the compositional changes will depend primarily upon the heats of vaporization ΔH_{vap} or sublimation ΔH_{sub} of the unbound materials entrapped within the emersed layer. Obviously, excess water, unadsorbed gases, liquids, and sublimable solids will be removed readily in UHV. Water retained as part of the hydration sphere of the counter cations can survive the evacuation process if the hydration enthalpies ΔH_{hyd} are substantial.[81,126,127]

Adsorbed species with ΔH_{ads} well in excess of 40 kJ/mol are expected to be unaffected by exposure to ultrahigh vacuum. Counterions retained in the diffuse layer would also be stable in vacuum unless they undergo solvolysis reactions which would be enhanced at very low pressures; an example is provided by the hydrolysis of HCO_3^-:

$$HCO_3^- \rightarrow OH^- + CO_{2(g)} \qquad (22)$$

Strongly chemisorbed species, such as iodine at the noble-metal electrodes, are expected to form stable well-ordered adlattices in solution[68,128–130] that would not reconstruct in vacuum since there is no reason to expect that surface diffusion is less activated in solution than in vacuum. Similarly, the surface coverages and molecular structures of chemisorbed molecules are expected to withstand the evacuation process. One example is provided by 3-pyridylhydroquinone (Py-H$_2$Q), which is chemisorbed on a Pt(111) electrode surface through the N heteroatom.[131] In such a mode of surface attachment, the diphenol group is pendant and able to undergo the following reversible quinone/hydroquinone redox reaction:

$$Pt(111)\text{-}Py\text{-}Q + 2e^- + 2H^+ \leftrightarrow Pt(111)\text{-}Py\text{-}H_2Q \qquad (23)$$

where Q and H$_2$Q, respectively, represent the pendant quinone and hydroquinone groups. Figure 20 shows cyclic voltammetric curves that correspond to Reaction (23) for chemisorbed Py-H$_2$Q before and after a one-hour exposure to UHV. It is clear from the data presented that the reversible electrochemical reactivity of the chemisorbed layer has not been affected by the prolonged exposure to UHV.

Perturbations can also arise from the surface characterization method itself. For example, the extraction of ΔH_{ads} information from TPD is based

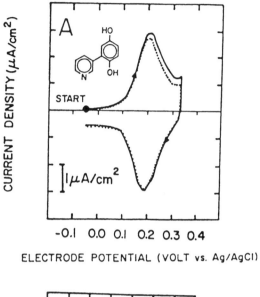

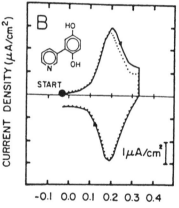

Figure 20. Cyclic voltammetry of 3-pyridylhydroquinone at Pt(111). (**A**) *Solid curve*: First scan: *Dotted curve*: After 1 hour in UHV. (**B**) *Solid curve*: First scan. *Dotted curve*: Second scan. (Reproduced from reference 131, with permission.)

upon the measurement of the desorption energy E_d; implicit in the technique is the requirement for complete desorption. Hence, by its very nature, TPD is a totally destructive technique. On the other hand, surface analytical methods based upon electron and optical spectroscopies are not intended to damage the surface layers; nevertheless, beam damage is common in these methods.

Unless exceedingly high photon fluxes are used, optical methods are non-deleterious relative to particle-based techniques.

Several surface processes are known to be stimulated by electron impact. Examples are binding-site conversions, dissociative chemisorption, and particle desorption.[17] Such processes take place even at very low electron power densities (minimal sample heating), which indicate that surface thermal effects are insignificant. The possibility of stimulation by momentum transfer can be assessed by noting that the maximum kinetic energy ΔE transferred to a particle of mass M upon collision with an electron of mass m_e and kinetic energy E_e is relatively small[17]:

$$\Delta E \approx 4E_e m_e/M \qquad (24)$$

As an example, the maximum energy imparted by an electron of 300-eV energy would be 0.3 eV an adsorbed H atom; in comparison, with adsorption enthalpy of strongly bound hydrogen is greater than 2 eV. Hence, for low-energy electrons, momentum transfer events cannot cause significant structural and compositional changes in the emersed layer. It is now accepted that electron-stimulated reactions occur mainly via electronic excitations. These excitations can lead to bond dissociation and form the basis of the surface spectroscopic technique known as *electron-stimulated desorption ion angular distribution* (ESDIAD).[132]

Pendant functional groups not directly bonded to the substrate surface, such as the diphenol moiety in Py-H_2Q, are most prone to electron-stimulated desorption. In other instances, electron irradiation can induce surface displacement reactions that involve species present as residual gas in the analysis chamber. It has been reported, for example, that prolonged electron-beam irradiation during LEED and/or AES caused the following reactions at a well-ordered adlattice formed at a Pt(111) electrode surface from aqueous Ba(CN)$_2$[126,127]:

$$\text{Ba(CN)}_2 \xrightarrow[\text{H}_2\text{O}]{e^- \text{ beam}} \text{Ba(OH)}_2 + 2\text{HCN}_{(g)} \qquad (25)$$

It is true that structural and compositional changes in the emersed double layer can be induced by electron irradiation. However, it is essential to note that the electron-stimulated alterations would be detrimental only if the postanalysis layers are to be used for further electrochemical experiments. In those rare instances when additional experiments have to be performed, it is a simple matter to regenerate the surface to exactly the point just prior the surface analysis. If desired, beam damage can be assessed by repeated analysis over a period of time followed by extrapolation of the data to zero time.

IV. CASE STUDIES

Over the past two decades, close to a thousand papers have appeared that describe a wide range of electrochemical phenomena at monocrystalline and polycrystalline electrodes examined by UHV-based methods. Most of these studies have already been covered in extensive reviews.[37-39] In this section, examples will be limited to studies that have employed well-defined single-crystal surfaces.

UHV-EC investigations with single-crystal electrode surfaces can be broadly classified into three groups. The first place emphasis on the structure and constitution of the electrochemical double layer as functions of electrode potential and solution composition. The second centers on electrodeposition reactions; included in this category are extensive studies on hydrogen and oxygen adsorption at platinum electrodes. The third deals with the interfacial structure and reactivity of chemisorbed complex molecules.

A. Electrochemical Double Layer

Two general strategies have been adopted in UHV-EC studies of the electrical double layer. One, strictly a model approach, involves the synthesis of the double layer in UHV by sequential cryogenic adsorption of its constituents[133-138]; the temperature must be maintained below 160 K at all times in order to prevent the evaporation of unbound solvent. The other approach is based upon the structural and compositional analysis of the emersed layer; since surface characterization is done at ambient temperatures, excess water in the diffuse layer is pumped away.

The viability of the cryogenic coadsorption approach was tested by comparison of work-function changes $\Delta\varphi$ for UHV-synthesized and actual $Ag(110)-X-H_2O$ layers, where X denotes Cl^- or Br^-.[133-136] The results showed that there is good agreement between the UHV and in situ results, provided that coadsorbed water is present in the UHV-generated layer. The satisfactory agreement indicates that, at least under zero diffuse-layer charge conditions, microscopic-level information from the UHV simulation work is relevant to electrochemical systems. The requirement of solvation implies that the electronic properties of the unsolvated $Ag(110)-Br$ interface are quite different from those of the fully solvated $Ag(110)-Br-H_2O$ layer. Other more complex interfacial systems have also been modeled via cryogenic coadsorption. For example, UHV-synthesized $H_2O-HF-CO$ coadsorbed layers were studied at $Pt(111)$ and $Rh(111)$ surfaces by HREELS, LEED, TPD, and XPS.[137,138] In that work, the "control" of electrode potential was based upon coadsorption of H_2. The cryogenic coadsorption approach to the study of the electrochemical double layer offers two main advantages[137,138]: (i) the control of interfacial parameters for more precisely than can be achieved in solution and (ii) the detailed characterization of fully solvated species by a host of complementary surface-sensitive spectroscopic methods. Still, it must be realized that this approach provides only models, and its relevance to electrochemistry still has to be fully established.

The direct approach to the study of the electrochemical double layer involves the surface characterization of the electrolyte layer retained at the electrode surface as it is withdrawn from solution. Although the direct approach is more realistic than the UHV simulation strategy, it is applicable only to cases in which the compact layer consists of materials which, because they are either strongly adsorbed or in the solid state, remain on the surface when evacuated to UHV. For example, no adsorption information can be expected when the electrolyte is aqueous HF since both H_2O and HF will be completely desorbed at ambient temperatures in vacuum. It is also implied in this approach that those solvent molecules pumped away are inconsequential in the formation and preservation of the electrochemical double layer. This is not an unreasonable premise because the double layer composed of chemisorbed species is governed by strong chemical and electrostatic interactions and should only be minimally perturbed by physisorbed species.

UHV-EC studies of specifically adsorbed anions have been carried out at well-defined Pt(111),[81,126,127,139-143] Pt(100),[144,145] stepped Pt(s) [6(111) × (111)],[146] Cu(111),[110,147] Ag(111),[148] Au(111),[149a] and Pd(111)[59-61,89,149b] electrodes; all these studies have been with aqueous solvents. The anions studies include monatomic species such as Cl^-,[11,140,144,145,148] Br^-,[139,144,145,148] I^-,[59-61,89,128,129,141,148] and SH^-[142,149b] and polyatomic species such as CN^-,[81,126,127] SCN^-,[127] and SO_4^{2-}.[146,150] All of these anions yield surface coverages and well-ordered structures that depend upon the solution pH and the applied potential.

B. Underpotential Electrodeposition

The deposition of monolayer quantities of one metal onto another generally occurs at potentials positive of that for bulk deposition because of preferential interactions between the substrate and the foreign-metal electrodeposit. The underpotential deposition (UPD) process is strongly influenced by the structure and composition of the substrate; hence, UPD research is one in which UHV-EC methods have been extremely valuable. The published literature on UHV-EC studies of UPD can be categorized according to whether the experiments were used to correlate the substrate structure with the electrodeposition voltammograms or to determine the resultant interfacial properties of the adatom-modified substrate. Investigations devoted to structure–voltammetry correlations help establish reference states which new experiments can be calibrated against; those focused on post deposition characterization yield important information concerning the electrocatalytic selectivity of the mixed-metal interfaces.

The first applications of LEED and AES in electrochemistry involved the correlation of the surface crystallographic orientation with the underpotential hydrogen deposition at Pt electrodes.[41,151-154] Those studies were motivated by earlier work with polycrystalline Pt electrodes whose cyclic voltammograms showed two hydrogen deposition peaks; the peaks were identified simply as weakly and strongly bound states of adsorbed hydrogen. Subsequent studies with single-crystalline but uncharacterized Pt(111), Pt(110), and Pt(100) electrodes associated the weakly bound hydrogen peak with the Pt(110) electrode and associated the strongly bound hydrogen with the Pt(100) electrode.[40] These studies, however, were not judged to be definitive because of the lack of a rigorous control of substrate structure and the use of multiple surface oxidation–reduction cycles to generate clean, but structurally disordered, surfaces. Nevertheless, they provided the impetus for further adaptation of UHV-based surface structural tools of interfacial electrochemistry.

Later studies based upon non-IHV-prepared single-crystal surfaces led to the discovery of new voltammetric features for Pt(111) in the form of highly reversible pseudocapacitance peaks at potentials *well positive* of the usual hydrogen deposition peaks.[42,53-55,155-157] These peaks can be seen when the Pt(111) voltammograms in H_2SO_4 and $HClO_4$ are compared for a flame-annealed sample with those for a UHV-prepared (but electrochemically cycled) electrode. Verification studies employing UHV-EC instrumentation equipped with improved vacuum-to-electrochemistry transfer technology were able to reproduce the new voltammetric results. Extensive follow-up *in situ* and UHV-EC work then ensued which clarified several aspects of this exceedingly surface-sensitive reaction.[53-55,150,156,157]

While studies of hydrogen chemisorption at Pt single-crystal electrode surfaces have been extensive, investigations on the formation of underpotential states of oxygen are not as numerous. Electrochemical experiments have been performed only with Pt(100) and Pt(111) electrodes.[30,158-160] Gas-phase and solution-state reactions with oxygenous species have been carried out at stainless steel single crystals.[161-163] The occurrence of place-exchange during anodic film formation has been studied via LEED spot-profile analysis.[30,160] This irreversible place-exchange reaction accounts for the common observation that the electrode surface loses its single crystallinity even after minimal surface oxidation.

The literature on monolayer metal deposits is extensive. Most of the work pertains to the geometric, electronic, and catalytic properties of foreign metals vapor-deposited in UHV onto single-crystal substrates; a compilation of the adlattice structures of such metal adlattices has been published.[164] Studies of foreign metal monolayers deposited electrochemically have been performed primarily with polycrystalline substrates. The first UHV-based investigation of electrochemically deposited admetals employed XPS to determine the core-level shifts of submonolayer Cu and Ag on polycrystalline Pt.[165,166] Quite a few investigations have been reported of admetal deposits on single-crystal surfaces, but only a minority of these involve UHV-EC experiments.

The first UHV-EC work on electrodeposition at well-defined electrode surfaces involved Ag at an iodine-coated Pt(111) electrode.[167,168] The iodine pretreatment was done in UHV to form a protective $Pt(111)(\sqrt{7} \times \sqrt{7})$ $R19.1°$–I adlattice before immersion into a solution containing dilute Ag^+ in 1 M $HClO_4$. Subsequent studies included Ag electrodeposition on I-coated $Pt(100)^{169}$ and stepped $Pt(s)[6(111) \times (111)],^{146}$ Cu on I-pretreated $Pt(111),^{170}$ and Pb on I-covered $Pt(111).^{171}$ Sn^{172} and Pb^{173} deposition onto iodine-free Pt(111) in Br^- or Cl^{-174} solutions has also been studied. Although the Pt substrate was not pretreated with I, the presence of halide ions in the plating solution led to specific adsorption of anions prior to the deposition process.

Electrodeposition from solutions free of surface-active anions have been studied. These investigations, carried out in ClO_4^- or F^- electrolyte, include UPD of Cu on $Pt(111),^{175-177}$ Ti, Pb, Bi, and Cu on $Ag(111),^{178}$ and Pb on the three basal planes of Ag.[108] Invariably, the underpotentially deposited films showed unique adlattice geometries that were dependent upon the substrate orientation and the admetal coverage.

The atomic layer epitaxy (ALE) approach to deposition of a compound film, based upon the alternate layer-by-layer deposition of the elements of the compound, has recently been adopted in the electrochemical synthesis of compound semiconductors. This electrochemical analogue, referred to as *electrochemical atomic layer epitaxy* (ECALE),[83,179,180] takes advantage of the fact that only monolayer quantities are produced by underpotential deposition. The UPD-based epitaxial growth of CdTe on Au(111) has been monitored by LEED and AES.[83]

C. Molecular Adsorption

The capability to prepare single-crystal surfaces by thermal treatment at ambient pressures[42,155–157,181] has fostered the proliferation of non-UHV studies of the adsorption of molecules at monocrystalline electrodes. The detail of information obtained from such *in situ* work, however, falls short of that provided by UHV-EC experiments. As one example, although *in situ* IRAS has provided much information about the structure-sensitivity of the chemisorption and anodic oxidation of CO, its sensitivity is too low to permit meaningful investigations with other molecules even as simple as ethylene.

1. Solvent–Electrode Interactions

The nature of the interactions between the solvent and the electrode surface has significant ramifications in electrochemical surface science. For instance, the use of strongly surface-active solvents would severely repress electrocatalytic processes that rely on a direct interaction between the reactant and the metal surface. The bonding of water to metal surfaces is an important issue in aqueous electrochemistry. In models suggested to explain the potential dependence of double-layer capacity, the existence has been postulated of monomeric and clustered water molecules, both of which are able to adopt two opposite dipolar orientations.[181–184] The studies of water adsorption on single-crystal electrodes are all based upon vapor deposition in UHV usually at cryogenic temperatures because water is not adsorbed on clean metal surfaces at ambient temperatures. Of significant interest to electrochemistry is the observation that, on Ni, Pt, Ag, Cu, and Pd, water is dissociatively chemisorbed if the surface contains submonolayer coverages of oxygen.[185] The reaction is thought to occur by hydrogen abstraction. This reaction is very metal-specific because at other noble metals such as Ru(001), adsorbed oxygen is inactive towards water dissociation.[185]

Nonaqueous solvents commonly used in electrochemistry include acetonitrile, dimethylformamide, p-dioxane, sulfolane, dimethylsulfoxide, pyridine, acetic acid, propylene carbonate, liquid ammonia, and dichloromethane.[186] Work involving such materials can be categorized according to whether the electrode is allowed to interact with the nonaqueous solvent by (i) vapor dosing in vacuum, (ii) exposure to aqueous solutions containing small quantities of nonaqueous-solvent material, or (iii) immersion in pure nonaqueous solvent. UHV-EC work of the latter type using single-crystal electrodes has not been pursued, although an XPS study of polycrystalline Li thin-film electrodes immersed in neat acetonitrile has been reported.[112] Studies that employed aqueous solutions containing small amounts of nonaqueous solvent are more abundant; however, such investigations are usually classified under electrode–solute, rather than electrode–solvent, interactions. Except for one case, all UHV-based adsorption studies with nonaqueous-solvent compounds (carboxylates, ammonia, and N-heteroaromatics) were carried out purely in the context of gas–solid surface science.[14,187] The intent of the one exception[188] was to use the reactions between the solvent vapor and the metal surface as models for the electrochemical analogues; for better simulation of solution conditions, vapor dosing was up to 0.3 torr, approaching the vapor pressures of the liquid solvents.

2. Group IB Electrodes

Most organic compounds are only weakly adsorbed on Cu, Ag, and Au electrode surfaces; hence, unless the adsorbate itself is a solid or when adsorption is carried out at cryogenic temperatures, meaningful UHV-EC experiments with the coinage metals are limited. One study, which took advantage of the strong interaction of the −SH functional group with the coinage metals, used HREELS, LEED, AES, and voltammetry to determine the influence of the location of the N heteroatom on the adsorption properties of the isomers 2-mercaptopyridine and 4-mercaptopyridine at Ag(111) in aqueous HF.[189] The subject compounds were postulated to undergo isomerization upon oxidative adsorption through the −SH moiety.

3. Group VIII Electrodes

The abundance of studies of organic molecular adsorption at electrode surfaces involves the platinum metals. This is of course not surprising because these metals are well known for their electrocatalytic activities and an immense body of work has already been amassed for these materials in their polycrystalline states.[190-195] Surface electrochemical studies of metal–organic compounds at single-crystal electrodes can be broadly classified according to whether the work was done with CO (and related small molecules) or with more complex molecules. The former are more numerous, although a vast majority of such studies have been carrried out without UHV-based surface characterrization. Work with well-defined surfaces have been limited to (a) LEED of CO adlattices on Pt(III)[196] and Pd(111)[197] and (b) HREELS, LEED, TPD, and XPS of mixed H_2O–HF–CO layers generated in UHV by cryogenic adsorption at Pt(111) and Rh(111) surfaces.[137,138] An impressive amount of detailed information on a wide variety of complex organic compounds chemisorbed at well-defined Pt(111) and Pt(100) electrode surfaces has been furnished by LEED, AES, TPD, and HREELS.[29,82,198] Electrocatalytic reactivity studies which accompanied these investigations were limited to anodic oxidation reactions, and only correlations between the mode of adsorbate bonding and extent of anodic oxidation were attempted.

a. Carbon Monoxide. Much of what is known about the structure and reactivity of CO chemisorbed at single-crystal electrodes, and their dependencies on surface crystallographic orientation, electrode potential, and adsorbate coverage are based almost entirely upon *in situ* IRAS measurements.[181] Only two UHV-EC studies on CO have been reported. One made use of a well-defined Pt(111) surface and sought to correlate anodic peak potentials with observed LEED structures.[196] The other, based upon LEED, AES, TPD,

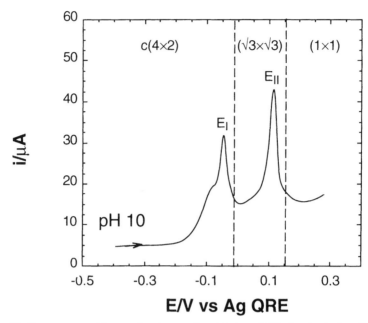

Figure 21. Current–potential curve for the anodic oxidation of CO chemisorbed on Pd(111), initially in the Pd(111)-c(4 × 2)–CO structure. The supporting electrolyte consisted of 0.1 M NaF and 0.1 mM NaOH. The shoulder in peak I is due to polycrystalline edge effects. Also indicated are the potential/coverage-dependent structures derived from the LEED patterns. The potential sweep rate was 2 mV/sec. (Reproduced from reference 197b, with permission.)

voltammetry, and coulometry, examined the chemisorption of CO at well-defined and anodically disordered Pd(111).[197] In the latter study, it was shown CO adsorption from solution onto a UHV-prepared Pd(111) surface yielded an ordered adlattice, Pd(111)-$c(4 \times 2)$–CO, in which the CO molecules occupy twofold hollow sites. Two CO-to-CO_2 anodic oxidation peaks were observed (Figure 21): The first peak is partial oxidation of the $c(4 \times 2)$ layer in which CO is chemisorbed only on twofold sites, followed by an adlattice reconstruction to form a Pd(111)$(\sqrt{3} \times \sqrt{3})R30°$–CO structure; the second peak is due to complete oxidative desorption of the Pd(111)$(\sqrt{3} \times \sqrt{3})R30°$–CO adlayer to yield a clean and well-ordered Pd(111) single-crystal surface.[197b] At the surface disordered by extensive anodic oxidation, chemisorption of CO occurred spontaneously but no ordered CO adlayers were produced; the CO molecules are thought to reside on atop sites.[197a]

b.   Other Organic Compounds. UHV-EC investigations have been undertaken to understand the nature of the chemical interactions between the organic molecule and the electrode surface as a function of interfacial parameters, such as pH and electrode potential, and also to correlate the mode of attachment with the reversible and/or catalytic electrochemistry of these materials.

The differences between gas-phase and solution-state chemisorption and catalytic hydrogenation of ethylene have been documented[199]: Variations in the structures of ethylene chemisorbed at the solid–solution and gas–solid interfaces lead to different reaction pathways. In solution, ethylene chemisorption occurs molecularly through its π-electron system, whereas chemisorption in UHV is accompanied by molecular rearrangements to form a surface ethylidyne species. In electrocatalytic hydrogenation, ethylene is reduced on the Pt surface by adsorbed H atoms; in gas-phase hydrogenation, H atoms must be transferred from the Pt surface through a layer of irreversibly adsorbed ethylidyne to ethylene adsorbed on top of the ethylidyne layer. Other studies[200–202] compared the electrocatalytic hydrogenation of ethylene at polycrystalline and well-defined Pt(111) and Pt(100) single crystals. Further work with alkenes[205–209] has focused on the effects of hydrocarbon chain length and the presence of weakly surface-active substituents such as carboxylates[210] and alcohols.[211] These studies showed that (i) the primary mode of surface coordination of terminal alkenes, alkenols, and alkenoic acids is through the π-electron system of the olefinic double bond and (ii) the pendant alkyl chain is always extended outward on top of the propylene moiety. This type of coordination is the same for the alkenols and alkenoic acids. Under favorable circumstances, intermolecular hydrogen bonding may occur within the alkenol layer,[211] or the carboxylate group may interact directly with the metal surface.[210] From coulometric measurements, it was concluded that electrocatalytic oxidation of the chemisorbed higher alkenes is limited largely to the olefinic anchor. Electrochemical oxidation of the lower alkenols (such as allyl alcohol) proceeds to completion, yielding only CO_2 and H_2O. Evidently, only groups that are in close proximity to the electrode surface undergo anodic oxidation.[210,211]

Early srudies with smooth polycrystalline based upon Pt thin-layer electrochemical techniques and *ex situ* IRAS indicated that aromatic compounds such as 1,4-dihydroxybenzene are chemisorbed in discrete, nonrandom orientations that depend upon interfacial factors such as temperature, concentration, and electrolyte coadsorption.[212–217] Experiments implemented with well-defined Pt(111) electrodes support the earlier findings, although the conditions at which the multiple orientational transitions occur are different for the polycrystalline and single-crystal electrodes.[29,81] The electrocatalytic oxidation of multiply oriented aromatic molecules has been shown to be strongly dependent on their initial adsorbed orientations.[218–220] For example, flat-adsorbed hydroquinone is oxidized completely to CO_2, while oxidation of the edge-oriented chemisorbed species is less extensive.

Sulfur-containing compounds investigated included thiophenol, penta-fluorothiophenol, 2,3,5,6-tetrafluorothiophenol, 2,3,4,5-tetrafluorothiophenol, 2,5-dihydroxythiophenol, 2,5-dihydroxy-4-methylbenzyl mercaptan, and benzyl mercaptan; chemisorption of these compounds occurs oxidatively through the sulfur group with loss of the sulfhydryl hydrogen.[89,193] The tethered diphenolic moieties in the adsorbed dihydroxythiophenols show reversible quinone/diphenol redox chemistry.

S-heterocyclic compounds studied were thiophene, bithiophene, and their carboxylate and methyl derivatives.[221] Experimental evidence indicates that these compounds are bound exclusively through the S heteroatom, although the chemisorption process is accompanied by desulfurization reactions; the extent of self-desulfurization increases as the adsorption potential is made more positive. The electropolymerization of 3-methylthiophene at clean Pt(111) and monomer-treated Pt(111) pretreated has been studied, and the properties of the two types of polymer film were compared.[221] In terms of the HREELS spectra, two major differences were noted which were attributed to changes in the physical nature of the polymer film, such as swelling or losses in reflectivity, and/or to excitation of phonon modes in the polymer.

The chemisorption of pyridine,[222] bipyridine,[223] multinitrogen hetero-aromatic compounds,[224] and their derivatives has been examined as a function of isomerism and substituents. Pyridine forms a well-ordered layer of admolecules chemisorbed through the N heteroatom in a tilted vertical orientation. The derivatives are coordinated similarly unless the ring nitrogen is sterically hindered such as in 2,6-dimethylpyridine, where chemisorption is in the flat orientation via the π-system of the aromatic ring. Pyrazine, pyrimidine, and pyridazine are chemisorbed through only one N heteroatom in a tilted-vertical orientation.[224] For their derivatives, adsorption occurs through the least hindered ring nitrogen. Carboxylate substituents located in positions *ortho* or *meta* to the N heteroatom interact with the Pt(111) surface at positive potentials, behavior similar to that shown by the corresponding pyridine carboxylates. The chemisrobed layers were disordered as indicated by the absence of LEED patterns and were observed to be electrochemically unreactive. The adsorption behavior of the bipyridyls is also sensitive to steric hindrance at the positions *ortho* to the N heteroatom.[223]

The mode of chemisorption at well-defined Pt(111) of L-dopa, L-tyrosine, L-cysteine, L-phenylalanine, alanine, and dopamine has been studied.[225,226] Except for L-phenylalanine, chemisorption occurs preferentially through the –SH moiety or the atomatic ring. This is as expected from what is known about the relative surface activities of various functional groups at polycrystalline Pt: It has been empirically determined via competitive chemisorption experiments that the strength of adsorption decreases in the following order:

$-SH$ > hetero N > quinone/diphenol ring > $C{=}C$ ⩾ benzene ring ⩾ amine N (pH 7) > $-OH$ > $-CO{=}O$.[216]

ACKNOWLEDGMENTS

In addition to reviewing the entire manuscript with Professor A. Wieckowski (University of Illinois); Professor D. A. Harrington (University of Victoria) re-wrote the section on work function measurements. Dr. R. J. Barriga (Texas A&M University) compiled the bibliography. MPS wishes to acknowledge the National Science Foundation (Presidential Young Investigator program, DMR-8958440) and the Robert A. Welch Foundation for research support. JLS acknowledges the financial assistance of the Office of the Chief of Naval Research (Grant N00014-91-J-1919).

REFERENCES

1. E. Yeager, *Surf. Sci.*, **101**, 1 (1980).

2. R. Parsons, *J. Electroanal. Chem.*, **150**, 51 (1983).

3. C. B. Duke, *Surf. Sci.*, **101**, 624 (1980).

4. E. Yeager, *J. Electroanal. Chem.*, **150**, 679 (1983).

5. T. E. Furtak, K. L. Kliewer, and D. W. Lynch, eds., *Non-Traditional Approaches to the Study of the Solid–Electrolyte Interface*, North-Holland, Amsterdam (1980).

6. W. N. Hansen, D. M. Kolb, and D. W. Lynch, eds., *Electronic and Molecular Structure of Electrode–Electrolyte Interfaces*, Elsevier, Amsterdam (1983).

7. J. D. E. McIntyre and M. J. Weaver, eds., *The Chemistry and Physics of Electrocatalysis*, The Electrochemical Society, Pennington (1984).

8. L. R. Faulkner, *In Situ Characterization of Electrochemical Processes*, National Academy Press, Washington, D.C. (1987).

9. M. P. Soriaga, ed., *Electrochemical Surface Science*, American Chemical Society, Washington, D.C. (1988).

10. E. Yeager and J. Kuta, in *Physical Chemistry, An Advanced Treatise*, Vol. IXA, H. Eyring, D. Henderson and W. Jost, eds., Academic, New York (1970).

11. A. J. Bard and L. R. Faulkner, *Electrochemical Methods*, Wiley, New York (1980).

12. M. P. Soriaga and A. T. Hubbard, *J. Electroanal. Chem.*, **167**, 79 (1984).

13. R. E. White, J. O'M. Bockris, B. E. Conway, and E. Yeager, *Comprehensive Treatise of Electrochemistry*, Vol. VII, Plenum, New York (1984).

14. G. A. Somorjai, *Chemistry in Two Dimensions: Surfaces*, Cornell University, Ithaca (1981).

15. T. H. Rhodin and G. Ertl, eds., *The Nature of the Surface Chemical Bond*, North-Holland, New York (1979).

16. G. Ertl and J. Kuppers, *Low Energy Electrons and Surface Chemistry*, VCH Publishers, New York (1985).

17. D. P. Woodruff and T. A. Delchar, *Modern Techniques of Surface Science*, Cambridge University Press, New York (1986).
18. A. Zangwill, *Physics at Surfaces*, Cambridge University Press, New York (1988).
19. M. A. Van Hove, S. W. Wang, D. F. Ogletree, and G. A. Somorjai, *Adv. Quantum Chem.*, **20**, 1 (1989).
20. E. L. Muetterties, *Bull. Chim. Belg.*, **84**, 959 (1975).
21. E. L. Muetterties, *Bull. Chim. Belg.*, **85**, 451 (1976).
22. E. Shustorovich, R. C. Baetzold, and E. L. Muetterties, *J. Phys. Chem.*, **87**, 1100 (1983).
23. R. Hoffman, *Solids and Surfaces: A Chemist's View of Bonding in Extended Structures*, VCH Publishers, New York (1988).
24. M. R. Albert and J. T. Yates, Jr., *A Surface Scientist's Guide to Organometallic Chemistry*, American Chemical Society, Washington, D.C. (1987).
25. A. B. Anderson, *J. Electroanal. Chem.*, **280**, 37 (1990).
26. A. B. Anderson, R. Kotz and E. Yeager, *Chem. Phys. Lett.*, **82**, 130 (1981).
27. S. P. Mehandru and A. B. Anderson, *J. Phys. Chem.*, **93**, 2044 (1989).
28. (a) A. T. Hubbard, *Accts. Chem. Res.* **13**, 177 (1980). (b) A. T. Hubbard, M. P. Soriaga, and J. L. Stickney, in *New Directions in Chemical Analysis*, B. Shapiro, ed., Texas A&M University Press, College Station, TX (1985).
29. A. T. Hubbard, *Chem. Rev.*, **88**, 633 (1988).
30. P. N. Ross, in *Chemistry and Physics of Solid Surfaces*, R. Vanselow and R. Howe, eds., Springer-Verlag, New York (1982).
31. P. N. Ross and F. T. Wagner, in *Advances in Electrochemistry and Electrochemical Engineering*, Vol. XIII, H. Gerischer and C. W. Tobias, eds., Wiley-Interscience, New York (1984).
32. E. Yeager, A. Homa, B. D. Cahan, and D. Scherson, *J. Vac. Sci. Technol.*, **20**, 628 (1982).
33. A. S. Homa, E. Yeager, and B. D. Cahan, *J. Electroanal. Chem.*, **150**, 181 (1983).
34. D. M. Kolb, *Z. Phys. Chem. Neue Folge*, **154**, 179 (1987).
35. D. M. Kolb, *Ber. Bunsenges. Phys. Chem.*, **92**, 1175 (1988).
36. M. P. Soriaga, *Prog. Surf. Sci.*, **39**, 325 (1992).
37. J. Augustynski and L. Balsenc, in "*Modern Aspects of Electrochemistry*, Vol. XIV, J. O'M. Bockris and B. E. Conway, eds., Plenum, New York (1979).
38. P. M. A. Sherwood, in "*Contemporary Topics in Analytical and Clinical Chemistry*, D. M. Hercules, G. M. Hieftje, L. R. Snyder, and M. E. Evenson, eds., Plenum, New York (1982).
39. P. M. A. Sherwood, *Chem. Soc. Rev.*, **14**, 1 (1985).
40. G. G. Will, *J. Electrochem. Soc.*, **112**, 451 (1965).
41. A. T. Hubbard, R. P. Ishikawa and J. Katekaru, *J. Electroanal. Chem.*, **86**, 271 (1978).
42. J. Clavilier, *J. Electroanal. Chem.*, **107**, 205 (1980).

43. T. Solomun, B. C. Schardt, S. D. Rosasco, A. Wieckowski, J. L. Stickney, and A. T. Hubbard, *J. Electroanal. Chem.*, **176**, 309 (1984).

44. R. Adzic, A. Tripkovic and V. Vesovic, *J. Electroanal. Chem.*, **204**, 329 (1986).

45. A. Rodes, J. M. Orts, J. M. Feliu, A. Aldaz and J. Clavilier, *J. Electroanal. Chem.*, **281**, 199 (1990).

46. R. Adzic, in *Modern Aspects of Electrochemistry*, Vol. XXI, R. E. White, J. O'M. Bockris, and B. E. Conway, eds., Plenum, New York (1990).

47. (a) C. A. Kittel, *Introduction to Solid-State Physics*, Wiley, New York (1971). (b) L. J. Clarke, *Surface Crystallography: An Introduction to Low Energy Electron Diffraction*, Wiley, New York (1985).

48. E. A. Wood, *Crystal Orientation Manual*, Columbia University Press, New York (1963).

49. B. Kaischew and B. Mutaftschiew, *Z. Phys. Chem.*, **204**, 334 (1955).

50. P. O. Nilsson and D. E. Eastman, *Phys. Scr.*, **8**, 113 (1973).

51. M. S. Zei, Y. Nakai, G. Lehmpfuhl, and D. M. Kolb, *J. Electroanal. Chem.*, **150**, 201 (1983).

52. C. E. D. Chidsey, D. N. Loiacono, T. Sleator, and S. Nakahara, *Surf. Sci.*, **200**, 45 (1988).

53. J. Clavilier, D. Armand, S. G. Sun, and M. Petit, *J. Electroanal. Chem.*, **205**, 267 (1986).

54. D. Aberdam, R. Durand, R. Faure, and F. El-Omar, *Surf. Sci.*, **171**, 303 (1986).

55. J. Clavilier, in *Electrochemical Surface Science*, M. P. Soriaga, ed., American Chemical Society, Washington, D.C. (1988).

56. D. M. Kolb, G. Lempfuhl, and M. S. Zei, *J. Electroanal. Chem.*, **179**, 289 (1984).

57. D. J. Trevor, C. E. D. Chidsey, and D. N. Loiacono, *Phys. Rev. Lett.*, **62**, 929 (1989).

58. E. Holland-Moritz, J. Gordon II, K. Kanazawa, and R. Sonnenfeld, *Langmuir*, **7**, 1981 (1991).

59. J. F. Rodriguez, M. E. Bothwell, G. J. Cali, and M. P. Soriaga, *J. Am. Chem. Soc.*, **112**, 7392 (1990).

60. G. J. Cali, G. M. Berry, M. E. Bothwell, and M. P. Soriaga, *J. Electroanal. Chem.*, **297**, 523 (1991).

61. M. E. Bothwell, G. J. Cali, G. M. Berry and M. P. Soriaga, *Surf. Sci.*, **249**, L322 (1991).

62. A. Bewick and B. Thomas, *J. Electroanal. Chem.*, **65**, 911 (1975).

63. J. L. Stickney, I. Villegas, and C. B. Ehlers, *J. Am. Chem. Soc.*, **111**, 6473 (1989).

64. H. H. Farrell and G. A. Somorjai, *Adv. Chem. Phys.*, **20**, 215 (1971).

65. M. A. van Hove and S. Y. Tong, *Surface Crystallography by LEED*, Springer-Verlag, New York (1979).

66. A. Wieckowski, B. C. Schardt, S. D. Rosasco, J. L. Stickney, and A. T. Hubbard, *Surf. Sci.*, **146**, 115 (1984).

67. A. W. Adamson, *Physical Chemistry of Surfaces*, Wiley, New York (1990).

68. J. F. Rodriguez, T. Mebrahtu, and M. P. Soriaga, *J. Electroanal. Chem.*, **264**, 291 (1989).

69. (a) H. D. Hagstrum and G. E. Becker, *J. Vac. Sci. Technol.*, **14**, 369 (1977). (b) J. H. Neave, B. A. Joyce, P. J. Dobson, and N. Norton, *Appl. Phys.*, **A31**, 1 (1983).

70. H. H. Madden, *Surf. Sci.*, **126**, 80 (1982).

71. D. R. Jennison, *J. Vac. Sci. Technol.*, **20**, 548 (1982).

72. R. R. Rye and J. E. Houston, *Accts. Chem. Res.*, **17**, 41 (1984).

73. K. W. Nebesny and N. R. Armstrong, *Langmuir*, **1**, 469 (1985).

74. K. W. Nebesny and N. R. Armstrong, *J. Electron Spectrosc.*, **37**, 355 (1986) 355.

75. M. C. Burrell and N. R. Armstrong, *Appl. Surf. Sci.*, **17**, 53 (1983).

76. R. E. Weber and W. T. Peria, *J. Appl. Phys.*, **38**, 4355 (1967).

77. J. M. Morabito, *Surf. Sci.*, **49**, 318 (1975).

78. W. A. Coghlan and R. E. Clausing, *At. Data*, **5**, 317 (1973).

79. J. Houston, *Surf. Sci.*, **38**, 283 (1973).

80. J. A. Schoeffel and A. T. Hubbard, *Anal. Chem.*, **49**, 2330 (1977).

81. B. C. Schardt, J. L. Stickney, D. A. Stern, D. G. Frank, J. Y. Katekaru, S. D. Rosasco, G. N. Salaita, M. P. Soriaga and A. T. Hubbard, *Inorg. Chem.*, **24**, 1419 (1985).

82. G. N. Salaita and A. T. Hubbard, in *Molecular Design of Electrode Surfaces*, R. W. Murray, ed., Wiley, New York (1989).

83. D. W. Suggs, I. Villegas, B. W. Gregory, and J. L. Stickney, *Mater. Res. Soc. Symp. Proc.*, **222**, 283 (1991).

84. M. Gryzinski, *Phys. Rev.*, **A138**, 305 (1965).

85. T. A. Carlson, *Photoelectron and Auger Spectroscopy*, Plenum, New York (1975).

86. C. S. Fadley, in *Electron Spectroscopy II*, C. R. Bundle and A. D. Baker, eds., Academic, New York (1978).

87. K. Siegbahn, *ESCA, Atomic Molecular and Solid State Structure Studies by Means of Electron Spectroscopy*, Almqvist-Wiksell Boktryckeri, Sweden (1967).

88. M. P. Seah, in *Practical Surface Analysis by Auger and X-Ray Photoelectron* Spectroscopy, D. Briggs and M. P. Seah, eds., Wiley, New York (1983).

89. M. P. Soriaga, *Chem. Rev.*, **90**, 771 (1990).

90. H. Ibach and D. A. Mills, *Electron Energy Loss Spectroscopy*, Academic, New York (1982).

91. N. R. Avery, in *Vibrational Spectroscopy of Molecules on Surfaces*, J. T. Yates, Jr., and T. E. Madey, eds., Plenum, New York (1987).

92. F. M. Hoffman, *Surf. Sci. Rep.*, **3**, 2 (1983).

93. B. E. Hayden, in *Vibrational Spectroscopy of Molecules on Surfaces*, J. T. Yates, Jr., and T. E. Madey, eds., Plenum, New York (1987).

94. H. A. Pearce and N. Sheppard, *Surf. Sci.*, **59**, 205 (1976).

95. A. Bewick and B. S. Pons, in *Advances in Infrared and Raman Spectroscopy*, R. J. H. Hester and R. E. Clark, eds., Wiley-Hayden, London (1985).

96. B. J. Barner, M. J. Green, E. I. Saez, and R. M. Corn, *Anal. Chem.*, **63**, 55 (1991).

97. S. Pons and A. Bewick, *Langmuir*, **1**, 141 (1985).

98. H. G. Tompkins, in *Methods of Surface Analysis*, A. Czanderna, ed., Elsevier, New York (1975).

99. N. Sheppard and T. T. Nguyen, in *Advances in Infrared and Raman Spectroscopy*, R. E. Hester and R. J. H. Clark, eds., Wiley-Hayden, London (1978).

100. W. Ho, R. R. Willis, and E. W. Plummer, *Phys. Rev. Lett.*, **40**, 1463 (1978).

101. S. C. Chang and M. J. Weaver, *J. Phys. Chem.*, **95**, 3391 (1991).

102. J. K. Foley and S. Pons, *Anal. Chem.*, **57**, 945A (1985).

103. H. Froitzheim, in *Topics in Current Physics*, Vol. IV, H. Ibach, ed., Springer-Verlag, New York (1977).

104. W. N. Hansen, *J. Electroanal. Chem.*, **150**, 133 (1983).

105. W. N. Hansen and G. J. Hansen, in *Electrochemical Surface Science*, M. P., Soriaga, ed., American Chemical Society, Washington, D.C. (1988).

106. P. A. Redhead, *Vacuum*, **12**, 203 (1962).

107. H. A. Strobel and W. R. Heineman, *Chemical Instrumentation*, Wiley, New York (1989).

108. M. E. Hanson and E. Yeager, in *Electrochemical Surface Science*, M. P. Soriaga, ed., American Chemical Society, Washington, D.C. (1988).

109. J. L. Stickney, C. B. Ehlers, and B. W. Gregory, in *Electrochemical Surface Science*, M. P. Soriaga, ed., American Chemical Society, Washington, D.C. (1988).

110. T. Mebrahtu, J. F. Rodriguez, M. E. Bothwell, I. F. Cheng, D. R. Lawson, J. R. McBride, C. R. Martin, and M. P. Soriaga, *J. Electroanal. Chem.*, **267**, 351 (1989).

111. T. Solomun, W. Richtering, and H. Gerischer, *Ber. Bunsenges. Phys. Chem.*, **91**, 412 (1987).

112. K. R. Zavadil and N. R. Armstrong, *J. Electrochem. Soc.*, **137**, 2371 (1990).

113. L.-W. H. Leung, T. W. Gregg, and D. W. Goodman, *Rev. Sci. Instrum.*, **62**, 1857 (1991).

114. B. E. Conway, H. Angerstein-Kozlowska, W. B. A. Sharp, and E. E. Criddle, *Anal. Chem.*, **45**, 1331 (1973).

115. W. Hansen and D. M. Kolb, *J. Electroanal. Chem.*, **100**, 493 (1979).

116. D. Rath and D. M. Kolb, *Surf. Sci.*, **109**, 641 (1981).

117. G. J. Hansen and W. N. Hansen, *J. Electroanal. Chem.*, **150**, 193 (1983).

118. W. N. Hansen. *Surf. Sci.*, **101**, 109 (1980).

119. J. D. E. McIntyre, *Adv. Electrochem. Electrochem. Eng.*, **9**, 61 (1973).

120. W. Hansen, D. M. Kolb, D. Rath, and R. Wille, *J. Electroanal. Chem.*, **110**, 369 (1980).

121. O. Hofman, K. Doblhofer, and H. Gerischer, *J. Electroanal. Chem.*, **161**, 337 (1984).

122. G. J. Hansen and W. N. Hansen, *Ber. Bunsenges. Phys. Chem.*, **91**, 317 (1987).

123. J. E. Pemberton, R. L. Sobocinksi, and M. A. Bryant, *J. Am. Chem. Soc.*, **112**, 6177 (1990).

124. J. E. Pemberton and R. L. Sobocinksi, *J. Electroanal. Chem.*, **318**, 157 (1991).

125. E. R. Koetz, H. Neff, and R. H. Muller, *J. Electroanal. Chem.*, **215**, 331 (1986).

126. S. D. Rosasco, J. L. Stickney, G. N. Salaita, D. G. Frank, J. Y. Katekaru, B. C. Schardt, M. P. Soriaga, D. A. Stern, and A. T. Hubbard, *J. Electroanal. Chem.*, **188**, 95 (1985).

127. D. G. Frank, J. Y. Katekaru, S. D. Rosasco, G. N. Salaita, B. C. Schardt, M. P. Soriaga, J. L. Stickney, and A. T. Hubbard, *Langmuir*, **1**, 587 (1985).

128. T. E. Felter and A. T. Hubbard, *J. Electroanal. Chem.*, **100**, 473 (1979).

129. G. A. Garwood, Jr., and A. T. Hubbard, *Surf. Sci.*, **92**, 617 (1980).

130. B. C. Schardt, S. L. Yau, and F. Rinaldi, *Science*, **243**, 1050 (1989).

131. D. A. Stern, L. Laguren-Davidson, F. Lu, C. H. Lin, D. G. Frank, G. N. Salaita, N. Walton, J. Y. Gui, D. C. Zapien, and A. T. Hubbard, *J. Am. Chem. Soc.*, **111**, 877 (1989).

132. T. E. Madey and F. P. Netzer, *Surf. Sci.*, **117**, 549 (1980).

133. J. K. Sass, K. Kretzschmar, and S. Holloway, *Vacuum*, **31**, 483 (1981).

134. J. K. Sass, *Vacuum*, **33**, 741 (1983).

135. J. K. Sass and K. Bange, in *Electrochemical Surface Science*, M. P. Soriaga, ed., American Chemical Society, Washington, D.C. (1988).

136. J. K. Sass, K. Bange, R. Dohl, E. Plitz, and R. Unwin, *Ber. Bunsenges. Phys. Chem.*, **88**, 354 (1984).

137. F. T. Wagner, S. J. Schmieg, and T. E. Moylan, *Surf. Sci.*, **195**, 403 (1988).

138. F. T. Wagner and T. E. Moylan, in *Electrochemical Surface Science*, M. P. Soriaga, ed., American Chemical Society, Washington, D.C. (1988).

139. D. A. Stern, H. Baltruschat, M. Martinez, J. L. Stickney, D. Song, S. K. Lewis, D. G. Frank, and A. T. Hubbard, *J. Electroanal. Chem.*, **217**, 101 (1987).

140. G. N. Salaita, D. A. Stern, F. Lu, H. Baltruschat, B. C. Schardt, J. L. Stickney, M. P. Soriaga, D. G. Frank, and A. T. Hubbard, *Langmuir*, **2**, 828 (1986).

141. F. Lu, G. N. Salaita, H. Baltruschat, and A. T. Hubbard, *J. Electroanal. Chem.*, **222**, 305 (1987).

142. N. Batina, J. W. McCarger, L. Laguren-Davidson, C.-H. Lin, and A. T. Hubbard, *Langmuir*, **5**, 123 (1989).

143. J. L. Stickney, S. D. Rosasco, G. N. Salaita, and A. T. Hubbard, *Langmuir*, **1**, 66 (1985).

144. G. A. Garwood and A. T. Hubbard, *Surf. Sci.*, **112**, 281 (1982).

145. G. A. Garwood and A. T. Hubbard, *Surf. Sci.*, **121**, 396 (1982).

146. T. Solomun, A. Wieckowski, S. D. Rosasco, and A. T. Hubbard, *Surf. Sci.*, **147**, 241 (1984).

147. C. B. Ehlers and J. L. Stickney, *Surf. Sci.*, **239**, 85 (1990).

148. G. N. Salaita, F. Lu, L. Laguren-Davidson, and A. T. Hubbard, *J. Electroanal. Chem.*, **229**, 1 (1987).

149. (a) B. G. Bravo, S. L. Michelhaugh, M. P. Soriaga, I. Villegas, D. W. Suggs, and J. L. Stickney, *J. Phys. Chem.*, **95**, 5245 (1991). (b) T. Mebrahtu, M. E. Bothwell, J. E. Harris, G. J. Cali, and M. P. Soriaga, *J. Electroanal. Chem.*, **300**, 487 (1991).

150. P. N. Ross, in *Electrochemical Surface Science*, M. P. Soriaga, ed., American Chemical Society, Washington, D.C. (1988).

151. W. E. O'Grady, M. Y. C. Woo, P. L. Hagans, and E. Yeager, *J. Vac. Sci. Technol.*, **14**, 365 (1977).

152. K. Yamamoto, D. M. Kolb, R. Kotz, and G. Lempfuhl, *J. Electroanal. Chem.*, **96**, 233 (1979).

153. P. N. Ross, *J. Electroanal. Chem.*, **126**, 67 (1979).

154. A. S. Homa, E. Yeager, and B. D. Cahan, *Electroanal. Chem.*, **150**, 181 (1983).

155. J. Clavilier, A. Rodes, K. El Achi, and M. A. Zamakhchari, *J. Chim. Phys.*, **88**, 1291 (1991).

156. P. N. Ross, *J. Chim. Phys.*, **88**, 1353 (1991).

157. G. Jerkiewicz and B. E. Conway, *J. Chim. Phys.*, **88**, 1291 (1991).

158. F. T. Wagner and P. N. Ross, *Surf. Sci.*, **160**, 305 (1985).

159. J. Horkans, B. D. Cahan, and E. Yeager, *Surf. Sci.*, **37**, 559 (1973).

160. F. T. Wagner and P. N. Ross, *J. Electroanal. Chem.*, **150**, 141 (1985).

161. J. B. Lumsden, G. A. Garwood, and A. T. Hubbard, *Surf. Sci.*, **121**, 1524 (1982).

162. D. A. Harrington, A. Wieckowski, S. D. Rosasco, B. C. Schardt, G. N. Salaita, J. B. Lumsden, and A. T. Hubbard, *Corrosion Sci.*, **25**, 849 (1985).

163. D. A. Harrington, A. Wieckowski, S. D. Rosasco, B. C. Schardt, G. N. Salaita, and A. T. Hubbard, *Langmuir*, **1**, 232 (1985).

164. J. P. Biberian and G. A. Somorjai, *J. Vac. Sci. Technol.*, **16**, 2073 (1979).

165. J. S. Hammond and N. Winograd, *J. Electroanal. Chem.*, **80**, 123 (1977).

166. J. S. Hammond and N. Winograd, *J. Electrochem. Soc.*, **124**, 826 (1977).

167. A. T. Hubbard, J. L. Stickney, S. D. Rosasco, M. P. Soriaga, and D. Song, *J. Electroanal. Chem.*, **150**, 165 (1983).

168. J. L. Stickney, S. D. Rosasco, M. P. Soriaga, D. Song, and A. T. Hubbard, *Surf. Sci.*, **130**, 326 (1983).

169. J. L. Stickney, S. D. Rosasco, B. C. Schardt, and A. T. Hubbard, *J. Phys. Chem.*, **88**, 251 (1984).

170. J. L. Stickney, S. D. Rosasco, and A. T. Hubbard, *J. Electrochem. Soc.*, **131**, 260 (1984).

171. J. L. Stickney, D. A. Stern, B. C. Schardt, D. C. Zapien, A. Wieckowski, and A. T. Hubbard, *J. Elctroanal. Chem.*, **213**, 293 (1986).

172. J. L. Stickney, B. C. Schardt, D. A. Stern, A. Wieckowski, and A. T. Hubbard, *J. Electrochem. Soc.*, **133**, 648 (1986).

173. B. C. Schardt, J. L. Stickney, D. A. Stern, A. Wieckowski, D. C. Zapien, and A. T. Hubbard, *Langmuir*, **3**, 239 (1987).

174. B. C. Schardt, J. L. Stickney, D. A. Stern, A. Wieckowski, D. C. Zapien, and A. T. Hubbard, *Surf. Sci.*, **175**, 520 (1986).

175. P. C. Andricacos and P. N. Ross, *J. Electroanal. Chem.*, **167**, 301 (1984).

176. D. Aberdam, R. Durand, R. Faure, and F. El-Omar, *Surf. Sci.*, **162**, 782 (1985).

177. L.-W. H. Leung, T. W. Gregg, and D. W. Goodman, *Langmuir*, **7**, 3205 (1991).

178. L. Laguren-Davidson, F. Lu, G. N. Salaita, and A. T. Hubbard, *Langmuir*, **4**, 224 (1988).

179. B. W. Gregory and J. L. Stickney, *J. Electroanal. Chem.*, **300**, 543 (1991).

180. D. W. Suggs and J. L. Stickney, *J. Phys. Chem.*, **95**, 10056 (1991).

181. M. P. Soriaga, in *Frontiers of Electrochemistry II*, P. N. Ross and J. Lipkowski, eds., VCH, New York (1992).

182. J. O'M. Bockris and A. K. N. Reddy, *Modern Electrochemistry*, Plenum, New York (1970).

183. B. B. Damaskin and A. N. Frumkin, *Electrochim. Acta*, **19**, 173 (1974).

184. R. Parsons, *J. Electroanal. Chem.*, **59**, 229 (1975).

185. P. A. Thiel and T. E. Madey, *Surf. Sci. Rep.*, **7**, 211 (1987).

186. C. K. Mann and K. K. Barnes, *Electrochemical Reactions in Nonaqueous Solvents*, Marcel Dekker, New York (1970).

187. H. Ohtani, C. T. Kao, M. A. Van Hove, and G. A. Somorjai, *Prog. Surf. Sci.*, **23**, 155 (1987).

188. G. A. Garwood, Jr., and A. T. Hubbard, *Surf. Sci.*, **118**, 233 (1982).

189. J. Y. Gui, F. Lu, D. A. Stern, and A. T. Hubbard, *J. Electroanal. Chem.*, **292**, 245 (1990).

190. B. B. Damaskin, O. A. Petrii, and V. V. Batrakov, *Adsorption of Organic Compounds on Electrodes*, Plenum, New York (1971).

191. G. Horanyi, *J. Electroanal. Chem.*, **51**, 163 (1974).

192. R. D. Snell and A. G. Keenan, *Chem. Soc. Rev.*, **8**, 259 (1979).

193. A. Wieckowski, *Electrochim. Acta*, **26**, 1121 (1981).

194. J. L. Stickney, M. P. Soriaga, A. T. Hubbard, and S. E. Anderson, *J. Electroanal. Chem.*, **125**, 73 (1981).

195. S. L. Michelhaugh, C. Bhardwaj, G. J. Cali, B. G. Bravo, M. E. Bothwell, G. M. Berry, and M. P. Soriaga, *Corrosion*, **47**, 322 (1991).

196. D. Zurawski, M. Wasberg, and A. Wieckowski, *J. Phys. Chem.*, **94**, 2076 (1990).

197. (a) G. M. Berry, M. E. Bothwell, S. L. Michelhaugh, J. R. McBride, and M. P. Soriaga, *J. Chim. Phys.*, **88**, 1591 (1991). (b) G. M. Berry, J. R. McBride, J. A. Schimpf, and M. P. Soriaga. *J. Electroanal. Chem.*, **353**, 281 (1993).

198. A. T. Hubbard and J. Y. Gui, *J. Chim. Phys.*, **88**, 1547 (1991).

199. A. Wieckowski, S. D. Rosasco, G. N. Salaita, A. T. Hubbard, B. E. Bent, F. Zaera, D. Godbey, and G. A. Somorjai, *J. Am. Chem. Soc.*, **107**, 5910 (1985).

200. A. T. Hubbard, M. A. Young, and J. A. Schoeffel, *J. Electroanal. Chem.*, **114**, 273 (1980).

201. F. Zaera and G. A. Somorjai, *J. Am. Chem. Soc.*, **106**, 2288 (1984).

202. M. Hourani and A. Wieckowski, *Langmuir*, **6**, 379 (1990).

203. N. Batina, S. A. Chaffins, J. Y. Gui, F. Lu, J. W. McCargar, J. W. Rovang, D. A. Stern, and A. T. Hubbard, *J. Electroanal. Chem.*, **284**, 81 (1990).

204. N. Batina, J. W. McCargar, C. H. Lin, G. N. Salaita, B. E. Kahn, and A. T. Hubbard, *Electroanalysis*, **1**, 213 (1989).

205. J. Y. Gui, D. A. Stern, D. C. Zapien, G. N. Salaita, F. Lu, C. H. Lin, B. E. Kahn, and A. T. Hubbard, *J. Electroanal. Chem.*, **252**, 169 (1988).

206. B. E. Kahn, S. A. Chaffins, J. Y. Gui, F. Lu, D. A. Stern, and A. T. Hubbard, *Chem. Phys.*, **141**, 21 (1990).

207. G. N. Salaita, C. Lin, P. Gao, and A. T. Hubbard, *Arabian J. Sci. Eng.*, **15**, 319 (1990).

208. J. Y. Gui, L. Laguren-Davidson, C. H. Lin, F. Lu, G. N. Salaita, D. A. Stern, B. E. Kahn, and A. T. Hubbard, *Langmuir*, **5**, 819 (1989).

209. N. Batina, D. C. Zapien, F. Lu, C. H. Lin, McCargar, B. E. Kahn, J. Y. Gui, D. G. Frank, G. N. Salaita, D. A. Stern, and A. T. Hubbard, *Electrochim. Acta*, **34**, 1031 (1989).

210. S. A. Chaffins, J. Y. Gui, C. H. Lin, F. Lu, G. N. Salaita, D. A. Stern, B. E. Kahn, and A. T. Hubbard, *J. Electroanal. Chem.*, **284**, 67 (1990).

211. M. P. Soriaga and A. T. Hubbard, *J. Am. Chem. Soc.*, **104**, 2735 (1982).

212. M. P. Soriaga and A. T. Hubbard, *J. Am. Chem. Soc.*, **104**, 2742 (1982).

213. M. P. Soriaga and A. T. Hubbard, *J. Am. Chem. Soc.*, **104**, 3937 (1982).

214. M. P. Soriaga, P. H. Wilson, A. T. Hubbard, and C. S. Benton, *J. Electroanal. Chem.*, **142**, 317 (1982).

215. M. P. Soriaga, J. H. White, and A. T. Hubbard, *J. Phys. Chem.*, **87**, 3048 (1983).

216. M. P. Soriaga, E. Binamira-Soriaga, A. T. Hubbard, J. B. Benziger, and K. W. P. Pang, *Inorg. Chem.*, **24**, 65 (1985).

217. M. P. Soriaga, J. H. White, V. K. F. Chia, D. Song, P. O. Arrhenius, and A. T. Hubbard, *Inorg. Chem.*, **24**, 73 (1985).

218. M. P. Soriaga, J. L. Stickney, and A. T. Hubbard. *J. Mol. Catal.*, **21**, 211 (1983).

219. M. P. Soriaga, J. L. Stickney, and A. T. Hubbard, *J. Electroanal. Chem.*, **144**, 207 (1983).

220. M. P. Soriaga and A. T. Hubbard, *J. Phys. Chem.*, **88**, 1758 (1984).

221. N. Batina, B. E. Kahn, J. Y. Gui, F. Lu, J. W. McCargar, H. B. Mark, C. H. Lin, B. N. Salaita, H. Zimmer, D. A. Stern, and A. T. Hubbard, *Langmuir*, **5**, 588 (1989).

222. D. A. Stern, L. Laguren-Davidson, F. Lu, C. H. Lin, D. G. Frank, G. N. Salaita, N. Walton, J. Y. Gui, D. C. Zapien, and A. T. Hubbard, *J. Am. Chem. Soc.*, **111**, 877 (1989).

223. S. A. Chaffins, J. Y. Gui, B. E. Kahn, C. H. Lin, F. Lu, G. N. Salaita, D. A. Stern, D. C. Zapien, A. T. Hubbard, and C. M. Elliott, *Langmuir*, **6**, 951 (1990).

224. S. A. Chaffins, J. Y. Gui, C. H. Lin, F. Lu, G. N. Salaita, D. A. Stern, and A. T. Hubbard, *Langmuir*, **6**, 1273 (1990).

225. A. T. Hubbard, D. G. Frank, D. A. Stern, M. J. Tarlov, N. Batina, N. Walton, E. Wellner, and J. W. McCargar, in *Redox Chemistry and Interfacial Behavior of Biological Molecules*, G. Dryhurst and R. Niki, eds., Plenum, New York (1988).

226. D. A. Stern, N. Walton, J. W. McCargar, G. N. Salaita, L. Laguren-Davidson, F. Lu, C. H. Lin, J. Y. Gui, N. Batina, D. G. Frank, and A. T. Hubbard, *Langmuir*, **4**, 711 (1988).

CHAPTER

2

SPECTROSCOPIC METHODS FOR THE CHARACTERIZATION OF ELECTROCHEMICAL INTERFACES AND SURFACES

JEANNE E. PEMBERTON and SEAN D. GARVEY

Department of Chemistry
University of Arizona
Tucson, Arizona 85721

Modern Techniques in Electroanalysis, Edited by Petr Vanýsek, Chemical Analysis Series, Vol. 139.
ISBN 0-471-55514-2 © 1996 John Wiley & Sons, Inc.

I. INTRODUCTION

Electrochemical methods rely on the measurement of electrical parameters such as current, voltage, charge, and capacitance for the description of electrochemical systems. Although considerable information is available from the thermodynamic and kinetic measures provided by common electrochemical methods, such electrical signatures are generally devoid of detailed molecular content. Electrochemical methods also suffer from limited selectivity based on the similarity of $E^{0'}$ values of many systems and the breadth in potential of the current response in a common electrochemical experiment such as voltammetry. Finally, electrochemical measures are very susceptible to interference from electrical background signals that are the result of charging phenomena (the so-called nonfaradaic response) or electrochemical responses from interfering redox-active species in the system. The problems associated with this electrical background become most severe in the most demanding measurements in which the redox species of interest are present in very low concentrations.

Many of the limitations of purely electrochemical methods have been mitigated by the successful coupling of spectroscopic methods with electrochemical systems over the past several decades. Indeed, significant advances in the understanding of electrochemical phenomena have resulted from such spectroelectrochemical approaches. Electrochemical redox pathways and processes have been elucidated through definitive spectroscopic identification of reactants, products, intermediates, and adsorbed species. Moreover, the use of such spectroelectrochemical modes has been particularly fruitful because of the sensitivity of electron transfer events to the structural, electronic, and molecular aspects of the electrochemical interface. Given the important role that electrochemical phenomena play today in industries such as sensors, corrosion, battery technology, energy storage devices, and the electrosynthetic production of specialty chemicals, continued interest in fundamental elucidation of electrochemical processes is anticipated. Spectroelectrochemical methods will undoubtedly play an important role in such fundamental efforts.

II. OVERVIEW OF SPECTROELECTROCHEMICAL METHODS

A. Ultraviolet–Visible Spectroelectrochemical Methods

A large number of spectroscopic tools have been successfully used with electrochemical systems. These have been reviewed adequately in the literature[1-16] and are listed in Table 1. Early efforts to couple electrochemistry and

Table 1. **Spectroelectrochemical Methods**

Spectroelectrochemical method	Representative References
UV–vis transmission methods	1–19
Optically transparent electrodes (OTEs)	1, 2
Minigrid electrodes	1
Optically transparent thin layer electrochemical (OTTLE) cell	3
Long–pathlength methods (parallel absorption; diffractive spectroelectrochemistry)	14–16
Derivative cyclic voltabsorptometry (DCVA)	17–19
UV–vis reflectance methods	
Internal reflectance	20
External (specular) reflectance	21, 22
Ellipsometry	23–25
Fluorescence	16, 26
Radioactive labeling	27, 28
Photoacoustic and photothermal methods	29–34
NMR spectroscopy	35
EPR spectroscopy	36
Mass spectrometry	37, 38
X-ray methods	39–42
Absorption methods (EXAFS; x-ray standing waves)	
Diffraction methods	
Attenuated total reflection IR spectroscopy	43–50
External reflectance IR methods	51–72
Electrochemically modulated IR reflectance spectroscopy (EMIRS)	51, 52
Subtractively normalized interfacial FTIR spectroscopy (SNIFTIRS)	53–56
IR reflectance–absorbance spectroscopy (IRRAS)	57, 58
Polarization–modulated FT-IRRAS (PM-IRRAS)	59–61
Potential difference alteration IR spectroscopy (PDAIRS)	62
Single potential alteration IR spectroscopy (SPAIRS)	63–65
Raman spectroscopy methods	73–94
Resonance Raman spectroscopy	73–78
Surface enhanced Raman scattering (SERS)	79–93
Normal Raman spectroscopy	94
Surface plasmon polariton enhanced Raman scattering (SPPERS)	95, 96
Nonlinear spectroscopic methods	97–99
Second harmonic generation (SHG)	97, 98
Surface enhanced hyper-Raman scattering (SEHRS)	99

spectroscopy utilized ultraviolet–visible (UV–vis) transmission spectroscopy at optically transparent electrodes (OTEs) and minigrid electrodes[1,2,4–6] and in optically transparent thin-layer electrochemical (OTTLE) cell arrangements.[3–6] In these experiments, the incident light beam was passed through the electrochemical interface in a direction normal to the electrode surface. These configurations overcome the opaque nature of common bulk electrode materials through the use of either partially transparent, thin (10–500 mm) metal electrodes or metal wire mesh "minigrid" electrodes.

Although extremely useful for the characterization of the UV-vis spectroscopy of interfacial reactants, products or intermediates of redox reactions, the methods based on optically transparent electrode systems suffer from poor sensitivity as a result of the very small pathlength associated with such measurements. More recent developments designed to overcome this limitation are based on spectroscopic measurements in which the light beam is directed parallel to the electrode surface. This approach gave rise to the parallel absorption and diffractive spectroelectrochemistry methods of the early 1980s.[9,14–16]

One other development of note in UV–vis transmission spectroelectrochemical methods was the derivative cyclic voltabsorptometry (DCVA) method.[17–19] In this approach, the derivative of the absorbance as a function of potential is determined. When plotted as a function of potential, this derivative has the shape of the current–potential response for the redox species of interest. This result occurs because the absorbance, whose electrical analogue is the charge, is an integral measure of the amount of the absorbing redox species produced. The derivative of the charge-potential function yields the current–potential response. Similarly, the derivative of the absorbance-potential function gives rise to the optical analogue of the current–potential response. This method possesses the unique advantage that all of the signal measured results from the redox species of interest. Thus, because of the selectivity of the spectroscopic measurement relative to the electrochemical measurement, this technique is much less susceptible to interferences from nonfaradaic and unwanted faradaic processes.

Numerous reflectance spectroscopies in the UV–vis region have also been used to investigate electrochemical systems. Internal reflectance spectroscopy has been coupled with electrochemistry at either transparent semiconductor electrodes or at semitransparent thin metal film electrodes on internal reflection elements.[20] External reflectance methods[21,22] and ellipsometry[23–25] have been used to study electrochemical processes at bulk metal and semiconductor electrodes. Ellipsometry has been particularly useful for the study of electrode surface films and adsorbed layers.

B.   Spectroelectrochemical Methods Based on Emission, Resonance, and X-Ray Methods

Other spectroscopic tools which have been coupled with electrochemistry include fluorescence[16,26] and radioactive labeling.[27,28] The latter approach has been extremely successful in the quantification of adsorbed species, especially in cases in which the adsorbates of interest are not redox active or the electrochemical measurement of surface coverage suffers from interfering redox activity of other species. Photoacoustic and photothermal techniques have been employed to study electrochemical interfaces as well.[29-34]

Techniques which are commonly employed for the study of bulk solution systems such as nuclear magnetic resonance (NMR) spectroscopy,[35] electron spin resonance (ESR) spectroscopy,[36] and mass spectrometry[37,38] have been used *ex situ* for the characterization of products and intermediates of electrochemical reactions. In such studies, bulk electrolysis is allowed to proceed for a period of time *in situ*, and then the products are analyzed by these methods. These techniques generally provide information rich in molecular detail. However, after considerable excitement about these methods in the 1970s, the interest in these techniques has decreased, although advances in the application of mass spectrometry to electrochemistry have been reviewed recently.[38] Therefore, these techniques will not be considered further in this chapter.

Methods based on the use of x-rays have emerged as important spectroelectrochemical tools during the past decade as a result of the availability of synchrotron radiation sources of intense x-rays. Techniques based on the coupling of x-ray absorption and diffraction methods with electrochemical systems have been implemented and shown to provide important structural information about electrode surfaces and electrochemical interfaces.[39-42]

C. Vibrational Spectroelectrochemical Methods

Much of the development of spectroelectrochemical methods during the past decade has focused on coupling vibrational spectroscopic methods to electrochemical systems, particularly for the *in situ* assessment of electrode interfacial and adsorbed species structure. Vibrational tools are somewhat more desirable than their electronic spectroscopic counterparts when detailed information about molecular structure is desired. Techniques which appear to be most promising in this regard are infrared (IR) and Raman spectroscopies, and these are the main focus of this chapter.

The potential of these approaches lies in the extreme sensitivity of molecular vibrations to chemical environment. In terms of applications to *in situ* electrochemical systems, promising experimental approaches must meet several criteria: adequate sensitivity to detect species on the order of 10^{14} to 10^{15} molecules cm^{-2}, applicability to a wide range of experimental systems, and interfacial selectivity. This last criterion is critical for minimal spectral interference from the solvent and from species in bulk solution, which are generally at much higher concentration levels than those confined to the interfacial region. Without interfacial selectivity, adsorbate spectra are observed as only weak features, which may be shifted in position from those of the solution components due to interaction with the surface. These goals are being realized through recent application of IR and Raman spectroelectrochemical methodologies.

1. Infrared Spectroelectrochemical Methods

Long accepted as a routine vibrational spectroscopic tool, IR spectroscopy was initially identified as a desirable method for coupling with electrochemical systems, based on the sensitivity of functional groups to chemical environment. Early IR spectroelectrochemical approaches relied on internal reflection or attenuated total reflection (ATR) for the study of the electrochemical interfaces at Ge electrodes.[43-45] This electrode material was the only one found to be both conductive and IR transparent. Thus, this method was severely limited by the poor IR transmission characteristics of most commonly used electrode materials. Internal reflection approaches have been more recently coupled with metallic electrodes through the use of the Kretschmann ATR prism configuration in which a thin Ag film is deposited onto an Si ATR prism,[46,47] as well as through the use of Si multiple reflection elements coated with metals such as Pt, Fe, Ag, and Au.[48-50]

More recently, successful IR spectroelectrochemical probes have focused mainly on external reflection approaches and have led to a variety of techniques which differ only slightly in the acquisition protocol used to discriminate against absorption by the bulk electrolyte solution. Thus, electrochemists have the following IR spectroelectrochemical techniques. Electrochemically modulated IR spectroscopy (EMIRS) relies on rapid potential alteration during spectral acquisition on a dispersive spectrometer to result in a difference spectrum.[51,52] Subtractively normalized interfacial Fourier transform (FT) IR spectroscopy (SNIFTIRS),[53-56] potential difference IR spectroscopy (PDIRS),[57] and single potential alteration IR spectroscopy (SPAIRS)[58-60] similarly rely on the potential-induced difference in interfacial spectral response measured using an FTIR spectrometer. Infrared reflectance–absorbance spectroscopy (IRRAS) generates spectra of interfacial and surface species through reliance on the reflectivity differences of incident light of alternating polarization.[61-65] Collectively, these techniques have facilitated the characterization of many electrochemical systems. This work has been reviewed extensively elsewhere.[66-71] Selected examples of recent work are highlighted below in this chapter.

2. *Raman Spectroelectrochemical Methods*

Raman spectroscopy provides several attractive features that address certain limitations of IR spectroelectrochemical techniques. First, Raman scattering in aqueous media is very weak. Therefore, many common aqeuous electrochemical systems are readily amenable to study with Raman spectroscopy. Moreover, the low-frequency region down to $\sim 5\text{--}10 \, \mathrm{cm}^{-1}$ is accessible in Raman spectroscopic investigations. Thus, despite the poor sensitivity normally inherent in light scattering methods, Raman spectroscopy has the potential to provide a wealth of information not readily obtainable with other methods.

Several enhancement approaches have been used to successfully overcome the inherent sensitivity limitations of Raman spectroscopy for the study of interfacial species in electrochemical systems. Van Duyne and coworkers[73-78] were the first to demonstrate the power of resonance Raman spectroscopy for the study of electrochemically generated species of a transient nature. Since this original work, other researchers have utilized resonance Raman spectroscopy to study monolayer and submonolayer quantities of adsorbed species in electrochemical systems.

A second enhancement approach, which allows for the investigation of the electrochemical double layer and which plays a prominent role in the work reported in this chapter, is surface enhanced Raman scattering or SERS. The origins of SERS are intimately connected with electrochemical systems and date back to pioneering work by Fleischmann et al.[79] In this 1973 paper, they reported the use of Raman spectroscopy to study Hg_2Cl_2, Hg_2Br_2, and HgO on a thin-film Hg electrode supported on a Pt substrate. They were able to obtain Raman spectra of quite reasonable signal-to-noise ratio (S/N) from thin films (two or more monolayers) of these mercurous halide and mercuric oxide species on relatively high surface area electrodes. The spectra resembled those of the respective bulk compounds, and although not particularly exciting from a surface chemistry perspective, these experiments did demonstrate the feasibility of using Raman spectroscopy as a probe of electrochemical interfaces.

These researchers later expanded this approach to the study of Ag electrodes using pyridine as a surface adsorbed probe molecule.[80] Their approach was to increase the number of surface pyridine molecules sampled in the Raman scattering experiment by increasing the surface area of the electrode by electrochemical roughening through multiple oxidation–reduction cycles in an aqueous Cl^- media. In their report of this work in 1974, they presented spectra of very good quality for adsorbed pyridine on these roughened Ag electrodes.

Recognizing that the spectra of Fleischmann et al. were of higher S/N than could be rationalized on the basis of an increase in surface area, Jeanmaire and

Van Duyne[81] and later Albrecht and Creighton[82] followed this original study with more careful investigations of this observation. Both groups independently recognized that some form of enhancement, responsible for a 10^5 to 10^6 increase in intensity relative to an an equivalent amount of pyridine in solution, was operable in these systems.

Since the first report of surface enhancement, a staggering amount of research has been undertaken with the intent of elucidating the mechanisms responsible for SERS and using this phenomenon to study a wide variety of solid–liquid, solid–gas, and solid–solid interfaces. A significant fraction of the research on SERS has been performed in electrochemical environments. Since the original observations at Ag by Van Duyne and Creighton, SERS has been extended to other metal electrodes as well, most notably Cu and Au. Indeed, thousands of papers in this area have been published in the last decade.

Electrochemical systems too numerous to detail here have been studied using SERS, although SERS studies in electrochemical systems have been reviewed.[83–93] However, the requirement for surface roughness in order to realize the largest enhancements and the limited applicability of SERS in terms of metals have been viewed unfavorably, especially in light of the recent tendency of surface electrochemists to focus more effort on well-defined single-crystal electrodes. The sensitivity problems associated with Raman scattering have been traditionally viewed as a fatal impediment to the success of *in situ* Raman experiments on nonenhancing polycrystalline and single-crystal electrode surfaces. However, in 1988, Shannon and Campion[94] published a study that began to erode this myth. In this study, Raman spectra of 4-cyanopyridine adsorbed at a nonenhancing polycrystalline Rh electrode were acquired *in situ*. Verification that the signal came from the surface confined 4-cyanopyridine species was provided by the observation that the spectrum could be observed only for light p-polarized with respect to the plane of incidence.

Recent work of Furtak is also particularly noteworthy in this regard. Taking advantage of surface plasmon polariton enhanced Raman scattering (SPPERS) in a Kretschmann geometry, Byahut and Furtak[95,96] were able to acquire Raman spectra on Ag(111) single-crystal electrode surfaces *in situ*. This development is a significant and exciting one, because it opens the door to *in situ* surface Raman investigations on well-defined electrode surfaces not available previously.

D. Nonlinear Optical Spectroelectrochemical Methods

Recent efforts have also resulted in the development of nonlinear spectroscopic probes of electrochemical systems in which the response is not linearly dependent on the incident field strength but some power thereof. Specifically, second harmonic generation (SHG) and hyper-Raman spectroscopy (HRS) are second-order nonlinear techniques in which the response depends on the square of the incident field that has been used to investigate electrochemical systems. The inherent sensitivity to the interface makes these nonlinear techniques particularly attractive for such studies.

SHG involves the production of light at twice the fundamental frequency at an interface where the inversion symmetry is broken. This technique was applid to the study of electrochemical interfaces as early as 1967,[97] but significant attention had to await the development of suitable pulsed laser sources which made the acquisition of the weak SHG signals routinely feasible. SHG has been subsequently applied to the study of a range of electrochemical surface and interfacial phenomena. This work has been reviewed extensively elsewhere.[98]

HRS is a three-photon process in which the energy of the hyper-Raman photon depends on the difference between two times the incident beam energy and the molecular vibrational energy. This method has been applied to the study of electrochemical interfaces in the surface enhanced mode (SEHRS) by Van Duyne and coworkers.[99] The interest in this phenomenon results from the vibrational selection rules which suggest that for highly symmetric molecules, vibrational modes that are both IR and Raman active will be observed in the hyper-Raman spectrum. Interest in this technique continues, and it is likely that useful molecular information about electrochemical systems will be forthcoming.

Obviously, all of the preceding work in these spectroelectrochemical areas cannot be reviewed in the space of this chapter. Rather, the intent is to give the reader a flavor of the power of certain methods as probes electrochemical systems. Due in part to the authors' bias in assessing the power of various spectroelectrochemical methods, the examples presented are drawn exclusively from recent advances in IR and Raman spectroelectrochemical methods.

III.   APPLICATIONS OF VIBRATIONAL SPECTROELECTROCEHMISTRY

A.   *In Situ* Infrared Studies of the Electro-oxidation of Formic Acid on Bimetallic Surfaces

Recent developments in IR spectroelectrochemistry have provided valuable molecular information about electrochemical interfaces. An example which nicely demonstrates the utility of such an approach is the IR study by Weaver and coworkers[100] of the electro-oxidation of formic acid on bimetallic electrodes. Among the electrochemical community, interest exists in understanding, at a molecular level, the mechanisms associated with electrocatalytic processes. Of particular interest in the electrocatalysis literature has been the use of bimetallic electrocatalysts to increase the electron transfer rate for a given system. Although numerous electrochemical studies on these bimetallic electrocatalysts for electro-oxidation processes have been reported,[101] more molecular information is needed before these systems will be completely understood. In order to aid in the development of a detailed chemical picture of these systems, vibrational studies on well-defined single-crystal surfaces would be desirable. Until recently, such IR spectroscopic studies were restricted to metal–ultrahigh-vacuum interfaces.[102] However, developments in IR spectroelectrochemical approaches have enabled studies on such systems to be performed *in situ* as well.

Weaver and coworkers[100] have recently used *in situ* FTIR spectroelectrochemistry to study the electro-oxidation of formic acid on unmodified Pt(100) surfaces and on Pt(100) surfaces modified with Bi overlayers. Due to the chemical irreversibility of the electrocatalytic reaction, SPAIRS was used. This approach allows the oxidation process to be monitored in real-time.

Figure 1 shows cyclic voltammograms on unmodified (dashed line) and Bi-modified (solid line) Pt(100) electrodes in a solution of 0.1 M formic acid in 0.1 M $HClO_4$. The Bi-modified Pt(100) electrode in this investigation had a Bi surface coverage (θ_{Bi}) of ~ 0.25 monolayer (ML). The voltammogram from the unmodified Pt(100) surface shows very little oxidation of formic acid to CO_2 on the forward sweep, even at 0.6 V, which is close to the onset potential of Pt surface oxidation. More oxidation is evident on the reverse sweep, however, presumably due to complete oxidation of surface-confined CO which poisons the oxidation reaction of formic acid as discussed below. Upon addition of Bi adatoms to the Pt(100) electrode surface, a substantial oxidation current is noted in the voltammogram for both the forward and reverse scans.

The SPAIR spectra from unmodified Pt(100) electrodes in this formic acid solution are shown in Figure 2. Each spectrum was obtained by acquiring 50 interferometer scans during a 2 mV/s potential sweep from -0.25 to 0.6 V. The potential reported by the side of each spectrum represents the average potential applied during the acquisition of that spectrum. Interfering bands due to the solvent were removed by subtracting the spectrum obtained at 0.6 V, where CO electro-oxidative removal is complete, from the spectrum at the potential of interest.

The bands observed in these spectra at 2030–$2060\,cm^{-1}$ and at $\sim 1850\,cm^{-1}$ are due to CO bound at terminal (atop) and twofold bridging sites, respectively.[103] The intensity of the band at $\sim 2035\,cm^{-1}$ associated with terminal CO increases significantly as the potential is made positive from -0.25 V to 0.3 V. At potentials positive of 0.3 V, the intensity of this band decreases and disappears entirely by 0.55 V. These results suggest that

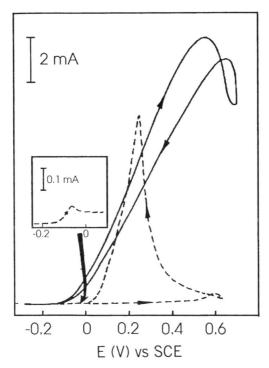

Figure 1. Cyclic voltammograms (50 mV/sec) for electro-oxidation of 0.1 M formic acid in 0.1 M HClO$_4$ on Pt(100) in the absence (dashed traces) and presence (solid trace) of Bi adlayer ($\theta_{Bi} = 0.25$). Electrode area is 0.80 cm^2. (Reprinted from reference 100, with permission.)

CO species bound at terminal sites form spontaneously from formic acid on Pt(100) and increase in coverage until a potential of ~ 0.3 V. Beyond this potential, these terminal CO species are removed from the surface by oxidation.

SPAIRS data were also acquired for formic acid oxiation on Pt(100) modified by various coverages of Bi. The Bi was pre-dosed onto the Pt(100) surface to a predetermined coverage. This coverage by Bi, θ_{Bi}, remained constant during the subsequent electrochemical and SPAIR measurements. The data are summarized in Figure 3. The solid circles represent the fractional CO coverage of terminal CO species determined from the intensity of the SPAIRS band at ~ 2035 cm^{-1} for the potential region between ~ 0.0 and ~ 0.25 V, where θ_{CO} coverage remains essentially constant on unmodified Pt(100) surfaces. The open circles represent the corresponding voltammetric current, i_{max}, at ~ 0.5–0.6 V obtained during the positive going potential sweep as a function of θ_{Bi}. The open triangles are the oxidative

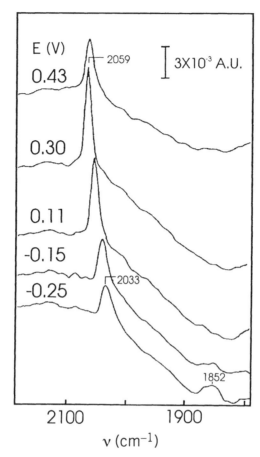

Figure 2. Single potential alteration infrared (SPAIR) spectra in the 1800–2150 cm^{-1} region for 0.1 M formic acid in 0.1 M HClO$_4$ on Pt(100). Spectral sequence was obtained during 2 mV/sec positive-going potential sweep from -0.25 V. Each spectrum was acquired by using 30 interferometer scans; the potentials indicated beside each spectrum are the average values during the data acquisition. A corresponding set of interferograms obtained upon completion of CO electro-oxidation was subtracted from each spectrum to remove solvent and other spectral interferences. (Reprinted from reference 100, with permission.)

current observed during the positive sweep at a potential of ~ 0.1 V, close to the onset of the electro-oxidation of formic acid, also as a function of θ_{Bi}. Figure 3 shows that i_{max} is the largest for surfaces on which there is a significant Bi coverage ($\theta_{Bi} > 0.25$) and the fractional coverage of terminal CO species is at a minimum.

To better understand the structure of the CO adlayer formed from the dissociative chemisorption of formic acid, the frequency of the band due to

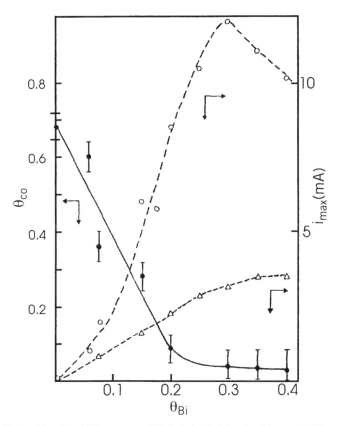

Figure 3. Plots of fractional CO coverage (filled circles, left-hand axis) at ~ 0.25 V versus SCE during electro-oxidation of 0.1 M formic acid in 0.1 M $HClO_4$ on Pt(100), and corresponding voltammetric peak current (open circles, right-hand axis) as a function of Bi coverage, θ_{Bi}. The open-triangle–dashed curve is the corresponding current at 0.1 V, also versus θ_{Bi}. Electrode area 0.80 cm^2. (Reprinted from reference 100, with permission.)

terminal CO, v_{CO}^t, was monitored for different coverages of CO adlayers on either unmodified or Bi-modified Pt(100). Table 2 compares the v_{CO}^t values obtained on Pt(100) at 0.15 V for various Bi adlayer coverages for CO produced spontaneously by dissociation of formic acid with those observed for different fractional CO coverages produced by adsorption of CO from CO-containing solutions. These data demonstrate that a maximum fractional CO coverage of ~ 0.7 can be achieved by the spontaneous dissociation of formic acid for $\theta_{Bi} = 0.0$.

The data for CO coverages labeled "dosing" in Table 2 were acquired from surfaces dosed with dilute ($\sim 10^{-5}$ M) CO solutions for the appropriate period

Table 2. Terminal C–O Stretching Frequencies, ν_{CO}, for Adsorbed CO Formed at 0.15 V versus SCE on Pt(100) in 0.1 M HClO$_4$ from 0.1 M Formic Acid, as Compared with Adsorption from Solution CO, in the Absence and Presence of Coadsorbed Bi

Environment	$\theta_{CO}{}^a$	$\theta_{Bi}{}^b$	ν_{CO}^t (cm^{-1})
0.1 M HCOOH	0.70	0.00	2059
Solution CO (dosing or stripping)^{c,d}	0.70	0.00	2060
0.1 M HCOOH	0.60	0.06	2050
Solution CO	0.60	0.24	2055
Solution CO (dosing)^c	0.60	0.00	2052
Solution CO (stripping)^d	0.60	0.00	2059
0.1 M HCOOH	0.35	0.08	2035
Solution CO (dosing)^c	0.35	0.00	2037
Solution CO (stripping)^d	0.35	0.00	2055

^{a}Coverage of CO as determined from integrated absorbance of the ν_{CO} band.
^{b}Coverage of irreversibly adsorbed bismuth as estimated from the voltammetric charge for its reversible two-electron oxidation.
^{c}Adlayer formed by dosing with dilute ($\sim 10^{-5}$ M) solution CO to coverage indicated (see reference 103).
^{d}Adlayer formed by saturation adsorption from solution CO followed by partial electro-oxidative stripping (see reference 103).
Source: From reference 100, with permission.

of time (1–10 min).[100] The data for CO coverages labeled "stripping" in Table 2 were acquired from surfaces formed by first dosing a saturated CO adlayer onto the surface and then partially electro-oxidizing that layer to the desired coverage.[100] For CO solution dosing on an unmodified Pt(100) surface at CO coverages below saturation ($\theta_{CO} = 0.7$), the frequency of ν_{CO}^t increases with increasing coverage of CO (from 0.35 to 0.6). However, a CO adlayer formed by stripping exhibits very little change in ν_{CO}^t with CO coverage. The significant dependence of ν_{CO}^t on θ_{CO} for the solution-dosed adlayers reflects the increasing contribution of adsorbate dipole–dipole coupling as the CO coverage increases. In other words, upon dosing larger coverages of CO onto the surface, larger and larger CO islands are formed, resulting in increasing dipole–dipole interaction among adjacent CO molecules. When the CO adlayer is formed by stripping, however, such a dramatic decrease in the ν_{CO}^t frequency at low CO coverages is not observed. These results indicate that upon stripping a saturated CO adlayer, the dipole–dipole coupling allows local CO coverages to remain high, suggesting that the remaining CO stays on the surface in islands.

The Bi-modified Pt(100) surface shows trends in the ν_{CO}^t frequency similar to those seen for the solution-dosed CO adlayer on unmodified Pt(100). The ν_{CO}^t frequency increases from 2035 cm^{-1} to 2050 cm^{-1} as the CO coverage increases from 0.35 to 0.6 for similar Bi coverages. At $\theta_{Bi} = 0.25$, where the oxidative currents are the largest, the v_{CO}^t frequency is still lower than that observed for a saturated CO adlayer. This behavior suggests the presence of smaller CO islands resulting in less dipole–dipole coupling. This conclusion is consistent with a previous finding by Chang and Weaver[104] that Bi adatoms are distributed uniformly across the Pt(100) surface. Such a structure would hinder the formation of large CO islands which promote extensive dipole–dipole coupling. Without the strong dipole–dipole coupling between adjacent CO molecules, the adsorbed CO molecules are more easily converted to CO_2 and desorbed from the surface.

B. Surface Enhanced Raman Spectroscopy of Carbon Electrode Surfaces

The advantages that accrue to the use of the SERS effect have allowed researchers to study numerous electrochemical systems in terms of the molecular structures of the electrolyte side of the interfacial region. However, few researchers have used SERS to study the microstructure of the electrode surface itself. Recently, Alsmeyer and McCreery[105] have used the SERS effect to study the microstructure of carbon electrode surfaces and related this information to electron transfer processes at these electrodes.

Carbon is a very popular electrode material. However, a complete understanding of the variables that affect electron transfer kinetics at carbon electrodes has not been developed. Much previous work has indicated that surface preparation and microstructure are of critical importance in dictating electrode kinetics at such surfaces. Parameters such as electron transfer rate, adsorption, and background current have been shown to depend strongly on surface preparation protocols.[106] In the late 1980s, McCreery and coworkers[107–109] demonstrated the feasibility of relating surface structure to electron-transfer activity using normal Raman spectroscopy. In particular, they found a direct relationship between the 1360 cm^{-1} "disorder" (D) graphite band and the electron-transfer kinetics for $Fe(CN)_6^{3-/4-}$ and dopamine. The appearance of this band has been taken to be indicative of the presence of edges of graphite domains.[110]

Although normal Raman spectroscopy is able to provide structural information about bulk carbon, the surface selectivity of this measurement is poor due to a sampling depth of ~ 30 nm for carbon probed with a visible wavelength laser at normal incidence.[107] This sampling depth decreases at more glancing angles, but it is still significant relative to the few tens of angstroms important in electron transfer processes. In recent work, Alsmeyer and McCreery[105] have utilized the properties of the SERS effect to reduce the sampling depth by the electrochemical deposition of a Ag adlayer on the carbon electrode. The optimum amount of Ag deposited on the carbon was determined by trial and error. Varying amounts of Ag were deposited and correlated with the SERS signal of the graphite bands at $1360\,cm^{-1}$ (D band) and $\sim 1600\,cm^{-1}$ (E_{2g} band) for mechanically polished glassy carbon (GC). The optimum SERS intensity was obtained following the deposition of $0.21\,\mu mol/cm^2$ or ~ 70 monolayers of Ag based on an equivalent monolayer coverage of $350\,\mu C/cm^2$.[111]

Polished GC is known to have a large density of defect sites (edges) which give rise to an intense D band. The inherent intensity of the D band from these GC surfaces in the absence of deposited Ag makes accurate evaluation of the usefulness of the SERS technique to selectively probe the surface interfacial region difficult. The true effectiveness of the SERS approach was shown more clearly in a similar study utilizing a more ordered carbon substrate, namely, highly oriented pyrolytic graphite (HOPG). The estimated sampling depths for HOPG using normal Raman spectroscopy and SERS are 20 nm and 2 nm, respectively. The relatively small sampling depth for the SERS experiment is a result of the electromagnetic enhancement profile of the Ag adatoms.

Figure 4A shows Raman spectra of HOPG which had been lightly polished with $0.05\,\mu m$ alumina in $0.1\,M$ $AgNO_3/0.1\,M$ $NaNO_3$ solution. The bottom spectrum (a) was taken before any Ag was deposited and shows the appearance of only the E_{2g} band at $1582\,cm^{-1}$ on the H_2O background. The observation

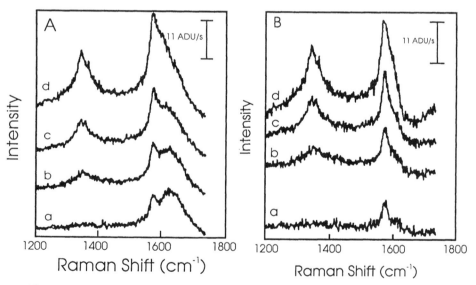

Figure 4. (**A**) *In situ* Raman spectra of lightly polished HOPG before and after Ag deposition for Ag coverages of (a) $0.0\,\mu mol/cm^2$, (b) $0.41\,\mu mol/cm^2$, (c) $0.12\,\mu mol/cm^2$, (d) $0.25\,\mu mol/cm^2$. (**B**) Same as **A** but with H_2O background subtracted. (Reprinted from reference 105, with permission.)

of this single band suggests that HOPG has long-range order, consistent with what is known about the microstructure of this substrate. As Ag is deposited (spectra b–d), both the D and E_{2g} bands become pronounced, and the D/E_{2g} ratio increases. This effect is seen more clearly in Figure 4B, in which the H_2O background has been subtracted. These result have been interpreted to indicate that the first few atomic layers of HOPG, those most important in electron transfer pocesses, possess a significant density of defect sites.

Although the surface selectivity of the SERS approach on these HOPG surfaces was demonstrated by this study, the authors note that the increase in the D band intensity upon deposition of Ag was large. Two theories are proposed to explain this observation. The first explanation is that the SERS enhancement length is short enough (i.e., <5 nm) to be more surface-selective than normal Raman as suggested by the HOPG example. The second theory is that the Ag preferentially deposits at defect or edge sites, as expected on the basis of other redox systems, and reinforces the surface selectivity of the SERS technique. Unfortunately, attempts to extract quantitative information about the density of surface defect sites from the SERS spectra may not be accurate. Thus, distinguishing between these two theories is not possible.

As mentioned above, the surface preparation of carbon electrodes can drastically affect electrode kinetics. Through the use of the SERS technique described above, Alsmeyer and McCreery[112] characterized carbon electrodes which were laser-activated and electrochemically pretreated. The SERS results were correlated with heterogeneous electron transfer rates for $Fe(CN)_6^{3-/4-}$. Figure 5A shows the normal Raman spectra of HOPG which had been irradiated in air at various power densities with a 9-nsec, 1064-nm pulsed Nd:YAG laser. The D band at $1360 \, cm^{-1}$ is initially observed at $40-50 \, MW/cm^2$, while spectra at laser powers of $60 \, MW/cm^2$ and greater show prominent D bands indicating severe disorder.

Figure 5B shows spectra of HOPG following laser irradiation and subsequent electrodeposition of ~ 70 monolayers of Ag. The Raman scattering due to H_2O has been subtracted from these spectra. Following irradiation at $40 \, MW/cm^2$, the D band in the SERS spectrum is more pronounced than in the normal Raman spectrum (Figure 5A). At $50 \, MW/cm^2$, the disorder (as indicated by the D band intensity) in the SERS spectrum is larger than observed at $80 \, MW/cm^2$ in the normal Raman spectrum. These results imply greater surface-selectivity in the SERS technique for the laser-induced lattice damage.

The electron transfer rate constant, k^0, which reflects interfacial processes, was determined for $Fe(CN)_6^{3-/4-}$ following laser irradiation of HOPG electrodes. The magnitude of k^0 is predicted to increase with increasing surface disorder. Figure 6 shows the log of k^0 plotted versus laser irradiation power density. On the same graph, the intensity ratios of the D/E_{2g} bands from the

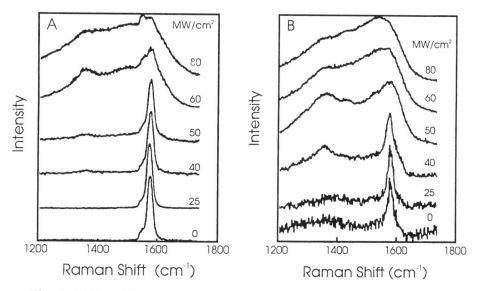

Figure 5. (A) Normal Raman spectra for HOPG obtained in air following laser activation at indicated laser power density. (B) Surface enhanced spectra of laser-activated HOPG obtained in solution following electrodeposition of 0.21 μmol/cm^2 Ag. Surfaces prepared as in **A** before immersion in Ag deposition solution. Solvent background was subtracted from Raman spectra. (Reprinted from reference 112, with permission.)

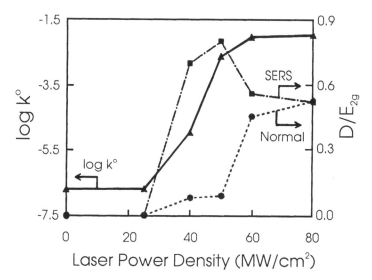

Figure 6. Integrated D/E$_{2g}$ intensity ratio from the spectra of Figures 5A and 5B, plotted with the k^0 for Fe(CN)$_6^{3-/4-}$ obtained following the same laser activation procedure. (Reprinted from reference 112, with permission.)

SERS and normal Raman spectra are plotted as a function of laser power. As shown in Figure 6, the magnifude of k^0 tracks the D/E_{2g} intensity ratio for the SERS experiment. The D/E_{2g} intensity ratio for the normal Raman spectra, however, lags behind k^0. Although HOPG shows overall long-range order even after 40–50 MW/cm^2 irradiation (as shown by the normal Raman spectra in Figure 5A), there is considerable surface disorder as evidenced by the SERS spectra and confirmed by the electron transfer rate constants. This study nicely demonstrates the utility of the SERS technique for providing surface structural information which is directly related to electron transfer processes.

C. Raman Spectroelectrochemical Studies in Nonaqueous Media

The majority of surface Raman studies reported have been performed in aqueous solutions where SERS measurements are relatively straightforward; in contrast, considerably fewer measurements have been reported in nonaqueous electrochemical environments. Given the importance of nonaqueous electrochemical systems such as Li-based battery systems, semiconductor photoelectrochemical cells, and electrosynthetic strategies based on nonaqueous electrochemical media, surface Raman spectroscopic studies are expected to contribute significant information about electrochemical pathways and processes in the interface and at electrode surfaces in such systems.

The surface Raman spectroelectrochemical studies reported to date in nonaqueous media can be separated into three categories. The first category contains experiments in which the surface Raman spectroscopic behavior of an electrode adsorbate from a nonaqueous electrochemical system is studied. The second category contains experiments in which the fundamental molecular structure of the nonaqueous electrochemical interface is probed through Raman spectroscopy of the solvent and/or electrolyte species in a nonaqueous electrochemical system. The third category contains experiments in which a film formed at a reactive electrode is investigated with Raman spectroscopy. The work done in all categories is reviewed below, and specific examples from categories two and three are discussed in-depth.

Experiments in the first category involved characterization of both organic and inorganic adsorbates from nonaqueous solutions. The first report of the successful use of nonaqueous solvents for SERS adsorbate studies appeared in 1983. In this work, the SERS of pyridine adsorbed on Ag electrodes from dimethylformamide (DMF) solutions containing tetrabutylammonium perchlorate (TBAP) was reported.[113] Potential-dependent changes in the pyridine spectra were interpreted in terms of changes in the electrode surface morphology as the potential was altered.

Several workers have reported SERS studies of adsorbed transition metal complexes based on Ru(II) and Fe(III) at electrodes from nonaqueous solutions. Stacy and Van Duyne[114] reported SERS and surface enhanced reson-ance Raman scattering (SERRS) studies of $Ru(bpy)_3^{2+}$ adsorption at Ag electrodes from acetonitrile (AN) solutions containing TBAP. The surface Raman intensities were shown to correlate with the double potential step chronocoulometric determination of adsorbate surface coverage. Virdee and Hester[115] used SERS to study $Ru(bpy)_3^{2+}$ and electrochemically generated $Ru(bpy)_3^+$ adsorbed at Ag electrodes from propylene carbonate (PC) solutions containing TBAP. Van Duyne et al.[116] later reported a SERS study of $Ru(bpy)_3^{2+}$, $Fe(bpy)_3^{2+}$ and $Fe(Phen)_3^{2+}$ adsorption on GaAs(100) and p-Si(100) through the use of electrochemically deposited Ag overlayers. In 1987, Wertz and coworkers[117] used SERS to study the adsorption of the free 2,2'-bipyridine ligand and $Ru(bpy)_3^{2+}$ adsorbed at Ag electrodes from DMF solutions containing tetraethylammonium perchlorate (TEAP). In a similar study of inorganic complex adsorption, Sanchez and Spiro[118] used SERS to study the adsorption of iron(III) protoporphyrin IX and its dimethyl ester at Ag electrodes from aqueous and AN solutions, resepctively.

Irish and coworkers[119] studied the adsorption of SCN^- at Ag electrodes from AN solutions of KSCN. Potential-dependent changes in the intensity and frequency of the $v_1(CN)$ band were interpreted to indicate adsorption of the SCN^- through the S atom with desorption from the electrode at increas-ingly negative potentials.

The SERS characterization of adsorption of organic molecules from nonaqueous solvents has also continued. Shin and Kim[120,121] have been particularly active in this area. They have reported SERS studies at Ag electrodes for the adsorption of pyridine, benzoic acid, and nitrobenzene from methanol (MeOH) solutions containing either KCl or LiCl. Their results suggest that the adsorption behavior of these molecules from MeOH is not significantly different than that from aqueous media. In further work, these investigators reported SERS studies of pyridine and benzene adsorbed at Ag electrodes from ethanol (EtOH) solutions of LiCl.[122] Their conclusions were similar to those of the previous study in MeOH in that the adsorption behavior observed was not significantly different in EtOH or water.

1. Surface Raman Spectroscopy of Li Electrodes Important in Batteries

Considerable interest in the molecular structure of the Li–electrolyte interface has resulted from the evolution of Li batteries.[123] At present, there is general agreement that a passive film which stabilizes the electrode forms at the Li surface in organic electrolytes. The structure and properties of this film, although of an undetermined chemical nature in most cases, are of critical importance in dictating the performance of the Li battery. These films are thought to arise from reduction of the solvent, electrolyte, or nucleophilic impurities such as water or oxygen.[123] Determination of the origin of this passive film *in situ* is predicted on having the ability to study clean Li surfaces. Due to the reactivity of Li in ambient atmospheres, this is difficult task. Results from Auger electron spectroscopy[124] and x-ray photoelectron spectroscopy[126] studies have shown that upon exposure of a clean Li surface to less than 100 Langmuirs of O_2, the surface is covered with much more than a monolayer of Li_2O. In typical dry box conditions, both oxygen concentration and exposure times are much greater.[126] Thus, the challenge to advancing our understanding of the interfacial chemistry of these systems has been in finding an *in situ* technique which allows study of a clean Li surface in contact with the desired organic electrolytes.

Irish and coworkers[127] have developed a technique involving the *in situ* cutting of Li to expose a fresh Li surface to study the formation of these films using surface Raman spectroscopy. Their studies have focused on film formation in two ethereal solvents, tetrahydrofuran (THF) and 2-methyl tetrahydrofuran (2-Me-THF), because of the promise that systems based on these solvents show for secondary Li batteries.[128]

An *in situ* study of Li in an anhydrous (water not detectable) 0.6 M $LiAsF_6$/THF electrolyte solution was conducted to identify the molecular species formed in the interfacial region. The Li electrode was allowed to sit in solution for 24 hr in a dry box. During this time, the solution viscosity increased, presumably due to the formation of a polymer species. When the Li electrode was freshly cut *in situ*, a brown film formed on the surface. The growth of the surface film was suppressed by the formation of the polymer. After transfer back to the dry box, the excess electrolyte was removed and the polymeric species was separated and dried on a microscope slide.

Figure 7 shows Raman spectra of the dried polymer, the Li electrode observed *in situ* at open circuit potential, and the 0.6 M $LiAsF_6$/THF electrolyte solution in the $v(C-H)$ region, respectively. The top spectrum (Figure 7a) is of the dried polymer and is similar to that found in the literature for polytetrahydrofuran (PTHF).[129,130] PTHF has been shown to exist in a planar zigzag conformation and possesses two types of methylene groups[124] referred to as types I and II:

$$[-O(CH_2)_I-(CH_2)_{II}-(CH_2)-(CH_2)]_n$$

Type I CH_2 groups neighbor an oxygen atom while type II CH_2 groups are positioned between methylene groups. The band at 2807 cm^{-1} in Figure 7a and 2799 cm^{-1} in Figure 7b has been assigned to a type I CH_2 group $v(C-H)$ mode.[124] This band is not observed in the spectrum of the pure electrolyte

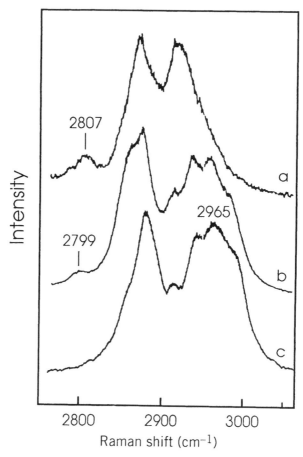

Figure 7. Raman spectra in the CH stretch region: (a) dried polymer, (b) Li electrode *in situ*, (c) 0.6 M LiAsF$_6$/THF electrolyte. (Reprinted from reference 127, with permission.)

(Figure 7c), suggesting that some type of amorphous PTHF forms at the electrode surface.

Figure 8 shows the 1150–1550 cm^{-1} region where the CH bending and wagging modes are found. The spectrum of dried PTHF (Figure 8a) shows two features at 1301 cm^{-1} and 1437 cm^{-1}, which are also observed in the spectrum from the Li electrode (Figure 8b) at 1293 cm^{-1} and 1437 cm^{-1} but not in the spectrum of the pure electrolyte (Figure 8c). The band at 1301 cm^{-1} has been assigned to a type I CH$_2$ group wagging mode, while the band at 1437 cm^{-1} has been left unassigned.

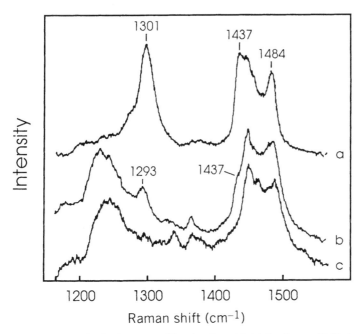

Figure 8. Raman spectra in the CH wag and bend regions: (a) dried polymer, (b) Li electrode *in situ*, (c) 0.6 M LiAsF$_6$/THF electrolyte. (Reprinted from reference 127, with permission.)

The polymerization of THF is known to take place by a cationic ring-opening mechanism.[131] This mechanism is predicted and observed to lead to disappearance of the ring breathing mode at 914 cm^{-1} as shown in Figure 9A(a). The *in situ* spectrum from the Li electrode (Figure 9A(b)], however, still shows the 914 cm^{-1} band. This band is sharper than that from the pure electrolyte [Figure 9A(c)], which has been attributed to monomeric electrolyte intermingled within the polymeric phase. Upon disappearance of the 914 cm^{-1} mode in the dried PTHF spectrum, a broad feature at 830 cm^{-1} is observed. This band has tentatively been assigned to a type II CH$_2$ rocking mode[132] and can be seen more clearly in Figure 9B, which is an expansion of Figure 9A. The 830 cm^{-1} band is also observed in the spectrum from the Li electrode but not in that of the pure electrolyte. Thus, the observation of this band confirms the presence of amorphous PTHF species at the electrode surface.

In addition to the prominent AsF$_6^-$ band at 679 cm^{-1}, the spectrum from the Li electrode in Figure 9B also shows a much weaker band at 704 cm^{-1}. Deng and Irish[133] have studied this band as a function of concentration in

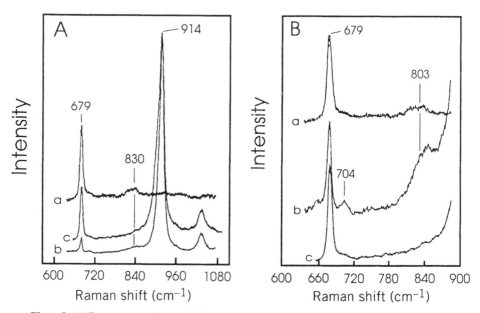

Figure 9. (A) Raman spectra in the THF ring stretching region: (a) dried polymer; (b) Li electrode *in-situ*; (c) 0.6 M LiAsF$_6$/THF electrolyte. (**B**) Expansion of the Raman spectra of **A** in the region 650–875 cm^{-1}: (a) dried polymer, (b) Li electrode *in situ*, (c) electrolyte. (Reprinted from reference 127, with permission.)

methyl acetate and attributed it to contact ion pairs of Li$^+$–AsF$_6^-$. The observation that this band is generally much weaker than that at 679 cm^{-1} in the spectrum from the Li electrode surface suggests that Li$^+$ and AsF$_6^-$ exist predominantly as mobile ions, and not contact ion pairs, throughout the polymer. Figure 10 shows the spectra of the pure electrolyte (a), dried PTHF (b), and dried LiAsF$_6$ powder (c). Both the pure electrolyte and dried PTHF show bands at identical frequencies, while the dried LiAsF$_6$ shows a shift in frequency for each of the bands. These data further support the premise that the Li$^+$ and AsF$_6^-$ exist as mobile ions in the polymer at the electrode interface.

Conductance measurements on the PTHF\LiAsF$_6$ solid were conducted using dc voltage and the four-point-probe method.[134] An initial conductance value of 10^{-4} S cm^{-1} was determined, indicative of ionic conducting material. Thus, the importance of the formation of this polymeric layer at the electrode surface in Li battery cells is obvious. This example demonstrates the power of surface Raman spectroscopy for determination of the molecular structure of such passivating films on Li surfaces in organic electrolytes.

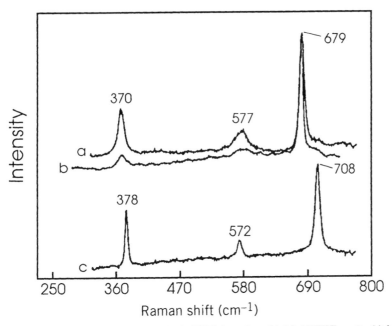

Figure 10. Raman spectra of (a) 0.6 M LiAsF$_6$/THF electrolyte, (b) dried PTHF *ex situ*, (c) dried LiAsF$_6$ powder. (Reprinted from reference 127, with permission.)

2. In Situ Studies of Interfacial Solvent Structure in Nonaqueous Electrochemical Systems

Irish et al.[135] were also the first to report the use of SERS for determination of interfacial structure at electrode surfaces in nonaqueous systems. These researchers reported a SERS study of Ag electrodes in AN solutions containing LiI. The ν(CN) region for interfacial acetonitrile molecules was monitored as a function of electrode potential. Bands attributed to interfacial AN molecules are observed at 2271 and 2300 cm^{-1}. These bands shift to lower frequencies as the potential is made more negative. Moreover, when Li$^+$ is replaced with Na$^+$ as the electrolyte cation, these two bands disappear at the expense of a band at 2262 cm^{-1}. Therefore, these bands are sensitive to the nature of the cation in solution. The frequencies of the bands at 2248 cm^{-1} in Li$^+$ and 2262 cm^{-1} in Na$^+$ are relatively insensitive to potential. On the basis of comparison with the Raman spectral behavior of electrolyte solutions of acetonitrile, these researchers assign the bands at 2271 and 2300 cm^{-1} in Li$^+$ and 2262 cm^{-1} in Na$^+$ to acetonitrile molecules at the electrode surface solvating cations in the outer Helmholtz plane. A second type of solvent molecule is also proposed to exist at the surface but to interact with other solvent molecules instead of cations. The molecular picture of the interface deduced by these researchers shows AN molecules weakly bonded to the Ag electrode surface through adsorbed I$^-$. Those molecules solvating cations are represented as solvent-separated ion pairs at the electrode surface.

The investigation of interfacial solvent structure at Ag electrodes in nonaqueous pyridine solutions containing tetrabutylammonium bromide (TBABr) was reported from this laboratory.[136] The potential dependence of the SERS pyridine frequencies, bandwidths and relative intensities was interpreted in terms of pyridine reorientation in the interface. In addition, a band at ~ 220 cm^{-1} was observed at certain potentials and attributed to the v(Ag–N) stretch for surface pyridine species.

Irish and coworkers[137] also reported studies at Ag electrodes in PC solutions containing LiI and NaI. The presence of a strongly specifically adsorbed anion such as I$^-$ or Br$^-$ was found to be needed in order to see a SERS response for interfacial PC molecules, suggesting that those PC molecules probed are those involved in weak interactions between C–H groups and the specifically adsorbed anion. Evidence for the photoelectrochemical reduction of PC to form insoluble carbonate salts with the cation of the supporting electrolyte was also presented. Moreover, the C=O moiety of the PC was observed to be very sensitive to chemical environment.

Additional systems which have been studied quite extensively in this laboratory are the alcohols (MeOH,[138–140] EtOH,[140] 1-propanol [PrOH],[139,140] 1-butanol [BuOH], 2-butanol [2-BuOH], iso-butanol [i-BuOH],[141] and 1-pentanol [PeOH][140]) as solvents at Ag and Au electrodes containing alkali halide supporting electrolytes. In all of these studies, elucidation of the interfacial solvent behavior (ion solvation, orientation) as a function of electrode potential has been the focus.

As a representative example of the SERS behavior of these interfacial alcohol systems at Ag electrodes, SERS spectra for BuOH containing 0.4 M LiBr at a Ag electrode in the v(C–H) frequency region as a function of electrode potential are shown in Figure 11. The top spectrum in this series shows the Raman scattering behavior of bulk BuOH in this region. The remaining spectra show the response of the interfacial BuOH molecules as a function of electrode potential. As can be seen, significant changes in relative intensity ratio of the vibrational modes occur with changes in electrode potential. Of particular interest under this envelope are the symmetric v(C–H) mode from the CH$_3$ group at 2873 cm^{-1}, and the corresponding antisymmetric v(C–H) mode of this group at 2963 cm^{-1}. The intensity ratio of these bands as a function of electrode potential has been determined by curvefitting and is shown in Figure 12. As can be seen, the ratio changes significantly with respect to the ratio observed in bulk (designated by the dashed line) as the potential

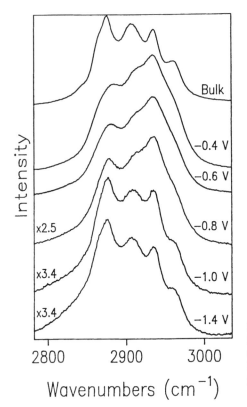

Figure 11. SERS spectra in $\nu(C–H)$ region from Ag electrode for 0.4 M LiBr/1-BuOH solution. Top spectrum represents bulk liquid; remaining spectra are potential-dependent spectra. (Reprinted from reference 141, with permission.)

changes. These changes are interpreted as being indicative of changes in the orientation of the BuOH molecules at the electrode surface in response to the changing metal surface charge at different electrode potentials.

Using the methyl group vibrations for an assessment of orientation as described elsewhere,[142] the proposed orientations of BuOH in three different potential regions relative to the potential of zero charge (PZC) are shown in Figure 13. In general, at potentials positive of the PZC, the alcohol molecules, including BuOH, exist in an orientation largely parallel to the electrode surface. As the potential is made more negative and approaches that of the PZC and goes beyond the PZC, the alkyl end of the alcohol molecule is driven away from the electrode surface to a more vertical orientation. This picture is consistent with the observation that the intensity ratio of the antisymmetric to the symmetric $\nu(C–H)$ modes decreases as the potential is made more negative.

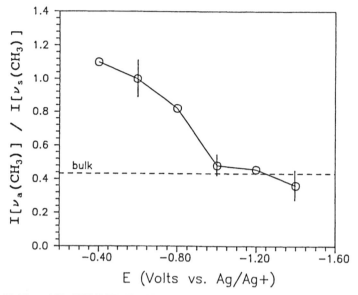

Figure 12. Plot of $I[\nu_a(CH_3)]/I[\nu_s(CH_3)]$ versus potential for 1-BuOH/0.4 M LiBr at an Ag electrode. (Reprinted from reference 141, with permission.)

This example demonstrates the utility of surface Raman scattering for the assessment of orientation of molecules at electrode surfaces.

Similar studies of the butanol isomers containing LiBr on Au electrodes have also been recently completed.[143] Quite significantly, the orientations deduced for these butanol isomers as a function of potential versus the PZC are *identical* to those proposed on Ag. This conclusion is noteworthy in that it supports the contention that the interfacial solvent species predominantly respond to the electrode surface charge and not the electrode material in cases in which the interaction between the solvent and the electrode surface is weak, as is presumed to be the case for these alcohols.

In addition to vibrational modes associated with the solvent molecules in this system, vibrational modes due to interfacial ions are also observed. Figure 14 shows the SERS spectra in the $\nu(Ag–Br)$ region as a function of electrode potential. At more positive potentials, a band at $156\,cm^{-1}$ is observed due to Br^- specifically adsorbed at the Ag electrode surface. As can be seen from the intensity of this band as a function of electrode potential, as the potential is made more negative and approaches the PZC, the Br^- is electrostatically repelled from the surface which results in a decrease in the intensity of this band. At potentials negative of the PZC, this band is completely absent from the spectra.

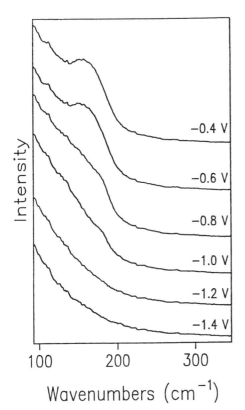

Figure 13. SERS spectra in v(Ag–Br) region from Ag electrode for 0.4 M LiBr/1-BuOH solution; potential-dependent spectra. (Reprinted from reference 141, with permission.)

Several reports describe SERS detection of H_2O in the interface at Ag electrodes in nonaqueous solvent systems. Vibrational bands in the v(O–H) region originate either from trace amounts of H_2O in these solvents, even after they are dried to the level of submillimolar concentrations of H_2O, or from H_2O intentionally added to the solvent. Irish et al.[135] observed vibrational modes due to H_2O, OH^-, and microcrystallites of LiOH at negative potentials (more negative than -0.8 V versus a W wire quasireference electrode) in SERS studies of acetonitrile solutions at Ag electrodes. In a later report, Irish and coworkers[137] identified similar interfacial species at Ag electrodes in PC electrolyte studies using SERS. Significantly, a SERS spectral band for microcrystalline NaOH was observed at negative potentials in propylene carbonate solutions containing NaI.

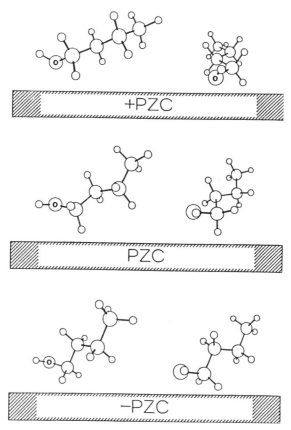

Figure 14. Proposed orientation of 1-BuOH at Ag electrode in 0.4 M LiBr/1-BuOH solution, front and side views, as a function of potential relative to the potential of zero charge. (Reprinted from reference 141, with permission.)

More recently, SERS in the ν(O–H) region was studied in H_2O/ethanol solutions containing NaBr at Ag electrodes.[144] Although perhaps somewhat counterintuitive, more intense SERS signals were detected in H_2O ν(O–H) region as the concentration of ethanol in the mixture increased. Moreover, two ν(O–H) bands were observed in the ethanol/H_2O solutions as compared to only one band in pure H_2O. The second band was attributed to H_2O species interacting directly with the electrode surface, presumably made possible by disruption of the H-bonded H_2O network by ethanol.

In a recent SERS investigation of interfacial solvent structure of butanol isomers at Ag electrodes in this laboratory,[145] distinct behavior in the ν(O–H) region where interfacial H_2O vibrations have been observed was noted. $v(O–H)$ bands originating from trace H_2O present in these solvents are observed which are quite sensitive to the anion and cation of the electrolyte. The behavior of these bands can be exploited to provide a wealth of information about the interfacial electrolyte structure in these systems. The nonaqueous environment of this H_2O helps to break the extensive H-bonded structure present at the interface in pure aqueous media resulting in better resolved $v(O–H)$ bands. Studies to investigate the interfacial H_2O and electrolyte behavior of LiCl, LiBr, LiI, and $LiClO_4$ and Ag electrodes in several isomers of butanol, including BuOH, i-BuOH, and 2-BuOH, have been performed.

Figure 15 shows spectra in the $v(O–H)$ region for this trace water in BuOH. Interestingly, six $v(O–H)$ bands are observed in the surface spectra depending on the electrode potential. These are assigned on the basis of previous assignments made for similar bands in aqueous solutions and the spectral response of crystalline LiOH. The intensities of these bands are a function of the nature of the anion in the LiX electrolyte, the particular isomer of butanol used as the solvent, and the electrode potential. A band at $\sim 3500\,cm^{-1}$ is assigned to H_2O hydrogen-bonded to specifically adsorbed anions at the electrode surface at potentials positive of the PZC. This band decreases in intensity as the PZC is reached, consistent with elimination of Br^- from the electrode surface. A band assigned to H_2O in the primary solvation shell of the Li^+ is observed at $\sim 3570\,cm^{-1}$ at potentials negative of the PZC. A band thought to be due to "free" H_2O not associated with specific ions, perhaps weakly held in the secondary solvation shell of Li^+, is observed between 3600 and $3650\,cm^{-1}$. Up to three bands are observed from OH^- formed at the electrode during the reduction of H_2O at negative potentials. At these potentials, a band observed at $3600\,cm^{-1}$ is assigned to OH^- weakly involved with Li^+, perhaps in a solvent-separated ion pair. A second OH^- band observed at $\sim 3630\,cm^{-1}$ is assigned to OH^- species adsorbed through the oxygen atom to the Ag electrode surface. The sharp band observed at $\sim 3660\,cm^{-1}$ is assigned to crystalline LiOH species which form due to the poor solubility of this salt in the isomers of butanol. This compound is either precipitated on the electrode surface or found as microcrystallites in the interfacial region.

In total, a very complete picture of the electrochemical interface in the LiBr–BuOH system at Ag electrodes has been deduced from these SERS studies. The proposed pictures of the complete interfacial structure at potentials positive of the PZC, in the vicinity of the PZC, and negative of the PZC are shown in Figure 16. Clearly, molecular pictures of this sort which are supported with spectroscopic information are much more complete than those determined on the basis of electrochemical data alone.

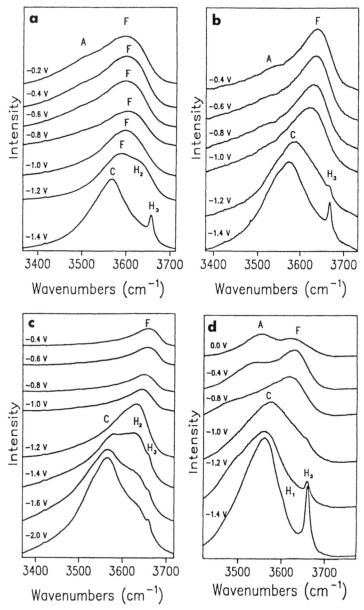

Figure 15. SERS spectra in $\nu(O-H)$ region from Ag electrode for 0.4 M LiX/1-BuOH solution: (a) $X = Cl^-$, (b) $X = Br^-$, (c) $X = I_-$, and (d) $X = ClO_4^-$. (Reprinted from reference 145, with permission.)

Positive of the PZC

At the PZC

Negative of the PZC

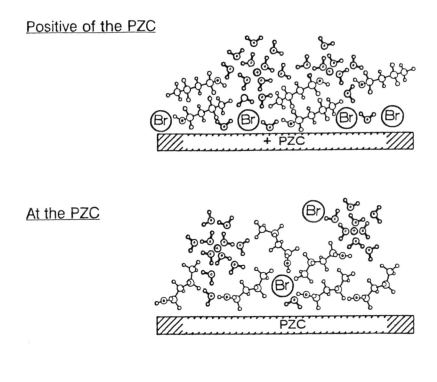

Figure 16. Schematic of interfacial structure at Ag electrode in 0.4 M LiBr/1-BuOH solution as a function of potential relative to the potential of zero charge.

3. Interfacial Structure at Electrodes Emersed from Electrochemical Systems

SERS spectra are generally acquired on rough Ag surfaces, but it is clearly desirable to be able to obtain similar spectra on much smoother surfaces. The extent to which such spectra can be acquired using the emersed electrochemical interface approach has been explored in this laboratory. In electrode

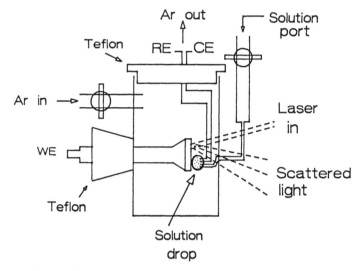

Figure 17. Schematic of a Raman spectroelectrochemical emersion cell. (Reprinted from reference 147, with permission.)

emersion, the first few layers of the solvent and supporting electrolyte in the interface are so strongly held at the electrode surface that upon careful removal of the electrode the bulk solution separates from the interfacial solution. Thus, upon emersion, the interfacial solution species should be retained intact at the electrode surface in their *in situ* orientations.

We have tested this hypotehsis using the alcohol solvent systems[139,146,147] noted above using the cell shown in Figure 17. This cell was designed with the electrode oriented vertically to facilitate drainage of bulk solvent from the electrode. Electrochemical contact was achieved in a drop of solvent separating the solvent-filled Luggin capillary tube and the lower portion of the working electrode. The counter and reference electrodes were also immersed in this drop. The working electrode was rotated at ~ 2 rpm between spectral acquisitions in order to reproducibly immerse/emerse the electrode. The laser beam enters the side of the cell and samples the emersed portion of the electrode.

Figures 18 and 19 show surface Raman spectra in the $v(C–H)$ of BuOH on Ag at -0.4, -1.0, and -1.4 V, and i-BuOH on Au at $+0.4$, 0.0, and -1.2 V, respectively, on *in situ* rough, emersed rough, and emersed smooth electrodes.[147] In all cases, the spectra acquired on the emersed rough and emersed smooth surfaces are strikingly similar to the spectra obtained *in situ* on rough surfaces. The changes in the relative intensities of different portions of the

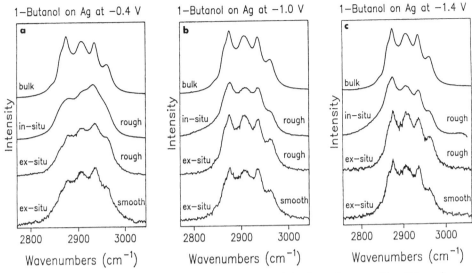

Figure 18. Raman spectra from bulk 1-BuOH, *in situ* rough Ag electrode, emersed (*ex situ*) rough Ag electrode, and emersed (*ex situ*) smooth Ag electrode; spectra from *in situ* and emersed surfaces are for 0.4 M LiBr/1-BuOH solution at (**a**) −0.4 V, (**b**) −1.0 V; and (**c**) −1.4 V. (Reprinted from reference 147, with permission.)

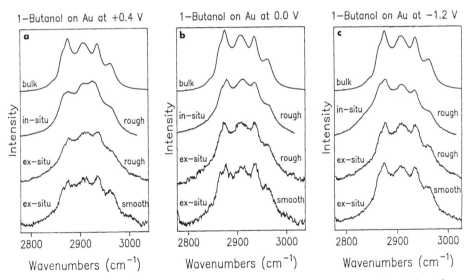

Figure 19. Raman spectra for bulk 1-BuOH, *in situ* rough Au electrode, emersed (*ex situ*) rough Au electrode, and emersed (*ex situ*) smooth Au electrode; spectra from *in situ* and emersed surfaces are for 0.4 M LiBr/1-BuOH solution at (**a**) 0.4 V, (**b**) 0.0 V, (**c**) −1.2 V.

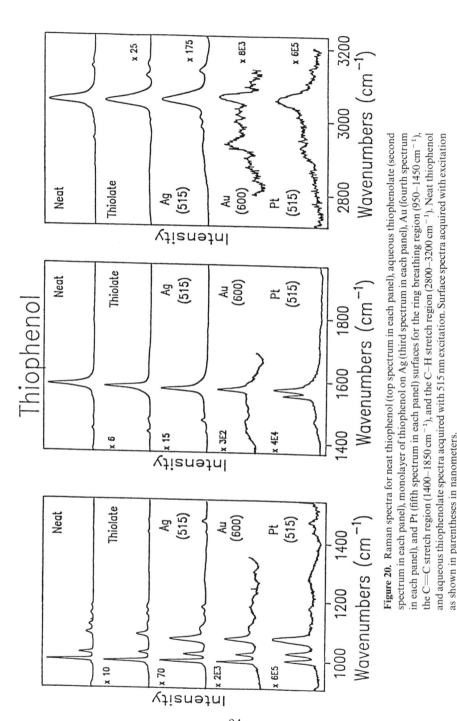

Figure 20. Raman spectra for neat thiophenol (top spectrum in each panel), aqueous thiophenolate (second spectrum in each panel), monolayer of thiophenol on Ag (third spectrum in each panel), Au (fourth spectrum in each panel), and Pt (fifth spectrum in each panel) surfaces for the ring breathing region (950–1450 cm^{-1}), the C=C stretch region (1400–1850 cm^{-1}), and the C–H stretch region (2800–3200 cm^{-1}). Neat thiophenol and aqueous thiophenolate spectra acquired with 515 nm excitation. Surface spectra acquired with excitation as shown in parentheses in nanometers.

v(C–H) envelopes are clearly visible in these spectra as a function of potential. Significantly, these ratios remain essentially unchanged after emersion of either the rough or smooth electrodes from all solvents studied. These observations collectively suggest that the solvents retain their potential-dependent *in situ* orientations upon emersion. These data validate the emersion approach and suggest that it may be used to achieve the necessary interfacial selectivity for systems in which significant surface enhancement is not operative.

Such an approach also allows extension of surface Raman methods to the study of metal surfaces which do not inherently support significant surface enhancement, such as Pt. We have pursued this concept in our laboratory with a particular interest in Pt based on its importance in electrochemical systems. Figure 20 shows the Raman spectra from a monolayer of thiophenol adsorbed on Ag, Au, and Pt surfaces.[148] These spectra were acquired from these metal surfaces in air after exposure to an ethanol solution of thiophenol. The spectra on Ag and Au were acquired with some surface enhancement at the excitation wavelengths indicated in the parentheses. However, Pt is not an enhancing metal at any excitation wavelength in the visible region of the spectrum. Thus, the spectrum shown for the monolayer of thiophenol on Pt acquired with 514.5 nm excitation is a truly unenhanced surface Raman spectrum. As can be seen, the spectrum from Pt, although much weaker than that from either Ag and Au, is one of excellent signal-to-noise ratio demonstrating the utility of this approach. Thus, emersed electrode approaches exhibit considerable promise for the study of electrochemical interfaces at metals which do not support surface enhancement.

IV.   THE FUTURE

Spectroelectrochemical methods are powerful tools for accessing detailed molecular information about electrochemical systems not available from measurement of electrical parameters alone. Improving technology has made many experiments feasible that were once thought to be virtually impossible. Hence, the quality of spectral information acquired from a myriad of electrochemical systems continues to improve. The future for application of these spectroelectrochemical methods to electrochemical systems of technological importance certainly looks very bright. Particularly attractive is the potential for continued advancement of the vibrational spectroelectrochemical methods. Developments in these areas will undoubtedly by driven by the need for chemical information of increasing levels of detail that can only be acquired with tools such as these that are inherently sensitive to molecular environment.

ACKNOWLEDGMENTS

The work described herein from the authors' laboratory has been generously supported by the National Science Foundation.

REFERENCES

1. T. Kuwana, R. K. Darlington, and D. W. Leedy, Electrochemical studies using conducting glass indicator electrodes, *Anal. Chem.*, **36**, 2023–2026 (1964).

2. T. Kuwana and N. Winograd, Spectroelectrochemistry at optically transparent electrodes, in *Electroanalytical Chemistry*, Vol. 7, A. J. Bard, ed., Marcel Dekker, New York (1974), pp. 1–64.

3. W. R. Heineman, B. J. Norris, and J. F. Goeltz, Measurement of enzyme $E^{0\prime}$ values by optically transparent thin layer electrochemical cells, *Anal. Chem.*, **47**, 79–84 (1975).

4. W. R. Heineman, Spectroelectrochemistry: Combination of optical and electrochemical techniques, *Anal. Chem.*, **50**, 390A–400A (1978).

5. W. R. Heineman, Spectroelectrochemistry: The combination of optical and electrochemical techniques, *J. Chem. Ed.*, **60**, 305–308 (1983).

6. W. R. Heineman, F. M. Hawkridge, and H. N. Blount, Spectroelectrochemistry at optically transparent electrodes. II. Electrodes under thin layer and semi-infinite diffusion conditions and indirect coulometric titrations, in *Electroanalytical Chemistry*, Vol. 13, A. J. Bard, ed., Marcel Dekker, New York (1984), pp. 1–113.

7. D. C. Johnson, M. D. Ryan, and G. S. Wilson, Dynamic electrochemistry: Methodology and applications, *Anal. Chem.*, **56**, 7R–20R (1984).

8. D. G. Jones, Photodiode array detectors in UV-vis spectroscopy. Part II, *Anal. Chem.*, **57**, 1207A–1214A (1985).

9. R. L. McCreery, Spectroelectrochemistry, in *Physical Methods of Chemistry*, Vol. 2: *Electrochemical Methods*, B. Rossiter and J. F. Hamilton, eds., Wiley, New York (1986), pp. 591–661.

10. D. C. Johnson, M. D. Ryan, and G. S. Wilson, Dynamic electrochemistry: Methodology and application, *Anal. Chem.*, **58**, 33R–49R (1986).

11. R. J. Gale, ed., *Spectroelectrochemistry: Theory and Practice*, Plenum Press, New York (1988).

12. M. D. Ryan and J. Q. Chambers, Dynamic electrochemistry: Methodology and applications, *Anal. Chem.*, **64**, 79R–116R (1992).

13. K. Rajeshwar, R. O. Lezna, and N. R. de Tacconi, Light in an electrochemical tunnel? Solving problems in electrochemistry via spectroscopy, *Anal. Chem.*, **64**, 429A–441A (1992).

14. R. Pruiksma and R. L. McCreery, Observation of electrochemical concentration profiles by absorption spectroelectrochemistry, *Anal. Chem.*, **51**, 2253–2257 (1979).

15. P. Rossi and R. L. McCreery, Diffractive spectroelectrochemistry: A sensitive probe of the electrochemical diffusion layer, *J. Electroanal. Chem.*, **151**, 47–64 (1983).

16. M. L. Simone, W. R. Heineman, and G. P. Kreishman, Long optical path electrochemical cell for absorption or fluorescence spectrometers, *Anal. Chem.*, **54**, 2382–2384 (1982).

17. E. E. Bancroft, J. S. Sidwell, and H. N. Blount, Derivative linear sweep and derivative cyclic voltabsorptometry. *Anal. Chem.*, **53**, 1390–1394 (1981).

18. E. E. Bancroft, H. N. Blount, and F. M. Hawkridge, Derivative cyclic voltabsorptometry of cytochrome c, *Biochem. Biophys. Res. Commun.*, **101**, 1331–1336 (1981).

19. E. E. Bancroft, H. N. Blount, and F. M. Hawkridge, Spectroelectrochemical determination of heterogeneous electron transfer kinetic parameters, in *Electrochemical and Spectrochemical Studies of Biological Redox Components*, ACS Symposium Series No. 201, K. Kadish, ed., American Chemical Society, Washington, DC (1982), pp. 23–49.

20. W. N. Hansen, Internal reflection spectroscopy in electrochemistry, in *Advances in Electrochemistry and Electrochemical Engineering*, Vol. 9, P. Delahay and C. W. Tobias, eds., Wiley, New York (1973), pp. 1–60.

21. J. D. E. McIntyre, Specular reflection spectroscopy of the electrode–solution interphase, in *Advances in Electrochemistry and Electrochemical Engineering*, Vol. 9, P. Delahay and C. W. Tobias, eds., Wiley, New York (1973), pp. 61–66.

22. D. M. Kolb, UV–visible reflectance spectroscopy, in *Spectroelectrochemistry: Theory and Practice*, R. J. Gale, ed., Plenum Press, New York (1988), pp. 87–188.

23. J. Kruger, Application of ellipsometry to electrochemistry, in *Advances in Electrochemistry and Electrochemical Engineering*, Vol. 9 P. Delahay and C. W. Tobias, eds., Wiley, New York (1973), pp. 228–280.

24. Ellipsometry: Principles and recent applications, in *Electroanalytical Chemistry*, Vol. 15, A. J. Bard, ed., Marcel Dekker, New York (1989), pp. 143–265.

25. R. H. Muller, Ellipsometry as an *in-situ* probe for the study of electrode processes, in *Techniques for Characterization of Electrodes and Electrochemical Processes*, R. Varma and J. R. Selman, eds., Wiley, New York (1991), pp. 31–125.

26. A. Yildiz, P. T. Kissinger, and C. N. Reilley, Evaluation of an improved thin layer electrode, *Anal. Chem.*, **40**, 1018–1024 (1968).

27. A. Wieckowski, *In-situ* surface electrochemistry: Radioactive labeling, in *Modern Aspects of Electrochemistry*, Vol. 21, R. E. White, J. O'M. Bockris, and B. E. Conway, eds., Plenum Press, New York (1990), pp. 65–119.

28. P. Zelenay and A. Wieckowski, Radioactive labeling: Toward characterization of well-defined electrodes, in *Electrochemical Interfaces: Modern Techniques for In-Situ Interface Characterization*, H. D. Abruna, ed., VCH Publishers, New York (1991), pp. 481–527.

29. G. H. Brilmyer and A. J. Bard, Application of photothermal spectroscopy to *in-situ* studies of films on metals and electrodes, *Anal. Chem.*, **52**, 685–691 (1980).

30. A. Fujishima, Y. Maeda, K. Honda, G. H. Brilmyer, and A. J. Bard, Simultaneous Determination of Quantum Efficiency and Energy Efficiency of Semiconductor Photoelectrochemical Cells by Photothermal Spectroscopy. *J. Electrochem. Soc.*, **127**, 840–846 (1980).

31. A. Fujishima, H. Masuda, K. Honda, and A. J. Bard, Measurement of gold electrode surface changes *in-situ* by laser photothermal spectroscopy, *Anal. Chem.*, **52**, 682–685 (1980).

32. R. E. Malpas, A. J. Bard, *In-situ* monitoring of electrochromic systems by piezoelectric detection photoacoustic spectroscopy of electrodes, *Anal. Chem.*, **52**, 109–112 (1980).

33. C. E. Vallett, Photoacoustic spectroscopy and the *in-situ* characterization of the electrochemical interface, in *Spectroscopic and Diffraction Techniques in Interfacial Electrochemistry*, C. Guttierez and C. Melendres, eds., NATO ASI Series, Vol. 320, Kluwer Academic Publishers, Dordrecht (1990), pp. 133–153.

34. J. D. Rudnicki, F. R. McLarnon, and E. J. Cairns, *In-situ* characterization of electrode processes by photothermal deflection spectroscopy, in *Techniques for Characterization of Electrodes and Electrochemical Processes*, R. Varma and J. R. Selman, eds., Wiley, New York (1991), pp. 127–166.

35. J. A. Richards and D. H. Evans, Flow cell for electrolysis within the probe of a nuclear magnetic resonance spectrometer, *Anal. Chem.*, **47**, 964–966 (1975).

36. T. M. McKinney, Electron spin resonance and electrochemistry, in *Electroanalytical Chemistry*, Vol. 10, A. J. Bard, ed., Marcel Dekker, New York (1977), pp. 97–278.

37. S. Bruckenstein and R. Rao Gadde, Use of a porous electrode for mass spectrometric determination of volatile electrode reaction products, *J. Am. Chem. Soc.*, **93**, 793–794 (1971).

38. B. Bittins-Cattaneo, E. Cattaneo, P. Konigshoven, and W. Vielstich, New developments in electrochemical mass spectrometry, in *Electroanalytical Chemistry*, Vol. 17, A. J. Bard, ed., Marcel Dekker, New York (1991), pp. 181–220.

39. J. Robinson, X-ray techniques, in *Spectroelectrochemistry: Theory and Practice*, R. J. Gale, ed., Plenum Press, New York (1988), pp. 9–40.

40. H. D. Abruna, X-ray absorption spectroscopy in the study of electrochemical systems, in *Electrochemical Interfaces: Modern Techniques for In-Situ Interface Characterization*, H. D. Abruna, ed., VCH Publishers, New York (1991), pp. 1–54.

41. J. H. White, Elucidation of structural aspects of electrode/electrolyte interfaces with x-ray standing waves, in *Electrochemical Interfaces: Modern Techniques for In-Situ Interface Characterization*, H. D. Abruna, ed., VCH Publishers, New York (1991), pp. 131–154.

42. M. F. Toney and O. R. Melroy, Surface X-ray scattering, in *Electrochemical Interfaces: Modern Techniques for In-Situ Interface Characterization*, H. D. Abruna, ed., VCH Publishers, New York (1991), pp. 55–129.

43. H. B. Mark and B. S. Pons, An *in-situ* spectrophotometric method for observing

the infrared spectra of species at the electrode surface during electrolysis, *Anal. Chem.*, **38**, 119–121 (1966).

44. D. R. Tallant and D. H. Evans, Applications of infrared internal reflectance spectrometry to studies of the electrochemical reduction of carbonyl compounds, *Anal. Chem.*, **41**, 835–838 (1965).

45. T. Higashiyama and T. Takenaka, Infrared attenuated total reflection spectra of adsorbed layers at the interface between a germanium electrode and an aqueous solution of sodium laurate, *J. Phys. Chem.*, **78**, 941–947 (1974).

46. A. Hatta, Y. Sasaki, and W. Suetaka, Polarization modulation infrared spectroscopic measurements of thiocyanate and cyanide at the silver electrode/aqueous electrolyte interface by means of Kretschmann's ATR prism configuration, *J. Electroanal. Chem.*, **215**, 93–102 (1986).

47. A. Hatta, Y. Chiba, and W. Suetaka, *In-situ* infrared measurements of thiocyanate at a silver/electrolyte interface by the excitation of surface plasmon polaritons, *Appl. Surf. Sci.*, **25**, 327–332 (1986).

48. H. Neugebauer, G. Nauer, N. Brinda-Kanopik, and R. Kellner, *In-situ* investigations of metal surfaces in electrolyte solutions, *Fresenius Z. Anal. Chem.*, **314**, 266–272 (1983).

49. M. C. Pham, F. Adami, P. C. Lacaze, J. P. Doucet, and J. E. Dubois, *In-situ* investigation of the film growth mechanism using multiple internal reflection Fourier transform infrared spectroscopy (MIRFTIRS). Part I. Behavior of iron and platinum electrodes in alkaline methanolic solutions, *J. Electroanal. Chem.*, **201**, 413–421 (1986).

50. D. B. Parry, J. M. Harris, and K. Ashley, Multiple internal reflection Fourier transform infrared spectroscopic studies of thiocyanate adsorption on silver and gold, *Langmuir*, **6**, 209–217 (1990).

51. A. Bewick, K. Kunimatsu, and B. S. Pons, Infrared spectroscopy of the electrode–electrolyte interphase, *Electrochim. Acta*, **25**, 465–468 (1980).

52. A. Bewick and K. Kunimatsu, Infrared spectroscopy of the electrode–electrolyte interphase, *Surf. Sci.*, **101**, 131–138 (1980).

53. T. Davidson, B. S. Pons, A. Bewick, and P. P. Schmidt, Vibrational spectroscopy of the electrode/electrolyte interface. Use of Fourier transform infrared spectroscopy, *J. Electroanal. Chem.*, **125**, 237–241 (1981).

54. S. Pons, T. Davidson, and A. Bewick, Vibrational spectroscopy of the electrode–solution interphase. 2. Use of Fourier transform spectroscopy for recording infrared spectra of ion radical intermediates, *J. Am. Chem. Soc.*, **105**, 1802–1805 (1983).

55. S. Pons, T. Davidson, and A. Bewick, Vibrational spectroscopy of the electrode–solution interface. Part III. Use of Fourier transform spectroscopy for observing double layer reorganization, *J. Electroanal. Chem.*, **140**, 211–216 (1982).

56. S. Pons, T. Davidson, A., and Bewick, Vibrational spectroscopy of the electrode–electrolyte interface. Part IV. Fourier transform infrared spectroscopy: Experimental considerations, *J. Electroanal. Chem.*, **160**, 63–71 (1984).

57. D. S. Corrigan and M. J. Weaver, Coverage-dependent orientation of adsorbates as probed by potential-difference infrared spectroscopy: Azide, cyanate, and thiocyanate at silver electrodes, *J. Phys. Chem.*, **90**, 5300–5306 (1986).

58. D. S. Corrigan, L. W. H. Leung, and M. J. Weaver, Single potential alteration surface infrared spectroscopy: Examination of adsorbed species involved in irreversible electrode reactions, *Anal. Chem.*, **59**, 2252–2256 (1987).

59. D. S. Corrigan and M. J. Weaver, Mechanism of formic acid, methanol, and carbon monoxide electro-oxidation at platinum electrodes as examined by single potential alteration infrared spectroscopy, *J. Electroanal. Chem.*, **241**, 143–162 (1988).

60. L. W. H. Leung and M. J. Weaver, Influence of adsorbed carbon monoxide on the electrocatalytic oxidation of simple organic molecules at platinum and palladium electrodes in acidic solution: A survey using real-time FTIR spectroscopy, *Langmuir*, **6**, 323–333 (1990).

61. A. Bewick, K. Kunimatsu, J. Robinson, and J. W. Russell, IR vibrational spectroscopy of species in the electrode–electrolyte solution interphase, *J. Electroanal. Chem.*, **119**, 175–185 (1981).

62. J. W. Russell, M. Severson, K. Scanlon, J. Overend, and A. Bewick, Infrared spectrum of CO adsorbed on a platinum electrode, *J. Phys. Chem.*, **87**, 293–297 (1983).

63. W. G. Golden, K. Kunimatsu, and H. Seki, Application of polarization modulated Fourier transform infrared reflection–absorption spectroscopy to the study of carbon monoxide adsorption and oxidation on a smooth platinum electrode, *J. Phys. Chem.*, **88**, 1275–1277 (1984).

64. K. Kunimatsu, H. Seki, and W. G. Golden, Polarization modulated FTIR spectra of cyanide adsorbed on a silver electrode, *Chem. Phys. Lett.*, **108**, 195–199 (1984).

65. K. Kunimatsu, W. G. Golden, H. Seki, and M. R. Philott, Carbon monoxide adsorption on a platinum electrode studied by polarization modulated FT-IRRAS, *Langmuir*, **1**, 245–250 (1985).

66. J. K. Foley and S. Pons, *in-situ* infrared spectroelectrochemistry, *Anal. Chem.*, **57**, 945A–956A (1985).

67. A. Bewick and S. Pons, Infrared spectroscopy of the electrode–electrolyte solution interface, in *Advances in Infrared and Raman Spectroscopy*, Vol. 12, R. J. H. Hester and R. E. Clark, eds., Wiley–Hayden, London (1985), pp. 1–63.

68. S. Pons, J. K. Foley, J. Russell, and M. Severson, Interfacial infrared vibrational spectroscopy, in *Modern Aspects of Electrochemistry*, Vol. 17, J. O'M. Bockris and E. Yeager, eds., Plenum Press, New York (1986), pp. 223–302.

69. J. K. Foley, C. Korzeniewski, J. L. Daschbach, and S. Pons Infrared vibrational spectroscopy of the electrode–solution interface, in *Electroanalytical Chemistry*, Vol. 14, A. J. Bard, ed., Marcel Dekker, New York (1986), pp. 309–440.

70. K. Ashley and S. Pons, Infrared spectroelectrochemistry, *Chem. Rev.*, **88**, 673–695 (1988).

71. B. Beden and C. Lamy, Infrared reflectance spectroscopy, in *Spectroelectro-*

chemistry: Theory and Practice, R. J. Gale, ed., Plenum Press, New York (1988), pp. 189–261.

72. S. M. Stole, D. D. Popenoe, and M. D. Porter, Infrared spectroelectrochemistry: A probe of the molecular architecture of the electrochemical interface, in *Electrochemical Interfaces: Modern Techniques for In-situ Interface Characterization*, H. D. Abruna, ed., VCH Publishers, New York (1991), pp. 339–410.

73. D. L. Jeanmaire, M. R. Suchanski, and R. P. Van Duyne, Resonance Raman spectroelectrochemistry. I. The tetracyanoethylene anion radical, *J. Am. Chem. Soc.*, **97**, 1699–1707 (1975).

74. D. L. Jeanmaire and R. P. Van Duyne, Resonance Raman spectroelectrochemistry. II. Scattering spectroscopy accompanying excitation of the lowest $^2B_{1u}$ excited state of the tetracyanoquinodimethane anion radical, *J. Am. Chem. Soc.*, **98**, 4029–4033 (1976).

75. D. L. Jeanmaire and R. P. Van Duyne, Resonance Raman spectroelectrochemistry. III. Tunable dye laser excitation spectroscopy of the lowest $^2B_{1u}$ excited state of the tetracyanoquinodimethane anion radical, *J. Am. Chem. Soc.*, **98**, 4034–4039 (1976).

76. M. R. Suchanski and R. P. Van Duyne, Resonance Raman spectroelectrochemistry. IV. The oxygen decay chemistry of the tetracyanoquinodimethane dianion, *J. Am. Chem. Soc.*, **98**, 250–252 (1976).

77. D. L. Jeanmaire and R. P. Van Duyne, Resonance Raman spectroelectrochemisty. V. Intensity transients on the millisecond timescale following double potential step initial of a diffusion controlled electrode reaction, *J. Electroanl. Chem.*, **66**, 235–247 (1975).

78. R. P. Van Duyne, M. R. Suchanski, J. M. Lakovits, A. R. Siedle, and R. D. Parks, Resonance Raman spectroelectrochemistry. 6. Ultraviolet laser excitation of the tetracyanoquinodimethane dianion, *J. Am. Chem. Soc.*, **101**, 2832–2837 (1979).

79. M. Fleischmann, P. J. Hendra, and A. J. McQuillan, Raman spectra from electrode surfaces, *J. Chem. Soc. Chem. Commun.*, **3**, 80–81 (1973).

80. M. Fleischmann, P. J. Hendra, and A. J. McQuillan, Raman spectra of pyridine adsorbed at a silver electrode, *Chem. Phys. Lett.*, **26**, 163–166 (1974).

81. D. L. Jeanmaire and R. P. Van Duyne, Surface Raman spectroelectrochemistry. Part I. Heterocyclic, aromatic, and aliphatic amines adsorbed on the anodized silver electrode, *J. Electroanal. Chem.*, **84**, 1–20 (1977).

82. M. G. Albrecht and J. A. Creighton, Anomalously intense Raman spectra of pyridine at a silver electrode, *J. Am. Chem. Soc.*, **99**, 5215–5217 (1977).

83. R. P. Cooney, M. M. Mahoney, and A. J. McQuillan, Raman spectroelectrochemistry, in *Advances in Infrared and Raman Spectroscopy*, Vol. 9, R. J. H. Clark and R. E. Hester, eds., Heyden, London (1982), pp. 188–278.

84. R. L. Birke, J. R. Lombardi, and L. A. Sanchez, Surface enhanced Raman spectroscopy. A new technique for studying interfacial phenomena, in *Electrochemical and Spectrochemical Studies of Biological Redox Components*,

ACS Symposium Series No. 201, K. Kadish, ed., American Chemical Society, Washington, DC (1982), pp. 69–107.

85. M. Fleischmann and I. R. Hill, Raman Spectroscopy, in *Comprehensive Treatise on Electrochemistry*, Vol. 8, R. E. White, J. O'M. Bockris, B. E. Conway, E. Yeager, eds., Plenum Press, New York (1984), pp. 373–432.

86. R. K. Chang and B. L. Laube, Surface enhanced Raman scattering and nonlinear optics applied to electrochemistry, *CRC Crit. Rev. Mater. Sci.*, **12**, 1–73 (1984).

87. S. Efrima, Surface enhanced Raman scattering, in *Modern Aspects of Electrochemistry*, Vol. 16, B. E. Conway, R. E. White, and J. O'M. Bockris, eds., Plenum Press, New York (1985), pp. 253–369.

88. R. K. Chang, Surface enhanced Raman scattering at electrodes: A status report, *Ber. Bunsenges. Phys. Chem.*, **91**, 296–305 (1987).

89. R. L. McCreery, Electronic and vibrational spectroscopy of electrochemical events, *Prog. Anal. Spectrosc.*, **11**, 141–178 (1988).

90. R. L. Birke and J. R. Lombardi, Surface enhanced Raman scattering, in *Spectro-electrochemistry: Theory and Practice*, R. J. Gale, ed., Plenum Press, New York (1988), pp. 263–348.

91. R. L. McCreery and R. T. Packard, Raman monitoring of dynamic electrochemical events, *Anal. Chem.*, **61**, 775A–789A (1989).

92. R. L. Garrell, Surface enhanced Raman spectroscopy, *Anal. Chem.*, **61**, 401A–411A (1989).

93. J. E. Pemberton, Surface enhanced Raman scattering, in *Electrochemical Interfaces: Modern Techniques for In-Situ Interface Characterization*, H. D. Abruna, ed., VCH Publishers, New York (1991), pp. 193–263.

94. C. Shannon and A. Campion, Unenhanced Raman scattering as an *in-situ* probe of the electrode–electrolyte interface: 4-Cyanopyridine adsorbed on a rhodium electrode, *J. Phys. Chem.*, **92**, 1385–1387 (1988).

95. S. Byahut and T. E. Furtak, A device for performing surface plasmon-polariton-assisted Raman scattering from adsorbates on single crystal silver surfaces, *Rev. Sci. Instrum.*, **61**, 27–32.

96. S. Byahut and T. E. Furtak, Direct comparison of the chemical properties of single crystal Ag(111) and electrochemically roughened Ag as substrates for surface Raman scattering, *Langmuir*, **7**, 508–513 (1991).

97. C. H. Lee, R. K. Chang, and N. Bloembergen, Nonlinear electroreflectance in silicon and silver, *Phys. Rev. Lett.*, **18**, 167–170 (1967).

98. G. L. Richmond, Investigation of electrochemical interfaces by nonlinear optical methods, in *Electrochemical Interfaces: Modern Techniques for In-situ Interface Characterization*, H. D. Abruna, ed., VCH Publishers, New York (1991), pp. 265–337.

99. J. T. Golab, J. R. Sprague, K. T. Carron, G. C. Schatz, and R. P. Van Duyne, A surface enhanced hyper-Raman scattering study of pyridine adsorbed on silver: Experiment and theory, *J. Chem. Phys.*, **88**, 7942–7951 (1988).

100. S. C. Chang, Y. Ho, and M. J. Weaver, Application of real-time infrared spectroscopy to electrocatalysis at bimetallic surfaces. I Electrooxidation of formic acid and methanol on bismuth-modified Pt(111) and Pt(100), *Surf. Sci.*, **265**, 81–94 (1992).

101. S. C. Chang and M. J. Weaver, *In-situ* infrared spectroscopy at single-crystal metal electrodes: An emerging link between electrochemical and ultrahigh-vacuum surface science, *J. Phys. Chem.*, **95**, 5391–4000 (1991).

102. For example, see R. Parsons and T. Vanderfoot, The oxidation of small organic molecules. A survey of recent fuel cell related research, *J. Electroanal. Chem.*, **257**, 9–45 (1988); B. Bittins-Cattaneo, E. Sento, W. Vielstich, and U. Linke, Study of the methanol adsorbates on Pt(100) and Pt(111) single crystal surfaces, *Electrochim. Acta*, **33**, 1499–1506 (1988).

103. S. C. Chang and M. J. Weaver, *In-situ* infrared spectroscopy of carbon monoxide adsorbed at ordered platinum (100) aqueous interfaces: Double-layer effects upon the adsorbate binding geometry, *J. Phys. Chem.*, **94**, 5095–5102 (1990).

104. S. C. Chang and M. J. Weaver, Influence of coabsorbed bismuth and copper on carbon monoxide adlayer structures at ordered low-index platinum-aqueous interfaces, *Surface Sci.*, **241**, 11–24 (1991).

105. Y. M. Alsmeyer and R. L. McCreery, Surface-enhanced Raman spectroscopy of carbon electrode surface following silver electrodeposition, *Anal. Chem.*, **63**, 1289–1293 (1991).

106. For example, see R. L. McCreery, Carbon electrodes: Structural effects on electron transfer kinetics, in *Electroanalytical Chemistry*, Vol. 17, A. J. Bard, ed., Marcel Dekker, New York (1991), pp. 221–374.

107. R. J. Bowling, R. T. Packard, and R. L. McCreery, Activation of highly ordered pyrolytic graphite for heterogeneous electron transfer: Relationship between electrochemical performance and carbon microstructure, *J. Am. Chem. Soc.*, **111**, 1217–1223 (1989).

108. R. J. Bowling, R. T. Packard, and R. L. McCreery, Mechanism of electrochemical activation of carbon electrodes: Role of graphite lattice defects, *Langmuir*, **5**, 683–688 (1989).

109. R. J. Rice and R. L. McCreery, Quantitative relationship between electron transfer rate and surface microstructure of laser-modified graphite electrodes, *Anal. Chem.*, **61**, 1637–1641 (1989).

110. Y. Wang, D. C. Alsmeyer, and R. L. McCreery, Raman spectroscopy of carbon materials: Structural basis of observed spectra, *Chem. Mater.*, **2**, 557–563 (1990).

111. J. E. Pemberton, SERS from thin Ag films electrochemically deposited onto Pt electrodes, *J. Electroanal. Chem.*, **167**, 317–323 (1984).

112. Y. W. Alsmeyer and R. L. McCreery, Surface enhanced Raman examination of carbon electrodes: Effects of laser activation and electrochemical pretreatment, *Langmuir*, **7**, 2370–2375 (1991).

113. K. Hutchinson, A. J. McQuillan, and R. E. Hester, Nonaqueous SERS: Pyridine

in *N,N'*-dimethylformamide solution at a silver electrode, *Chem. Phys. Lett.*, **98**, 27–31 (1983).

114. A. M. Stacy and R. P. Van Duyne, Surface enhanced Raman and resonance Raman spectroscopy in a nonaqueous electrochemical environment: Tris(2,2'-bipyridine)ruthenium(II) adsorbed on silver from acetonitrile, *Chem. Phys. Lett.*, **102**, 365–370 (1983).

115. H. R. Virdee and R. E. Hester, Surface enhanced Raman spectra of $Ru(bpy)_3^{2+}$. and electrochemically generated $Ru(bpy)_3^+$ on a silver electrode, *J. Phys. Chem.*, **88**, 451–455 (1988).

116. R. P. Van Duyne, J. P. Haushalter, M. Janik-Czachor, and N. Levinger, Surface enhanced resonance Raman spectroscopy of adsorbates on semiconductor electrode surfaces. 2. *In-Situ* studies of transition metal (Fe and Ru) complexes on Ag/GaAs and Ag/Si, *J. Phys. Chem.*, **89**, 4055–4061 (1985).

117. C. D. Tait, T. M. Vess, and D. W. Wertz, A nonaqueous SERS study of the 2,2'-bipyridine free ligand and the tris(bpy) Ru(II) complex reduction products, *Chem. Phys. Lett.*, **142**, 225–230 (1987).

118. L. A. Sanchez and T. G. Spiro, Surface enhanced Raman spectroscopy as a monitor of iron(III) protoporphyrin reduction at a Ag electrode in aqueous and acetonitrile solutions: Vibronic resonance enhancement amplified by surface enhancement, *J. Phys. Chem.*, **89**, 763–768 (1985).

119. D. A. Guzonas, G. F. Atkinson, and D. E. Irish, Surface enhanced Raman spectroscopy of the nonaqueous system silver/acetonitrile, KSCN, *Chem. Phys. Lett.*, **107**, 193–197 (1984).

120. G. S. Shin and J. J. Kim, Surface enhanced Raman scattering in a nonaqueous electrochemical cell: Pyridine, benzoic acid, and nitrobenzene in methanol, *Surf. Sci.*, **158**, 286–294 (1985).

121. J. J. Kim and G. S. Shin, Surface enhanced Raman scattering in nonaqueous electrochemical systems of alcoholic solvents, *Chem. Phys. Lett.*, **118**, 493–497 (1985).

122. G. S. Shin and J. J. Kim, Nonaqueous surface enhanced Raman scattering spectra of benzene, *Chem. Phys. Lett.*, **120**, 569–572 (1985).

123. E. Peled, Lithium stability and film formation in organic and inorganic electrolytes for lithium battery systems, in *Lithium Batteries*, J. P. Gabano, ed., Academic Press, New York (1983), pp. 43–72.

124. D. J. David, M. H. Fronig, T. N. Wittberg, and W. E. Modeman, Surface reactions of Li with the environment, *J. Electrochem. Soc.*, **126**, 306C (1979).

125. K. R. Zavadil and N. R. Armstrong, Surface chemical and electrochemical processes of well-characterized lithium thin films: Electrochemical reactions in SO_2/CH_3CN, *J. Electrochem. Soc.*, **137**, 2371–2378 (1990).

126. M. Odziemkowski and D. E. Irish, An electrochemical study of reactivity at the lithium electrolyte/Bare lithium metal interface. I. Purified electrolytes, *J. Electrochem. Soc.*, **139**, 3063–3074 (1992).

127. M. Odziemkowski, M. Krell, and D. E. Irish, A Raman microprobe *in-situ* and

ex-situ study of film formation at lithium/organic electrolyte interfaces, *J. Electrochem. Soc.*, **139**, 3052–3063 (1992).

128. For example, see G. E. Blomgren, Properties, structure, and conductivity of organic and inorganic electrolytes for lithium battery systems, in *Lithium Batteries*, J. P. Gabano, ed., Academic Press, New York (1983), pp. 13–41.

129. Y. Matsui, T. Kubota, H. Tadokoro, and Y. Mizumuto, Raman spectra of polyethers, *J. Polym. Sci. A1*, **3**, 2275–2288 (1965).

130. R. F. Schaufele, Advances in vibrational Raman scattering spectroscopy of polymers, *Trans. N.Y. Acad. Sci.*, **30**, 69–80 (1967).

131. P. Dreyfuss, Polymerization, in *Polytetrahydrofuran*, Gordon and Breach Publishers, New York (1983), pp. 17–74.

132. K. Imada, H. Tadokoro, A. Umehara, and S. Murahashi, Normal vibrations of the polymer molecules of helical conformation. VI Polytetrahydrofuran and deuterated polytetrahydrofuran. *J. Chem. Phys.*, **42**, 2807–2816 (1965).

133. Z. Deng and D. E. Irish, A Raman spectral study of solvation and ion association in the systems $LiAsF_6/CH_3CO_2CH_3$ and $LiAsF_6/HCO_2CH_3$, *Can. J. Chem.*, **69**, 1766–1773 (1991).

134. H. H. Weider, Resistivity and conductivity, in *Laboratory Notes on Electrical and Galvanomagnetic Measurements*, C. Laird, ed., Materials Science Monographs No. 2, Elsevier, Amsterdam (1979), pp. 1–38.

135. D. E. Irish, I. R. Hill, P. Archambault, and G. F. Atkinson, Investigations of electrode surfaces in acetonitrile solutions using surface-enhanced Raman spectroscopy, *J. Solution Chem.*, **14**, 221–243 (1985).

136. J. E. Pemberton, Surface enhanced Raman scattering at Ag electrodes in nonaqueous pyridine solutions, *Chem. Phys. Lett.*, **115**, 321–327 (1985).

137. I. R. Hill, D. E. Irish, and G. F. Atkinson, Investigations of silver electrode surfaces in propylene carbonate/alkali halide electrolytes by surface enhanced Raman scattering, *Langmuir*, **2**, 752–757 (1986).

138. R. L. Sobocinski and J. E. Pemberton, SERS investigation of interfacial methanol at silver electrodes, *Langmuir*, **6**, 43–50 (1990).

139. R. L. Sobocinski, M. A. Bryant, J. E. Pemberton, Surface Raman scattering of methanol, 1-propanol, 1-pentanol, and 1-butanethiol on *in-situ* and emersed silver electrodes, *J. Am. Chem. Soc.*, **112**, 6177–6183 (1990).

140. R. L. Sobocinski and J. E. Pemberton, Determination of alcohol solvent orientation and bonding at silver electrodes using surface enhanced Raman scattering: Methanol, ethanol, 1-propanol, and 1-pentanol, *Langmuir*, **8**, 2049–2063 (1992).

141. S. L. Joa and J. E. Pemberton, A surface enhanced Raman scattering investigation of interfacial structure at Ag electrodes in electrolyte solutions of the isomers of butanol, *Langmuir*, **8**, 2301–2310 (1992).

142. J. E. Pemberton, M. A. Bryant, and S. L. Joa, A simple method for determination of orientation of adsorbed organics of low symmetry using surface enhanced Raman scattering, *J. Phys. Chem.*, **96**, 3776–3782 (1992).

143. S. L. Joa and J. E. Pemberton, *J. Chem. Soc., Faraday Trans.* Determination of interfacial solvent structure in 1-butanol, 2-butanol, and iso-butanol at Au electrodes using surface enhanced Raman scattering: similarity to behavior at Ag electrodes, submitted.

144. B. W. Lee, M. Y. Kwon, and J. J. Kim, Surface enhanced Raman scattering of water molecules in water–ethanol solutions, *Chinese J. Phys.*, **28**, 1–8 (1990).

145. J. E. Pemberton and S. L. Joa, Water and electrolyte structure at Ag electrodes in nonaqueous butanol solutions using surface enhanced Raman scattering, *J. Electroanal. Chem.*, **378**, 149–158 (1994).

146. J. E. Pemberton and R. L. Sobocinski, Raman spectroscopy of the emersed Ag/alcohol electrochemical interface, *J. Electroanal. Chem.*, **318**, 157–169 (1991).

147. S. L. Joa and J. E. Pemberton, Surface Raman scattering of the butanol isomers on emersed Ag and Au electrodes, *J. Phys. Chem.*, **97**, 9420–9424 (1993).

148. M. A. Bryant, S. L. Joa, and J. E. Pemberton, Raman scattering from monolayer films of thiophenol and 4-mercaptopyridine at Pt surfaces, *Langmuir*, **8**, 753–756 (1992).

CHAPTER

3

SCANNING TUNNELING AND ATOMIC FORCE MICROSCOPY OF ELECTROCHEMICAL INTERFACES

MICHAEL D. WARD

Department of Chemical Engineering and Materials Science
University of Minnesota
Minneapolis, Minnnesota 55455

HENRY S. WHITE

Department of Chemistry
University of Utah
Salt Lake City, Utah 84112

I. INTRODUCTION

Electrochemical processes ultimately hinge on the atomic- and molecular-level interactions between an electrified surface and the components of a liquid electrolyte in contact with the surface. Consequently, measurements of

Modern Techniques in Electroanalysis, Edited by Petr Vanýsek, Chemical Analysis Series, Vol. 139.
ISBN 0-471-55514-2 © 1996 John Wiley & Sons, Inc.

interfacial structure and properties, including the electrode topography, the crystallographic orientation of the electrode, the double-layer structure, reactant concentration, and the presence and concentration of adsorbates and contaminants, are crucial in understanding the factors controlling interfacial electron-transfer reactions. Electrochemical processes conventionally are investigated by a variety of techniques that involve measurement of current response following a perturbation of the electrode potential.[1,2] However, these approaches generally provide only a macroscopic picture of behavior that are interpreted in terms of a statistical ensemble of atomic- or molecular-level events. Numerous ultrahigh vacuum (UHV) surface analysis techniques such as x-ray diffraction, low-energy electron diffraction, x-ray photoelectron spectroscopy, Auger electron spectroscopy, and infrared and Raman spectroscopy have provided atomic-level structural details of the electrode–electrolyte interface. While these techniques provide valuable characterizattion of adsorbates, elecctrode structure, and composition of the electrode–electrolyte interface,[3–11] the area of the sample under interrogation is at least several orders of magnitude larger than molecular dimensions. That is, these methods do not allow direct atomically resolved measurements of surface structure.

With the advent of scanning tunneling microscopy (STM) in 1982[12] and atomic force microscopy (AFM) in 1986,[13] there now exist experimental protocols that provide true atomic- and molecular-level characterization of electrode interfaces in real space. STM allows simultaneous measurement of electrode topography and electric current between an atomically sharp tip and the electrode, while AFM allows simultaneous measurement of electrode topography and interacton force between a small tip and the electrode. In principle, these capabilities allow investigations of electrode structure with angstrom resolution in the plane of the electrode and subangstrom resolution in the vertical direction. While both STM and AFM can be used for *ex situ* characterization of electrodes, they are uniquely capable of direct *in situ* characterization of electrode–electrolyte interfaces. STM and AFM therefore provide an opportunity to examine electrode processes at size scales that were unheard of only 10 years ago. Coincidentally, this development parallels trends in electrochemistry toward nanoscale electrodes.[14,15]

Both STM and AFM offer several real advantages in relationship to other surface analytical methods. The most obvious of these is that individual structures of atomic dimensions can be imaged. With the exception of high-resolution transmission electron microscopy, no other surface analytical method possesses this capability. Detailed structural information about individual molecular adsorbates, phase boundaries between domains of metal atoms, and atomic vacancies and steps are obtained readily using STM and, to a lesser extent, AFM. This is a particularly important feature for electrochemical studies, since such heterogeneities are often more important in controlling the electrode behaviour than the average structure of more ideally ordered regions of the surface. It is primarily for this reason that STM and AFM will be essential tools in future investigations of electrochemical interfaces. A second advantage of STM and AFM is that these microscopies provide *real-space* images. Provided that STM and AFM images are interpreted correctly in context of the electronic structure and force fields, real-space images can directly provide the positions of atoms and molecules at the interface. The unique capability distinguishes STM and AFM from diffraction techniques (e.g., LEED and SEXAFS) where structures are deduced by inverse Fourier transforms of reciprocal-space data. Except for the simplest structures, extraction of the true physical structure of an electrochemical interface from diffraction data is complex. In fact, the physical structure deduced from a diffraction method is not necessarily even a unique structure, but one that is only consistent with the diffraction pattern. Several detailed examples have recently appeared where *real-space* STM images have been shown to be consistent with LEED or SEXAFS data, but are entirely different from the physical structure deduced from the *reciprocal-space* data. Thus, within limitations discussed in later sections, STM and AFM images can be interpreted in a more direct fashion and without recourse to fitting data to models. Finally, STM and AFM are moderately rapid, structural probes that allow the dynamics of chemical reactions (e.g., adsorption of molecules and metal dissolution) to be observed in real time. In this manner, these techniques are complementary to *in situ* spectroscopic probes—for example, time-resolved, surface-enhanced Raman spectroscopy (SERS). This capability yields chemical information that far exceeds that obtained by static methods (electron spectroscopy) in which electrode structure is examined *post facto*.

The purpose of this review is to provide the reader with (1) a fundamental understanding of the two methods and (2) the basic principles of operation, including typical instrumental and experimental configurations. The advantages of *in situ* structural probes that provide real-space, atomically resolved images will be noted in comparison to techniques that provide averaged structural information. It should be emphasized that STM and AFM characterization, like other techniques, can occasionally produce artifacts that can lead to misinterpretation of data. Some of the difficulties in interpreting images will be discussed briefly in terms of current knowledge of the imaging mechanisms of STM and AFM. We will then present several applications of these microscopies to the characterization of electrode–electrolyte interfaces, with an emphasis on *in situ* characterization. Since space does not permit a full comprehensive review, the reader is referred to other review articles,[16–18] especially the recent text edited by Bonnell[19] that presents a pedagogical treatment of a broad range of topics of interest to STM and AFM users.

II. THEORY AND BASIC PRINCIPLES

STM and AFM have many features in common; in fact, the two methods frequently are performed on a common workstation and share much of the electronic and mechanical components. They differ, however, with respect to the features required for the measurements unique to each technique, namely, current measurement for the STM and force measurement for the AFM. In this section we will describe first the operation of the STM and follow with a description of the AFM, with emphasis on the components that distinguish it from the STM.

A. Scanning Tunneling Microscopy

The essential element of STM is a metallic tip, ideally terminating in a single atom, whose position with respect to a sample can be accurately controlled. When the tip is placed sufficiently close to a sample such that the atomic orbitals of the tip and sample overlap, generally within ~ 10 Å, there is a finite probability that electrons can tunnel across the gap between the tip and the sample. When a small voltage bias is applied between the tip and the sample, the resulting net tunneling current, i_T, can be described generally by Eq. (1), where $\rho_s(\mathbf{r}, E)$ is the combined density of states (DOS) of the sample (including the sample and any adsorbed impurities), $\rho_t(\mathbf{r}, E, eV)$ is the DOS of the tip, and $T(\mathbf{r}, E, eV)$ is the transmission probability for electrons crossing the tunneling gap.[20] The term $T(\mathbf{r}, E, eV)$ is strongly dependent upon distance due to the exponential decay of the electron wave function outside of the metal surface. Consequently, i_T decays exponentially with increasing distance between the tip and the sample according to Eq. (2), where W is the height of the sample surface potential barrier and s is the tip–sample separation. In addition, since the highest energy states of electrons have the largest spatial extension outside the metal surface, the tunneling current will result from electrons at, or slightly below, the Fermi level. Therefore, i_T is approximately proportional to the DOS of the metal and tip at a tip–sample separation s and the Fermi energy E_F.[21,22] For a constant DOS, i_T decreases approximately one order of magnitude for ~ 1 Å increase in the tip–sample distance. It is important to note, however, that the contribution from different orbitals may differ with distance from the electrode, with those near the center of the surface Brillouin zone dominating at large distances.

$$i_T = \int_0^{eV} \rho_s(\mathbf{r}, E)\rho_t(\mathbf{e}, E, eV)T(\mathbf{r}, E, eV)dE \qquad (1)$$

$$i_T \propto e^{-2s\sqrt{2(W-E)}} \qquad (2)$$

The STM is configured so that the sample is mounted to a stage affixed to a piezoelectric crystal capable of precisely controlling and measuring the tip height with respect to the sample (z), as well as the position of the tip in the xy plane of the sample (see Figure 1). The piezoelectric crystal experiences dimensional changes when an electric field is applied across its surfaces, with the direction of the deformation depending upon the orientation of the applied field. This deformation can be controlled to within < 0.1 Å for high-resolution scanners. Alternative configurations are emerging in which the z direction is controlled with a separate piezoelectric crystal. Decoupling the z control from the xy piezocontroller may have advantages in large-area scans where non-linearities can influence the z-axis deformation.

The ability to scan in the xy plane provides a spatial representation of the distribution $\rho(\mathbf{r}, E_F)$ over the sample surface. In practice, the STM can be operated in one of two modes. Operation with a constant applied current between the tip and the substrate forces the tip height to vary in order to maintain a constant $\rho(\mathbf{r}, E_F)$ as the tip is scanned across the surface. This is accomplished with a feedback mechanism that senses the change in i_T and adjusts the potential applied to the z-controller, V_z, so that the tip height is changed to maintain constant i_T. The data collected are $V_z(x, y)$, which can be directly converted, using the piezoelectric expansion coefficient, to a plot of apparent surface topography—that is, $z(x, y)$. Conversely, the STM can be operated so that the vertical position of the tip is fixed, and the variation in i_T is measured as the tip is rastered across the surface. The data in this mode reflect directly the spatial dependence of $\rho(\mathbf{r}, E_F)$. In either mode, the contrast observed in STM images (variation in z in constant current operaton or i_T in constant height operation) reflects the surface topography as well as the local surface DOS. It is important to remember that although the tunneling current decays exponentially with distance of the tip from states on the surface of the sample, it is possible that surface regions closer to the tip but with small DOS will afford minimal tunneling current, and vice versa.

While the majority of STM studies have been performed in air or vacuum, recent developments allow STM to be performed in electrolyte solutions as well. The electrochemical STM employs a bipotentiostat which is capable of setting the voltages E_{tip} and E_{sample} to fixed values with respect to a reference electrode contained in the solution. Consequently, the tip–sample bias potential required for tunneling is determined by the difference ($E_{\text{tip}} - E_{\text{sample}}$). In general the bias potential ($\leqslant 50\,\text{mV}$) is smaller than the potential difference between the sample and the reference electrode. In this manner, the sample remains under active potential control during STM imaging. In these experiments, it is critical to insulate the tip so that faradaic current at the tip is approximately an order of magnitude smaller than the tunneling current. Since tunneling currents in electrochemical STM are typically of the order of

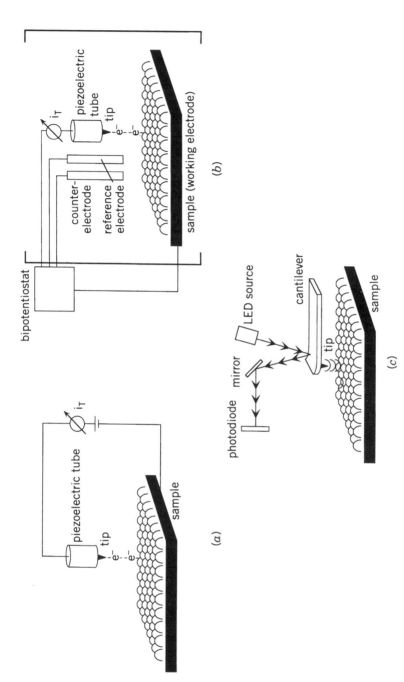

Figure 1. (**a**) Schematic representation of a scanning tunneling microscope. The scanning probe tip is scanned in the *xy* plane, and dependent variables of current (STM) or force (AFM) are measured. (**b**) Schematic representation of an electrochemical STM. A bipotentiostat controls the potential of both the tip and the sample with respect to a reference electrode. The tip is designed so that its area is small, minimizing faradaic current. (**c**) Schematic representation of an AFM. The displacement of the tip is calculated from the deflection of an LED beam off the cantilever supporting the tip to a photodiode.

112

1–10 nA, the faradaic current at the tip must be approximately 0.1 nA. This small value can be achieved if the exposed area of the tip is ∼ 5 μm^2. Large faradaic currents, such as that resulting from H^+ reduction at a Pt tip, greatly reduce the image resolution. Typically, the tip is insulated with glass or Apiezon wax so that its sides are insulated and tunneling only occurs through a small protrusion at the tip. Details of the tip preparation have been described elsewhere.[23]

The resolution of the STM is governed not only by the piezoelectric crystals and the sample rate, but also by the tip radius, the tip–sample distance, the nature of the states involved in tunneling, including those of the sample, molecular states in the fluid near the sample surface, and species adsorbed on the sample surface. The DOS of the tip is generally assumed to be constant at all voltages and to consist primarily of spherical s-states. However, larger tips may have contributions from higher angular quantum states. These considerations reveal that the STM tip should ideally be as small as possible, preferably a single atom.

B. Atomic Force Microscopy

The essential element of the AFM is a small tip, generally Si_3N_4, mounted on the end of a flexible cantilever. As the tip-sample distance is decreased, the tip experiences an attractive force which deflects the cantilever by an amount that depends upon the magnitude of the force and the spring constant of the cantilever. This deflection can be measured by several techniques, including the measurement of the deflection of a laser beam reflected from the cantilever, the variation in the capacitance of a capacitor detector, or the change in tunneling current between an STM tip and the cantilever. The z-deflection of the cantilever is then converted to force using Hooke's law, $F = -kz$, where k is the spring constant of the cantilever. As the tip is moved closer to the sample, the attractive force increases until the tip begins to experience repulsive forces. As the tip is withdrawn from the sample the force curve qualitatively retraces itself; however, hysteresis generally is observed owing to mechanical deformation of the tip and sample resulting from contact and capillary forces. Consequently, the "pull-back" force required to extract the tip from the sample generally is greater than the experienced moving toward the sample.

When the force measurement capability is employed while rastering the tip across the sample, the spatial dependence of the force can be measured. The electronic and mechanical controls for the AFM are essentially identical to those of the STM. Operation at constant force between the tip and the substrate forces the tip height to vary in order to maintain the constant force as the tip is scanned across the surface. This is accomplished with a feedback mechanism that senses the change in force and adjusts the potential applied to the z-controller so that the cantilever height is changed to maintain constant force. The data then reflect the tip deflection across the sample surface. Conversely, the AFM can be operated so that the vertical position of the cantilever is fixed, and the variation in force is measured as the tip is rastered across the surface. The data in this mode reflect the spatial dependence of the tip–sample forces. The AFM can be operated in either the "attractive" or "repulsive" mode, which simply signifies the operational tip–sample distance with respect to the force versus distance curve.

Recent developments have demonstrated that the AFM can be operated quite successfully in liquids. Operation in liquids eliminates capillary forces associated with films of atmospheric contaminants on the sample. Liquids also modify the attractive van der Waals forces due to changes in the Hamaker constant,[24,25] and in some cases the resulting forces can become respulsive. These properties conspire to alleviate damage to the surface caused by the tip and, in many cases, also enhance the resolution. This is particularly important, because imaging of atomic scale defects by AFM is otherwise difficult to accomplish. The ability of AFM to acquire topographical images in liquids allows *in situ* investigation of electrochemical interfaces under a wide variety of conditions, using a conventional potentiostatic circuit to control the potential of the sample.

C. Tip–Sample Interactions and Image Interpretation

The above description of STM and AFM emphasizes the operation and theory of these techniques under assumed ideal conditions. Although images of electrode–liquid interfaces displaying atomic corrugation have been obtained by both techniques, there are a number of fundamental questions associated with the imaging mechanisms that remain unresolved, especially in liquid electrolytes. The issues will become apparent in the examples of STM and AFM imaging presented below, especially in the description of tunneling spectroscopic data. However, a brief description of some nonideal behaviors commonly encountered in imaging will be presented here to alert the reader to the nuances and limitations of the techniques.

A key assumption of either STM or AFM is that the properties and geometry of the imaging tip remain absolutely constant during image acquisition. Before addressing how the tip properties might change during an experiment, it should be recognized that the tip geometry and properties are almost never characterized prior to, or, after experimentation. Claims of an atomically sharp tip are based solely on the ability to obtain images of features of atomic dimensions. As will be discussed below in context of the interactions between an AFM tip and sample, the appearance of corrugations with spacings on the order of atomic lattice dimensions can be interpreted in terms of models assuming a relatively blunt tip.

One type of variation in tip properties has already been addressed in the section on STM imaging. It was pointed out that the tunneling current is a function of the DOS of both the tip and the sample. Ideally, the DOS of a STM tip would have a weak dependence on energy, thereby allowing the measurement of the distribution of energy levels of the sample without a complex deconvolution of two contributions. Unfortunately, this is generally not the case, although it can be closely approximated in some studies, such as in the imaging of semiconductor electrodes where the DOS of the semiconducting electrode is much more highly structured than that of a sharp metal tip. Although these considerations are critical in tunneling spectroscopic (I–V) measurements (*vide infra*), most *in situ* STM images of metals and/or adsorbate structure are analyzed without critical examination of the dependence of the electronic structure of the tip. It is somewhat fortuitous for electrochemists that the apparent images of metals are not highly dependent on the sign or magnitude of the applied voltage. In electrolyte solutions, the situation may be more complex, however, because the distribution of solvent and ions around the tip depend upon of the tip bias. Under these conditions, it is reasonable to suggest that the density of states associated with the tip and surrounding electrolyte are potential dependent. It is thus possible (and highly likely in some cases) that the subtle chnges in the apparent structure of a metal–liquid interface assumed to be associated with potential-dependent property of the sample may in fact be due to changes in the electrolyte structure around the tip.

In addition to changes in the properties of the electronic structure of an STM tip, it is also likely that the tip undergoes mechanical deformations due to the electrostatic and adhesive forces between the tip and sample.[26] The following example demonstrates the magnitude of the stress induced by the electric field. Consider a parallel plate geometry for the tip and sample, separated by a vacuum. From a bias voltage of 1 V and a tip–sample separation of 10 Å, the local field-induced stress in the junction ($\sigma = 0.5\varepsilon_0 E^2$, where E is the electric field) is of the order of 10^9 N/m.[2] This estimate, which would be considerably larger for a calculation employing a sharp-tip geometry, is an order of magnitude larger than the maximum tensile stress of polycrystalline gold ($\sim 10^8$ N/m^2) and thus would be expected to cause plastic deformation of the tip (and sample). Indeed, from tunneling spectroscopic measrements of an Au–air–Au tunnel junction, we have observed that Au STM tips reproducibly undergo a stress-induced elongation, even to the point where tip fracture occurs.[26] Although more commonly employed materials for STM (e.g., Pt, Pt–Ir alloy, and W) have higher elastic modulii than does Au, it is likely that they also undergo deformations as the sample-to-tip bias is varied. Consequently, subtle potential-dependent features of images may in fact result from mechanical tip stretching, which can be manifest as a decrease or increase in the tip–sample distance and also in the electronic structure of the tip. The presence of an electrolyte in the *in-situ* experiment compounds the problem, since adsorbed ions and solvent dipoles may lower the interatomic metal–metal bond energy at the tip apex, making the STM tip even more susceptible to plastic deformation.

A second assumption that is universally implied in STM and AFM studies is that the tip does not strongly interact with or modify the surface under study (unless intentionally done so). There are a number of examples in the literature, however, that show that there is little justification for an ideal noninteracting tip. The situation is complicated by the fact that the mechanisms by which tip interactions occur will depend upon the specific system under study, requiring careful consideration of the forces within the tunnel junction. For example, consider recent *in situ* electrochemical STM investigations of the reconstruction of Au(111) and Au(100) electrodes, in which high-quality, atomically resolved, *in situ* STM images of the Au(111) surface in aqueous electrolytes revealed a potential-dependent reversible transition between the Au 1×1 and $\sqrt{3} \times 22$ surface phases.[27] Scanning the electrode potential to a value negative of the potential of zero charge (pzc) drove the 1×1 surface into a stable $\sqrt{3} \times 22$ phase. This reconstructed phase exhibited parallel pairs of corrugation lines (0.2 Å in height) oriented along the $(11\bar{2})$ direction with a pair-to-pair separation of ~ 63 Å. At potentials positive of the pzc, the $\sqrt{3} \times 22$ phase is converted back to the 1×1 structure. The interatomic bond distances in the $\sqrt{3} \times 22$ surface structure are compressed by $\sim 5\%$ relative to the 1×1 phase, suggesting that the observed transition is related to a potential-dependent tensile stress. In the original report of this behavior, the results were discussed in terms of the redistribution of charge at the electrode–electrolyte interface, without interference from the tip. However, an identical reversible transition between the $\sqrt{3} \times 22$ and 1×1 surface phases can also be induced solely by the electric field originating at the STM tip *even in the absence of an electrolyte.* For example, when a stationary STM tip was biased positive of the Au(111) sample in air, the parallel corrugation lines characteristic of the $\sqrt{3} \times 22$ phase appeared and quickly spread out over a larger region of the surface.[28] When the STM tip was biased negatively, the reconstructed phase was lifted, in accordance with dependence of the potential depenence in the *in situ* electrochemical experiment. An interesting feature of the measurements in air is that the excess, tip-induced, surface charge necessary to lift or induce the reconstruction is significantly smaller (a factor of 3–10) than the same quantity measured in the *in situ* electrochemical studies. This suggests that the field-induced stress caused by the STM tip is extremely efficient in initiating surface reconstructions. Clearly, in the case of the experiments in air, one cannot associate the surface rearrangements with a conventional double layer comprised of solvent and ions, as was done in interpreting the observations in the electrochemical environment. This obviously raises the question of the role of the STM tip in inducing the observed phase transitions in the *in situ* measurement. The STM tip is solution may modify the local double-layer structure and excess charge density, either inhibiting or accelerating the $\sqrt{3} \times 22 \rightarrow 1 \times 1$ transformation. Fortunately, the *in situ* STM results are in semiquantitative agreement with x-ray scattering measurements of the surface structure, indicating that the conventional electric fields within electrode–electrolyte interface are largely, but not necessarily exclusively, responsible for the observed reconstruction. However, this conclusion cannot be obtained solely from the original STM study.

The capability of AFM to obtain true atomically resolved images has been recently questioned by several authors (for a recent review of the subject, see Chapter 7 of reference 19). In the (repulsive) hard-contact imaging mode, it is predicted that the sample and the tip should both deform under the high local pressures, resulting in a considerably increase in the contact area. For a typical value of the force measured in AFM (1–100 nN), the local pressure on a single-atom contact of area $\sim 3 \times 10^{-20}\,m^2$ would be of the order of $10^{11}\,Pa$, considerably larger than the yield strength of the materials (tips and samples) used in the AFM experiment. It is thus likely that the area of the contact region between the tip and sample is increased well beyond a point atom contact. Pethica[29] and Landman et al.[30] have successfully used such blunt tip models to show that the *apparent* atomic corrugation observed in AFM images may actually result from a periodic variation in the forces between two well-ordered planes of atoms that comprise the surfaces of the sample and the tip. Such a model is in general accord with the fact that very few features of true atomic dimensions (an individual adatom or an atomic vacancy) have been reported in AFM, although such features are routinely observed in STM. Thus, the interpretation of contact-mode, high-resolution AFM images requires very special consideration of the influence of the tip in generating subtle features. On the other hand, the potential resolution of AFM imaging in the noncontact mode is determined by the shape of the force curve. Here, van der Waals and/or electrostatic forces dominate the interaction between the tip and sample.[31] These forces decay relatively slowly away from the surface. This factor and the fact that the mechanical instabilities prevent stable positioning of a tip close to the surface without jumping to direct contact limit the achievable resolution in the noncontact mode to a few nanometers. An unequivocal demonstration of true "atomic resolution" using AFM in *electrochemical* environments has not been presented at the time of this writing, although a recent report of atomic resolution of calcite crystals in aqueous media is encouraging in this respect.[32]

III. APPLICATIONS OF STM AND AFM IN ELECTROCHEMISTRY

An extensive compilation of electrochemical and nonelectrochemical STM and AFM studies was published in 1992,[18] and Bard and Fan[23] have more recently provided a complete tabulation of electrochemical STM and AFM experiments. In addition, STM investigations of molecular films of redox-active inorganic complexes have recently been described by Abruna et al.[33] Rather than duplicate these efforts, we have chosen to focus on a few selected examples in the electrochemical literature where STM and AFM have proven especially useful in understanding electrochemical phenomena at the atomic and molecular scale. The examples illustrate the power of being able to *visualize* the interface in real space. This capability has led to new atomic- and molecular-level models of the electrochemical interface, providing fresh insights into origin of numerous macroscopic behaviors—for example, electrochemical rates, double-layer capacitance, and crystal growth patterns.

A. Metal Electrodes

Investigations of highly oriented pyrolytic graphite (HOPG) and noble-metal electrodes in aqueous solutions have been the focus of STM and AFM studies. Much of the early work in this area was done to demonstrate the experimental capability of obtaining *in situ* STM and AFM images. More recently, these techniques have been applied successfully to detailed investigations of metal and alloy dissolution, electrocrystallization, surface adatom diffusion and trapping, roughening and oxide formation on metals and graphite, underpotential deposition of metal adlayers, adsorption of ions and carbon monoxide, and the structure of films comprising relatively large redox-active molecules or conductive polymers.[23]

The reconstruction of the surface layer of metal and semiconductor surfaces is a topic for which *in situ*, high-resolution STM is particularly well-suited. Such studies were first performed in the surface physics community, with the identification of the 7×7 reconstruction of Si serving as the prototype example of the unmatched capabilities of STM.[34] However, the possibility of *potential-dependent* surface reconstructions at metal electrode surfaces has long been a subject of discussion in electrochemistry. For instance, the capacitance of the (100) surface of an Au single-crystal electrode in electrolytes containing weakly adsorbing ions (e.g., perchloric acid solutions) exhibits considerable hysteresis as the electrode potential is slowly scanned between values positive and negative of the pzc. Since the interfacial capacitance reflects the magnitude and distribution of charge at the interface, it was postulated by Hamelin[35] that the reconstruction of the outermost layer of the electrode surface was responsible for the observed hysteresis. This postulate

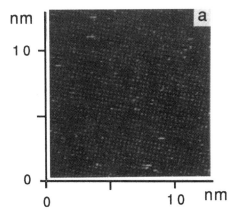

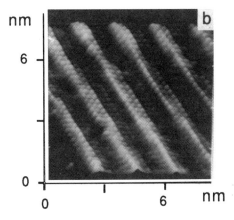

Figure 2. *In situ* STM images of Au(100) in 0.1 M perchloric acid. (a) Unfiltered image at −0.1 V versus SCE showing the (1 × 1) structure. (b) Height-shaded image at −0.3 V versus SCE showing the 5 × 26 reconstructed surface. (Reprinted from reference 37, with permission.)

was indeed supported by *ex situ* electron-diffraction measurements, but details of the potential-dependent positions of atoms, as well as numerous types of surface defects, were only first obtained by STM measurements by Gao et al.[36] and by Hamelin et al.[37] Figure 2 shows *in situ* STM images of a flame-annealed Au(100) surface immersed in a 0.1 M perchloric acid solution at two different electrode potentials. The image shown in Figure 2a is that of the electrode poised at a potential of −0.1 V versus saturated calomel electrode (SCE). At this potential, the STM image depicts the Au(100) surface to be predominantly unreconstructed, displaying the square-planar arrangement of atoms (interatomic spacing of 0.29 ± 0.2 nm), anticipated for the 1×1 Au structure. When the electrode potential is shifted just slightly more negative than -0.2 V, domains identified as a 5×27 surface phase are reported to gradually develop over a period of 5–10 minutes. Figure 2b depicts an atomically resolved STM image of one such domain obtained at -0.3 V versus SCE. This reconstruction of Au(100), as well as the $\sqrt{3} \times 22$ reconstructed phase on Au(111), can be lifted when the potential is scanned to positive potentials, as evidenced by the STM images which show that the atom arrangement returns to the original square-planar 1×1 structure of Figure 2a. Although the mechanism of the reconstruction is far from being resolved, the dependence of surface structure on the electrode potential clearly suggests that the atomic movement is due to stresses associated with the excess surface charge. In addition to demonstrating the ability to obtain atomically resolved images in an electrochemical environment, Hamelin et al.[37] have clearly demonstrated that the potential dependence of the reversible $1 \times 1 \rightarrow 5 \times 27$ reconstruction, as first observed by STM, is consistent with the observed hysteresis in the capacitance–potential curves.

One of the more interesting findings in the investigations of Hamelin et al.[37] is that flame-annealed Au(100) surfaces do not reconstruct if the sample is quenched in ultrapure H_2O and immediately transferred to the electrochemical cell. This finding is in distinct opposition to results obtained by Kolb and coworkers[38,39] where a flame-annealed Au sample was cooled in air and transferred to UHV for *ex situ* electron-diffraction investigations. The latter nonelectrochemical method yields an Au(100) surface that is reconstructed. Since no error is likely to have occurred in either determination of the surface structure, the natural conclusion gleaned from comparison of this set of results is that either the environment of the sample preparation (H_2O versus air) or that of the characterization facility (H_2O versus UHV) has a significant impact on the nature of the atomic arrangement of the interface. This point clearly emphasizes the need to perform *in situ* structural characterization of electrode surfaces under conditions that mimic, as close as possible, the relevant solution conditions.

The Au(111) surface undergoes the $1 \times 1 \rightarrow \sqrt{3} \times 22$ surface reconstruction that can be activated by variation of the surface excess charge (under potential control), by thermal methods, and by electrostatic forces originating at the STM tip during imaging in air. The details of the atomic arrangement in the Au $\sqrt{3} \times 22$ surface phase was initially observed in UHV–STM by Woll et al.[40] and Barth et al.[41] Later *in situ* electrochemical experiments by Gao et al.[27] and by Tao and Lindsey[42] demonstrated that the reaction could be activated by varying the electrode potential, in a manner analogous to the 5×5 reconstruction on Au(110) surfaces. Images of the 1×1 unreconstructed

Au surface display the characteristic hexagonal atomic arrangement expected from an ideally terminated Au(111) surface. The $\sqrt{3} \times 22$ reconstructed surface results from the insertion of one extra atom in a geometrical area that corresponds to ~ 22 Au atoms of the unreconstructed surface. This increased atom density results in a slight "buckling" of the surface, and STM images of the reconstructed Au surface exhibit parallel pairs of corrugation lines with a horizontal pair-to-pair separation of ~ 63 Å and vertical corrugation amplitude of ~ 0.2 Å.

Topographical features observed in STM images, such as the corrugation lines of the $\sqrt{3} \times 22$ reconstruction, are accompanied by local variations in the surface electron density profile. Thus, adsorbates may be expected to decorate the surface in a way which reflects these local variations in the electron density. This was first observed using STM in UHV during the vacuum evaporation of submonolayer quantities of Cu onto the $\sqrt{3} \times 22$ Au surface.[43] The Cu film exhibited a strong tendency to nucleate on the corrugation lines resulting a regular array of two-dimensional Cu film patches, each centered on one of the corrugation lines. A similar phenomena was observed in STM studies of the adsorption of $[(Ru(bpy)_2(bpy\text{-}(CH_2)_x\text{-}bpy)]^{2+}$ $(x = 5, 7,$ and $12)$ from dimethylformamide onto Au(111) surfaces, which occurs through electronic interactions of the pendent bipyridine (bpy) with the surface Au atoms.[44] On the reconstructed $\sqrt{3} \times 22$ surface, the molecules assemble with long-range order along the corrugation lines, resulting in an unusual structure in which the molecules are aligned in uniform rows separated by the characteristic ~ 63 Å spacing of the surface corrugations (Figure 3a). Conversely, on the unreconstructed surface (1×1), adsorption of $[(Ru(bpy)_2(bpy\text{-}(CH_2)_x\text{-}bpy)]^{2+}$ results in a completely random adsorption pattern (Figure 3b). Images of the molecular adsorbates obtained by STM in air are of significantly higher resolution than images obtained in DMF solution, for reasons that are probably associated with the translation and rotational motion of the Ru complex attached to the Au surface by the extended alkane chain. However, the results clearly demonstrate the role of the surface electronic structure in dictating the morphology of an electroactive film.

A second area of electrochemical research where STM and AFM have been extensively applied in the past three years is the underpotential deposition (UPD) of metal adatoms from solutions containing the corresponding metal ion. Examples include Ag on HOPF, Cu on Pt, Cu on Au, and Hg on Au. The interested reader should consult the tabulation of studies Bard and Fan for a more exhaustive listing.[23] The structural data obtained from *in situ* STM and AFM differ in several respects in comparison to what can be obtained by diffraction methods. First, the UPD adlayers frequently contain a number of atomic scale defects that can be easily imaged by STM. Second, the growth of the film can be monitored over a wide scale of dimensions, allowing a range of

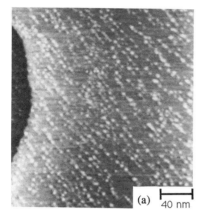 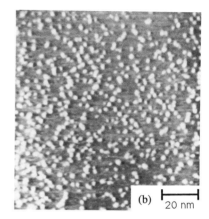

Figure 3. STM images (~ 150 × 150 nm) of adsorbed films of $Ru(bpy)_2(bpy-(CH_2)_5-bpy)^{2+}$ on (**a**) the $\sqrt{3} \times 22$ reconstructed Au(111) surface and (**b**) the unreconstructed Au(111) surface. The adsorbed films on the reconstructed surface display highly ordered patterns that are commensurate with the measured corrugation pattern of the substrate. (Reprinted from reference 44, with permission.)

structures ranging from individual adatoms to wide terraces to be routinely imaged. Third, it is possible to directly count the number of adatoms from analysis of STM images, allowing a comparison of true surface coverage with the electrochemical coverage, the latter being determined from coulometric techniques. Estimation of the surface coverage of adatoms (as well as molecular adsorbates) from STM and AFM measurements, however, suffers to a large extent from the lack of a general procedure for ensuring accurate sampling statistics. It is generally up to the individual investigator to determine a method for ensuring that the surface coverages deduced from images are truly representative of the entire surface. Explicit statements of the sampling procedure or experimenal statistics, however, rarely appear in the literature. Consequently, the degree to which a published image accurately represents the entire surface is frequently, and justifiably, the subject of debate.

Hachiya et al.'s *in situ* STM investigations of the UPD of Cu on Au(111) surfaces in 1 mM $CuSO_4/0.05$ M H_2SO_4 solutions are representative of this field of study.[45] Figure 4 shows a typical STM image of the Cu adlayer obtained while the Au substrate was biased at 0.15 V versus SCE at which UPD occurs. The image shows that the adlayer has a $\sqrt{3} \times \sqrt{3}/R30°$ structure, in agreement with other AFM and STM studies of the same system. An interesting feature of the image shown in Figure 4 is that phase boundaries are clearly observable between domains of the $\sqrt{3} \times \sqrt{3}/R30°$ Cu adlayer. In fact, the resolution of their electrochemical STM is sufficiently good that it is

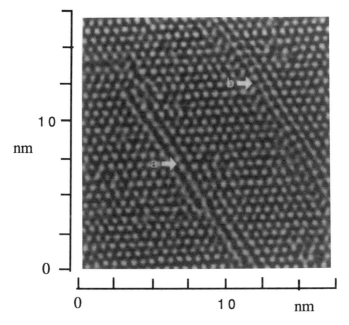

Figure 4. *In situ* STM image of a $14 \times 14\,nm^2$ region of the Cu adlayer on Au(111). Arrows indicate two different phase boundaries. (Reprinted from reference 45, with permission.)

possible to discern two different types of phase boundaries, labeled a and b in Figure 4 and schematically depicted using a ball model in Figure 5. In the type a boundary, two $\sqrt{3} \times \sqrt{3}/R30°$ domains are shifted by a half cell unit and a single row of Cu atoms is inserted between the domains. In the type b boundary, two domains are separated by a small mismatch of the adlayers at the boundary. The boundaries extend over several hundred angstroms and are essentially straight at the atomic level.

Hachiya et al.[45] compared the values of the surface coverages of Cu atoms evaluated by coulometry and from STM images. They found that the coulometrically measured value ($\sim 0.3\,mC/cm^2$) is twice as large as expected based on the observed $\sqrt{3} \times \sqrt{3}/R30°$ structure ($0.15\,mC/cm^2$) and suggested that the difference is due to a significant contribution of the charging current due to the adsorption and desorption of the electrolyte anions (SO_4^{2-} or HSO_4^{-}) during the UPD process. The difference is apparently real, since the authors report that the $\sqrt{3} \times \sqrt{3}/R30°$ structure is the only structure observed by STM in numerous experimental observations. Thus, the results suggest that the UPD of Cu involves the reduction of Cu^{2+}, with concurrent atom and anion adsorption. Although such a mechanism is conceptually familiar to electrochemists working in this area, it is important to note that an

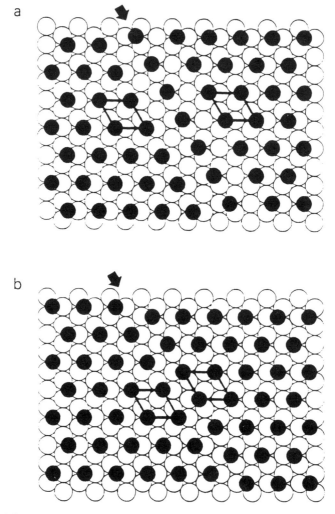

Figure 5. Schematic representations of the two different phase boundaries shown in Figure 4. The unit cells of the adlayer are shown. Unfilled and filled circles correspond to Au and Cu atoms, respectively. (Reprinted from reference 45, with permission.)

unequivocal demonstration of a real difference in the electrochemical coverage and true physical coverage hinges on *in situ* observations of the adlayer structure. Without such direct evidence as the STM provides, one can only speculate about the adlayer structure and coverage.

In our laboratories, we have recently developed a slightly different method using STM to evaluate the ion valancy of adsorbed halogen ions.[46] Atomically

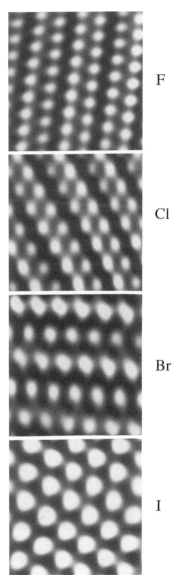

F

Cl

Br

I

Figure 6. Atomically resolved STM images (each $2.5 \times 2.5 \, \text{nm}^2$) of halogen adlayers on Ag(111).

smooth adlayers of fluorine, chlorine, bromine, and iodine on Ag(111) substrates can be prepared by flame-annealing and quenching a polycrystalline Ag sample in concentrated acid halide solutions. High-resolution STM images of these surfaces in air (Figure 6) show that the halogens form highly ordered layers on the Ag(111) surface. As suggested in the preceding paragraph, the degree of charge transfer between the adatoms and metal surface can be estimated from comparison of (a) the coulometric charge required to strip the atoms from the surface and (b) the charge associated with the atom surface densities observed in the STM images. The precision of this method for the Ag surfaces is low due to uncertainty in the real electrode area, differences in adlayer structures on different surface orientations of the polycrystalline Ag sample, and theoretical difficulties in separating faradaic and capacitive currents associated with charge transfer processes of surface confined species.[47]

However, the series of STM images of the halogen adlayers show that the relative *apparent* dimensions of the adatoms are consistent with expectations based on the relative sizes of halogen atoms (or ions); that is, the apparent size of I adatoms is roughly twice that of F atadoms, with Cl and Br having intermediate values. Since there is a comparatively large difference between the radius of a halogen atom and the corresponding halogen ion, measurements of the adatom radii provide an indirect method of determining the degree of charge transfer in the adlayer. With this goal in mind, we obtained the adatom radii directly from measurements of the variation in the tunnel current density measured along a line drawn through the center of the atom. This method is appealing because the data correspond directly to the apparent size of the atoms as defined by the image contrast. However, this measurement suffers from a number of intrinsic and instrumental effects. First, there is a significant variation in the apparent dimensions of neighboring adatoms, as is clearly evident in the images of Cl adlayers. The apparent differences in adatom dimensions may reflect variations in the degree of charge transfer between the metal and adatoms, resulting from coulombic forces within the film.[48] Second, the tunneling current distribution around individual adatoms appears often appears distorted, resulting in a nonspherical atom geometry (e.g., Figure 6, Br). The extent of this distortion is also tip-dependent. Third, without an appropriate theoretical model to describe the relationship between the tunneling current density and the adatom size, the choice of a cutoff in the tunneling current density at which the atom boundary is defined is subjective. The combined effects of these limitations make this method of estimating the adatom radii seemingly unreliable, and, indeed, statistical analyses of experimental data show that there are no significant differences in the adatom size.

A more precise and reliable method of determining atomic/ionic radii is based on the fact that the shortest nearest-neighbor bond distances observed

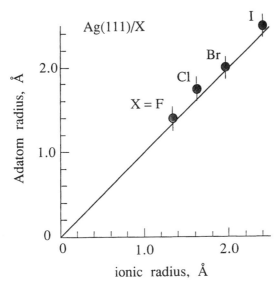

Figure 7. Plot of STM-measured adatom radius versus the literature value of halogen ions. The solid line has unity slope and zero intercept.

in each of the adlayer structures correspond to twice the theoretical maximum possible value of the adatom radius, assuming a hard-core repulsion between roughly spherical electron distributions.[49] Using the shortest nearest-neighbor bond distance measued for each adlayer reported above, the corresponding radii were calculated to be 2.30 Å for I, 2.05 Å for Br, 1.65 Å for Cl, and 1.40 ± 0.3 Å for F (based on approximately 25 determinations for each halogen species using different samples and tips). These distances are plotted in Figure 7 as a function of literature values of halide radii[50]: I, 2.20 Å; Br, 1.96 Å; Cl, 1.81 Å; F, 1.33 Å. Before discussing these data, we note that measurement of the nearest-neighbor adatom bond distances does not suffer from the same inherent limitations as discussed above and that the values obtained are very reproducible using different STM tips (better than ± 0.3 Å). The radii of any two halogen species based on this method are statistically different.

The radii of halogen adatoms, measured from the nearest-neighbor bond distance, are in excellent agreement with literature radii of the corresponding bare halide ion in the gas phase, as evident from comparison of the data with the solid line which has unity slope and a zero intercept.* As noted above, we

---

* Literature values of halide radii vary by a few hundredths of an angstrom between sources (e.g., see discussion in reference 17b). This variance has no effect on the linearity of the plot shown in Figure 3.

interpret these values of STM-measured radii as representing the theoretical *maximum* value of the adatom radius, since it is probable that the separation of the adatoms is not due solely to hard core interactions, but is effected by interactions with the substrate lattice. Indeed, it is most likely that some combination of the two interactions are responsible for the spatial arrangements of the atoms. However, the data in Figure 7 clearly suggest that each of the halogen adatoms are chemisorbed on the Ag substrate as ionic species, rather than as halogen atoms. In particular, the radius measured for each halogen adatom is at least 50% larger than anticipated if the chemisorbed species were not ionized, but rather existed as neutral atoms (atomic radius[51]: I, 1.28 Å; Br, 1.11 Å; Cl, 0.99 Å; F, 0.64Å). Based on these findings, we conclude that the structures we have imaged are best described as Ag halide monolayers.

B.   Electrocatalytic Systems

The structure of UPD layers plays a key role in electrocatalysis, but the use of AFM and STM to investigate this relationship has not been extensively employed. An exception is Chen and Gewirth's *in situ* AFM investigation of the electrocatalytic reduction of H_2O_2 using Au(111) substrates on which Bi was underpotential deposited to various degrees of coverage.[52] The UPD of Bi on Au(111) is well known to catalyze the reduction of H_2O_2, but the process is particularly interesting because the electrocatalytic activity is strongly dependent on the surface coverage of Bi. *Intermediate* Bi coverages show a significantly larger activity than either low or full monolayer coverages. Figure 8 shows images acquired (a) at potentials positive of the UPD potential for Bi deposition (i.e., bare gold), (b) at intermediate Bi coverages, corresponding to potentials between 250 and 190 mV versus SCE where the electrocatalytic effect is high, and (c) at full Bi coverage at potentials negative of 190 mV, where the catalyttic activity for H_2O_2 decreases significantly. The UPD adlayer at intermediate coverages has an open hexagonal structure, whereas the adlayer structure at high coverages corresponds to a rectangular, more closed-pack film. Proposed models of the structures imaged at the two coverages are shown in Figure 8d.

Chen and Gewirth[52] have suggested that the differences in the Bi adlayer structures, imaged at intermediate and full monolayer coverages, may be responsible for the decrease in catalytic activity as the Bi coverage is decreased. The hypothesis is based on the assumption that H_2O_2 binds end-on to Au during the reduction process, forming an Au–O–O complex. The open structure of the Bi layer at intermediate coverages would allow the Au–peroxide complex to interact, sterically or electronically, with Bi adatoms possibly forming a bimetallic complex. Such a complex may enhance the

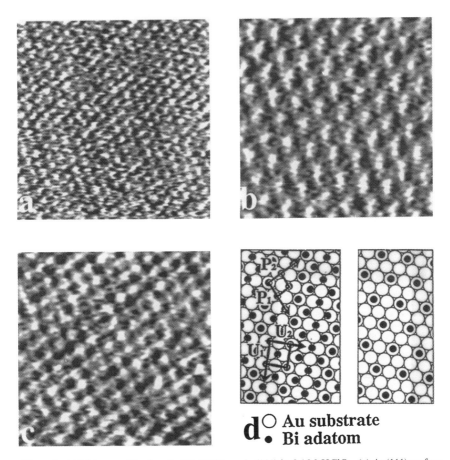

Figure 8. AFM images (5×5 nm) of Bi UPD on Au(111) in 0.1 M $HClO_4$. (**a**) Au(111) surface found positive of Bi UPD peaks. Atom–atom distance in 0.29 nm. (**b**) (2×2)-Bi adlattice found at 200 mV versus $E_{Bi}^{3+/0}$. Atom–atom distance is 0.57 ± 0.02 nm. (**c**) Uniaxially commensurate, rectangular Bi adlattice found at 100 mV. Atom–atom distance is 0.34 ± 0.02 nm. (**d**) Schematic of Bi structures. **Left**: Rectangular lattice where **P** and **U** are primitive and nonprimitive unit cell vectors, respectively. **Right**: (2×2)-Bi adlattice showing open Au and Bi sites. The Bi adatoms are larger than Au; they are shown smaller here for clarity. (Reprinted from reference 52, with permission.)

cleavage of the O–O bond, thereby catalyzing the reduction of H_2O_2 to H_2O. In contrast, a full monolayer of Bi results in a surface where the Au atoms are essentially blocked, preventing the formation of an Au–O–O complex. These results are particularly promising and suggest that it may be possible to image complex molecule–surface structures *during* the course of electrochemical reactions.

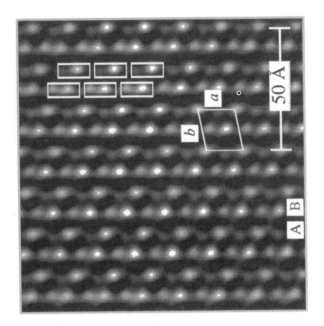

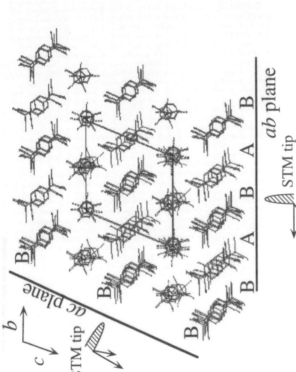

Figure 9. Left: A view of $[(\eta\text{-}C_5Me_5)_2Ru_2(\eta^6,\eta^6\text{-}[2_2]\text{-}(1,4)cyclophane)]^{2+}$ $[TCNQ]_4^{2-}$ looking down the stacking axis. The (001) face (ab plane) contains layers of TCNQ molecules, organized into two crystallographically independent TCNQ anion stacks, repeating ...ABAB... along the [010] direction. Each stack is composed of $(TCNQ)_4^{2-}$ tetramer units. **Right**: STM image of the (001) face of **1**. The rectangular boxes are included as a visual aid to depict the tetrameric repeat unit in the two crystallographically independent TCNQ stacks.

130

C. STM and AFM of Conducting Molecular Crystals

The unique abilities of STM and AFM have been exploited in the characterization of specific crystal planes of conducting molecular crystals, specifically the class of compounds referred to low-dimensional charge-transfer solids. These compounds have been examined because of their unique electronic properties, which stem from intermolecular interactions in the solid state. They have been developed with the premise that physical and electronic properties can be "molecularly engineered." One-dimensional solids share a common motif in which planar open-shell molecules stack upon each other, forming quasi-one-dimensional chains. As a consequence of the electronic interaction between molecules in these stacks, these materials can exhibit electrical conductivity and superconductivity. The aim of most STM and AFM studies of these materials has been to examine the electronic structure on a local molecular level. However, STM and AFM studies have also been prompted by the relevance of these materials to electrochemistry. For example, electrodes fabricated from low-dimensional conductors such as (tetrathiafulvalence)(tetracyanoquinodimethane) and related compounds have been reported to be catalytic for electrochemical oxidation of several biologically significant molecules, including NADH, flavoenzymes, and glucose.[53] Microelectrodes have also been fabricated from (tetrathiafulvalene)(tetracyanoquinodimethane).[54] There also exists a need to understand the synthesis of these materials, which is typically accomplished by electrocrystallization methods.[55] The initial stages of electrocrystallization involve electron transfer between an inert metal electrode and a solution redox species, resulting in the formation of a conducting crystalline salt on the electrode. After this stage, however, crystallization involves electron transfer at the *conducting crystal interface*. Characterization of the surface DOS and topography of specific surfaces of these crystals is instrumental in understanding the electron transfer events that underlie these processes. In addition, the capabilities of STM and AFM also allow for *in situ* characterization of nanoscale topographic features such as steps and kinks, which influence nucleation and crystal growth. The real time capability of these methods also allows characterization of the dynamic nucleation and growth events that take place on crystal faces.

Several *ex situ* STM and AFM studies of conducting molecular crystals have appeared recently, primarily addressing the electronic structure and topography of specific crystal planes of these materials (Figure 9 and 10, Table 1). For example, we have demonstrated that the (001) face of the low-dimensional superconductor, $(TMTSF)_2ClO_4$ (TMTSF = tetramethyltetra-selenafulvalene), reveals a DOS that is consistent with its structure. The presence of ClO_4^- that decorate this face was responsible for the observed

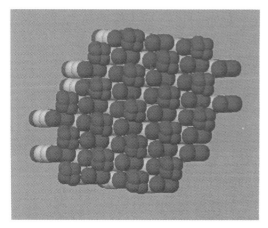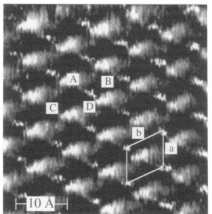

Figure 10. Left: Space-filling representation of the (001) face of $(TMTSF)_2ClO_4$, illustrating the TMTSF stacks aligned along the [100] direction, with ClO_4^- anions decorating the surface. **Right**: STM image of the (001) face of $(TMTSF)_2ClO_4$. Region A corresponds to TMTSF molecules, region B corresponds to the insulating ClO_4^- anions that obscure every other TMTSF molecule in the stack, and region C corresponds to states associated with Se–Se contacts between the stacks.

Table 1. Conducting Molecular Crystals Examined with STM or AFM

Crystal	Reference	Crystal Face Examined
TTF-TCNQ	56–58	(001)
$[(\eta-C_5Me_5)_2Ru_2(\eta^6,\eta^6-[2_2]-$ $(1,4)cyclophane)^{2+}[TCNQ]_4^{2-}$	59	(010), (001)
$(TMTSF)_2ClO_4$	60, 61	(001)
$(TMTSF)_2PF_6$	60, 61	(001)
$\alpha-(BEDT-TTF)_2I_3$	62	(001)
$\beta-(BEDT-TTF)_2I_3$	62	(001)
$\alpha-(BEDT-TTF)_2IBr_2$	62	(001)
$\beta-(BEDT-TTF)_2IBr_2$	62	(001)
$\beta-(BEDT-TTF)_2AuI_2$	62	(001)
$\kappa-(BEDT-TTF)_2Cu(NCS)_2$	62–64	(100)
$(TEA)(TCNQ)_2$	65	(010)
$(4-EP)(TCNQ)_2$	66	(001)
$(BEDT-TTF)_2Ag_{1.6}(SCN)_2$	67	(001)
$(BEDT-TTF)_2KHg(SCN)_4$	68	(010)

tunneling current contrast. Molecules on the surfaces of these crystals can be imaged readily, but obtaining atomic resolution is somewhat more difficult. Likewise, STM images of the (010) and (001) faces of the semiconducting charge-transfer salt $[(\eta-C_5Me_5)_2Ru_2(\eta^6,\eta^61[2_2]-(1,4)cyclophane)]^{2+}-[TCNQ]_4^{2-}$ revealed a DOS consistent with the molecular motif on these faces. These studies are representative of others, in that the DOS inferred from the tunneling current contrast is generally in agreement with that expected from the molecular arrangement on the respective-faces and the distribution of electronic charge, as deduced from the Hückel molecular orbitals of the crystal constituents.

Recently, we demonstrated in our laboratories that AFM could be used to obtain real-time images of (a) crystal growth of organic conductors on electrode surfaces and (b) the growth modes of specific faces of organic conductors.[69] These studies focused on the growth orientation during growth and on the role of topographic features in crystal growth. For example, slow electrochemical oxidation of tetrathiafulvalene in ethanol solution containing n-$Bu_4N^+Br^-$ resulted in the growth of $TTFBr_{0.7}$ crystals on a graphite electrode that could be imaged by AFM (Figure 11). The crystals were needle-shaped with approximate dimensions $1\ \mu m \times 10\ \mu m$, and they grew in an oriented fashion with the [100] axis and (010) face parallel to the electrode surface. The AFM was used as a goniometer to measure the angles subtended by the different crystal faces. Interestingly, the faces identified for these crystals were not observable on the macroscopic crystals. This suggested that at these dimensions, surface energies may have played an important role in the morphology of the crystals. The AFM also revealed that crystal growth on two of the observed faces involved the motion of ledges along the [100] direction.

Dynamic behavior was also observed in studies of the crystal growth and etching of $(TMTSF)_2ClO_4$ crystals.[70] A previously grown mature crystal was mounted in the AFM and connected to a potentiostat as a working electrode. Initial images of the crystal revealed a rough (001) surface consisting of numerous anisotropic (001) terraces oriented along the [100] direction (Figure 12). This roughness diminished when the crystal was immersed in propylene carbonate or ethanol solutions containing TMTSF and n-$Bu_4N^+ClO_4^-$, indicating a spontaneous transformation to a lower-energy surface. When a slightly anodic potential was applied to the crystal, TMTSF was oxidized and crystallization was observed on the (001) face. The crystal growth occurred by fast extension of 13-Å-high terraces (one unit cell height) along the [100] direction, leading to the topography observed in the initial crystal. Application of a cathodic potential to the $(TMTSF)_2ClO_4$ crystal resulted in dissolution, with AFM verifying that dissolution occurred by layer-by-layer recession of 13-Å-high (001) terraces that were bounded by [100] ledges (Figure 13). It was also shown that when dissolution was very fast relative to the AFM tip scan rate, etching of these terraces resulted in triangular features when the [100] axis of the crystal was aligned with the vertical scan direction of the AFM tip (Figure 14).

This behavior was due to the time-dependent positions of the terrace ledges during [100] ledge motion; as the AFM tip moved vertically it detected the

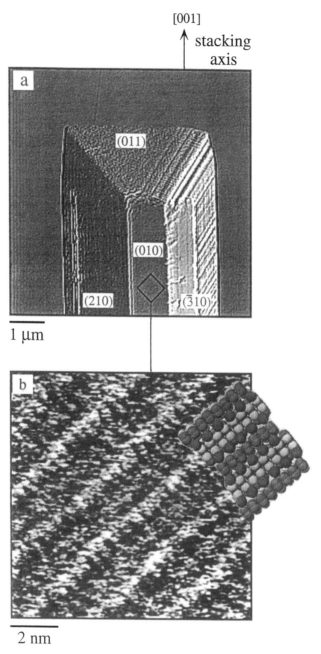

Figure 11. (a) AFM image of a $(TTF)Br_{0.76}$ crystal growing by electrocrystallization on a highly oriented pyrolytic graphite electrode. The [001] direction of the crystal is parallel to the basal plane of the graphite electrode. (b) Raw AFM data of the (010) face and the molecular packing of the (010) face.

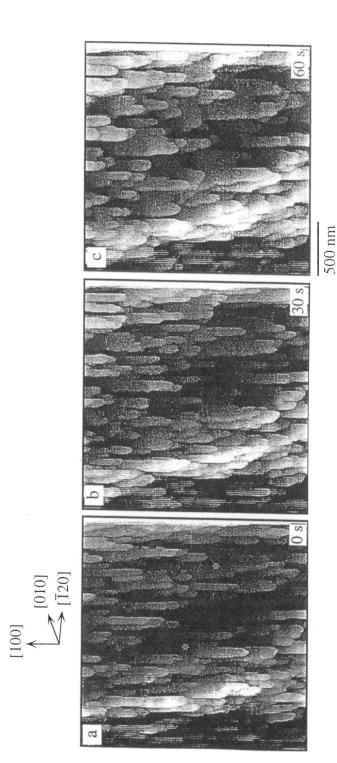

[100]

[010]

[120]

500 nm

Figure 12. *In situ* AFM of the growth on the (001) face of $(TMTSF)_2ClO_4$ in a solution of propylene carbonate containing 1 mM TMTSF and 0.1 M n-Bu$_4$NClO$_4$ at an applied potential of 700 mV versus SCE: (a) 0 sec, (b) 30 sec, and (c) 60 secs. The step heights range from 13 to 26 Å, which corresponds to one or two molecular layers. The AFM data were obtained in the constant height mode and are unfiltered. The regions marked with an asterisk are located on growing terraces.

135

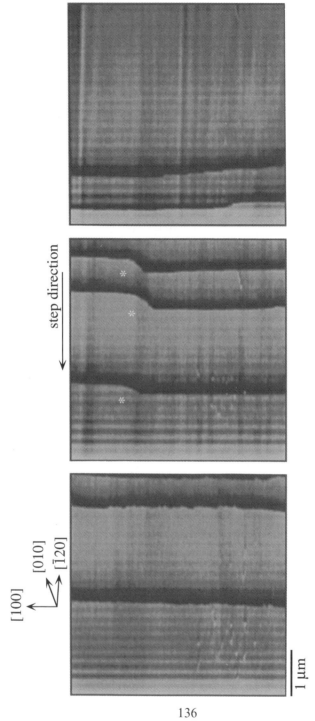

Figure 13. *In situ* AFM of the (001) face of (TMTSF)$_2$ClO$_4$ showing dissolution under potential control. During dissolution (*middle*), [100] ledges recede along [010]. The ledge motion can be arrested by increasing the electrochemical potential to $E > E_0$ (*left* and *right*). The step heights are 13 Å. The AFM data were obtained in the constant height mode and are unfiltered.

[100]
[010]
[$\bar{1}$20]

step direction

1 μm

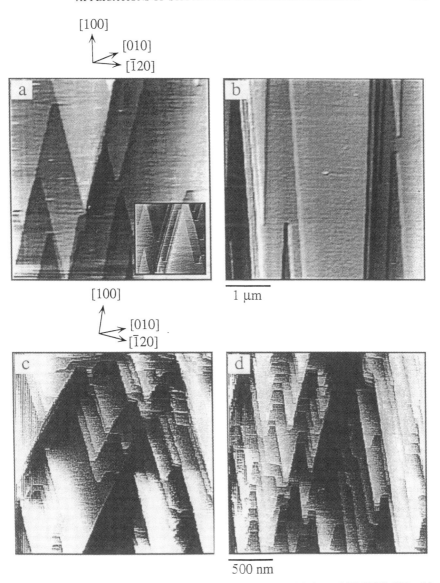

Figure 14. *In situ* AFM images of fast ledge motion on the (001) face of $(TMTSF)_2ClO_4$. (a) $E_{applied} = 0.42$ V (versus SCE), $v_t = 200$ nm/sec in the downward directiion, 0.1 M n-$Bu_4^+ClO_4^-$ ethanol (**inset**: scanning in the upward direction). (b) $E_{applied} = 0.42$ V, $v_t = 381$ nm/sec in the downward direction, 0.1 M n-$Bu_4^+ClO_4^-$/ethanol. (c) $E_{applied} = 0.35$ V, $v_t = 105$ nm/sec in the downward direction, 0.1 M n-$Bu_4^+ClO_4^-$/propylene carbonate. (d) $E_{applied} = 0.500$ V, $v_t = 105$ nm/sec in the downward direction, 0.1 M n-$Bu_4^+ClO_4^-$/propylene carbonate. The step heights correspond to 13 Å in all images. The AFM data were obtained in the constant height mode and are unfiltered.

terrace ledge at different positions along the x-axis. The triangular features were oriented along the vertical direction, the features inverting with each successive vertical scan. Two different orientations of the triangular features were observed in each image; triangles pointing in the tip scan direction were higher than adjacent terraces, whereas triangles pointing opposite the tip scan direction were lower than adjacent terraces. These observations were a consequence of the relative motions of companion [100] ledges. If two [100] ledges on a given layer recede *toward* each other, they eventually annihilated one another and disappeared from the surface. In this case, the triangles pointed in the same direction as the vertical AFM scanning direction because the terrace had a greater width at the beginning of the scan than at later times, and the triangle was higher than adjacent terraces. The apex of the triangle represented the time at which that terrace completely disappeared. However, for two [100] ledges on a given layer that originated within the frame and receded *away* from each other, the triangle pointed in the direction *opposite* the vertical AFM scanning direction because the newly exposed terrace, which was lower than the adjacent terraces, had a greater width at the end of the scan. The aspect ratio of the triangles depended upon the relative rates of etching and vertical AFM tip velocity. For a given etching rate, faster scan velocities resulted in narrower triangles as the tip velocity became comparable to the ledge motion rate. Similarly, for a given scan rate, increasing the rate of etching by applying a more cathodic potential gave wider triangles as the etching rate exceeded the tip velocity to a greater extent. These studies illustrated that AFM was valuable for investigating crystal growth and etching modes that could be correlated with molecular structure and the strength of intermolecular bonding.

D. Tunneling Spectroscopy

Tunneling spectroscopy has been employed in electrochemistry for investigating the electronic structure of semiconductor photoelectrodes (e.g., Si, TiO_2) and semiconducting metal oxides films. In a typical experiment, which can be performed in air, vacuum, or liquids, the STM tip is positioned several angstroms above the surface of a conductive sample. Depending on the polarity of the bias voltage applied between the tip and sample, electrons can tunnel from (filled) tip states near the Fermi level into (empty) sample states, or vice-versa. As noted in Section II.A, the dependence of tunneling current, I, on the voltage applied between the tip and the sample, V, is a complex function of the DOS of (a) both the sample ($\rho_s(\mathbf{r}, E, V)$) and tip ($\rho_t(\mathbf{r}, E, V)$) and (b) the transmission coefficient ($T(\mathbf{r}, E, V)$). Generally, one is interested in obtaining the density of states of the sample, $\rho_s(\mathbf{r}, E, V)$, which can be correlated with macroscopic observables (e.g., electron-transfer rates or the electrode photovoltage). Extracting $\rho_s(\mathbf{r}, E, V)$ from the I–V curves can be simplified by plotting the raw data as $(dI/dV)/(I/V)$ versus V spectra. The quantity (dI/dV) is the differential conductance of the tunnel junction, $(dI/dV)/(I/V)$ is the same quantity normalized to the total or integral conductance (I/V), and represents an estimate of the surface density of states (SDOS). The latter method of plotting tunneling data is preferred since it has been experimentally and theoretically shown in UHV studies to yield spectra for highly doped inorganic semiconductors (e.g., Si) that are essentially independent of the tip–substrate separation.[71] This type of analysis has been applied to measure the energy position electronic states of photoelectrodes such as TiO_2[72], p- and n-Si,[73] Fe_2O_3,[74] and FeS_2.[75]

Ex situ tunneling spectroscopy measurements have been used in investigations of the pitting corrosion of Ti electrodes.[76] The chemical stability of Ti is due to the formation of a thin (~ 20 Å) TiO_2 film on the surface which protects the underlying metal from corrosive environments. In the presence of aqueous solutions containing Br^- or I^-, localized breakdown of the film occurs at microscopic surface sites, resulting in rapid growth of a corrosion pit. Tunneling spectroscopy of the Ti surface provides a means of measuring the local electronic properties of the surface which can then be correlated with the regions where pitting occurs.

Figure 15 shows representative I–V curves obtained for (a) single crystals of TiO_2 (001), (b) TiO_2 (110), (c) a mechanically polished Ti rod on which an ~ 160-Å-thick TiO_2 film was anodically grown, and (d) a mechanically polished rod. The I–V curves for TiO_2 single crystals and thick anodically grown oxide films show highly rectifying behavior, in agreement with expectations for a large band gap semiconductor. Tunneling current at positive biases (tip versus substrate) results from tunneling of electrons from the valence band of TiO_2 (comprised of O 2p levels) to the metal tip. Conversely, tunneling current at negative bias (tip versus substrate) results from electron tunneling from the metal tip into the TiO_2 conduction band (comprised of Ti 3d levels). The corresponding $(dI/dV)/(I/V)$ versus V spectra are presented in Figure 16. The $(dI/dV)/(I/V)$ versus V plots for the anodically grown oxide film and for single-crystal electrodes show a large increase in the SDOS at bias voltages of ~ -0.5 V below the Fermi level (0 V bias) and ~ 1 V above it. Between these values the differential conductance is one to two orders of magnitude smaller, indicating a relatively small number of surface electronic states. This region corresponds to the bandgap of TiO_2. In contrast to this nearly ideal behavior, the $(dI/dV)/(I/V)$ versus V curve for the native oxide on the mechanically polished sample (Figure 16d) shows a relatively constant SDOS over a 5-eV range with no indication of an increase in state density at biases corresponding to the conduction or valence band edges.

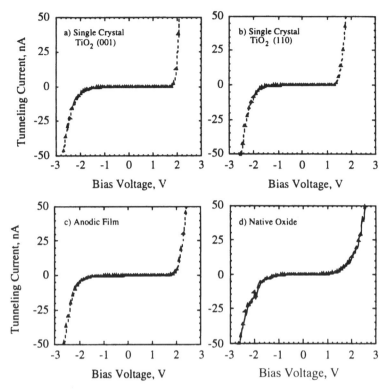

Figure 15. STM-measured current (I) versus voltage (V) curves for (**a**) single-crystal (001) TiO$_2$, (**b**) single-crystal (110) TiO$_2$, (**c**) 160-Å-thick anodically grown TiO$_2$ film, and (**d**) native TiO$_2$ film. (Reprinted from reference 76, with permission.)

The SDOS plots indicate that the thicker, anodically grown oxide displays more ideal semiconducting behavior, in agreement with previous photo-electrochemical measurements. The reason(s) that a well-defined bandgap is not observed for the thin native TiO$_2$ film cannot be discerned from the SDOS data alone. It has been suggested that the less structured SDOS observed for the native film may be a consequence of tunneling directly across the thin native oxide (rather than into the states of the film) or due to a large defect density in the native TiO$_2$ film resulting from oxygen vacancies and/or interstitial Ti^{3+}.

Of particular relevance to the pitting corrosion of Ti is the finding that I–V curves obtained for the native oxide at different locations on the same sample occasionally show rectifying behavior essentially identical to that observed for the anodically grown thick oxide layers. Thus, it is clear that the electronic

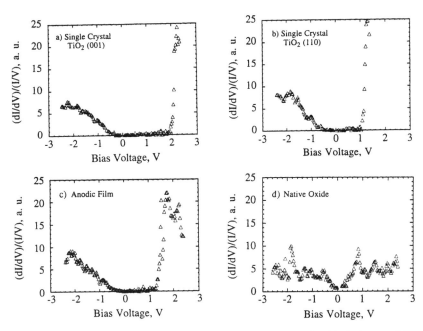

Figure 16. Surface density of states, $(dI/dV)/(I/V)$, versus voltage V for (**a**) single-crystal (001) TiO_2, (**b**) single-crystal (110) TiO_2, (**c**) 160-Å-thick anodically grown TiO_2 film, and (**d**) native TiO_2 film (~ 20 Å thick). (Reprinted from reference 76, with permission.)

properties of the native oxide film are spatially heterogeneous, with microscopic regions displaying either metallic or semiconductive behavior. The results suggest that the rate of electron-transfer reactions might also be spatially nonuniform. Recent experiments using scanning electrochemical microscopy (SECM) have demonstrated that the oxidation of Br^- at thin TiO_2 films is indeed highly nonuniform and that the more metallic-like sites on the Ti surface act as precursor sites for oxide breakdown.[77]

In situ tunneling spectroscopy was demonstrated initially by Tomita and Itaya for TiO_2 electrodes and hydrogen-terminated *p*- and *n*-type Si electrodes.[78] For example, Figure 17a shows the *I–V* response obtained by positioning the STM tip at a constant height above an *n*-Si electrode immersed in sulfuric acid. The potential of the *n*-Si electrode and tip were initially set a potentials of -0.5 and 0.45 versus SCE, respectively. The potential of the *n*-Si electrode was then scanned toward positive potentials at a rate of 1 V/sec (keeping the tip potential constant) and the *I–V* recorded. As seen in Figure 17a, the current is relatively constant at electrode potentials more negative than the flat-band position, E_{FB} (approximately equal to the conduction band edge of the *n*-Si electrode), indicating that tunnel junction resistance is nearly

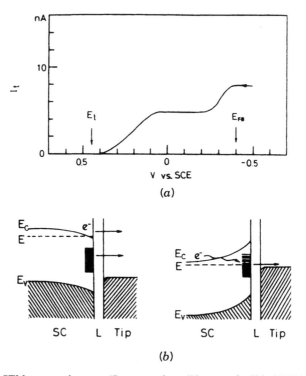

Figure 17. (a) STM-measured current (I) versus voltage (V) curve of n-Si in 0.05 M H_2SO_4. The electrode potential of the tip was held at 0.45 versus SCE. The electrode potential of n-Si was scanned from 0.5 V versus SCE at a rate of 1 V/sec. (b) Schematic of the interfacial energetics for the n-Si–electrolyte–tip tunnel junction. E, E_c, and E_v are the electrode potential, the edges of the conduction, and valence band edges, respectively. (Reprinted from reference 73, with permission.)

constant in this potential region. The current flowing in this potential region corresponds to electrons tunneling from the conduction band, at energies near the Fermi level of the semiconductor, into unoccupied states of the metal tip. When the n-Si electrode potential was increased to potentials positive of the E_{FB}, a decrease in I was observed that is due to a decrease in $\rho_s(\mathbf{r}, E, V)$ within the bandgap region. However, the current does not decrease to near-zero levels until the potential is increased to a value well within the bandgap (~ 0.4 V versus SCE), suggesting the presence of a relatively wide set of surface states that participate in the tunneling process. Figure 17b shows schematic diagrams of the tunnel junction and the direction of current flow for various electrode biases. As depicted in Figure 17b, the involvement of surface states in the tunneling process must allow for the assumption that the conduction band states are electronically coupled with the surface states. *In situ*, atomically

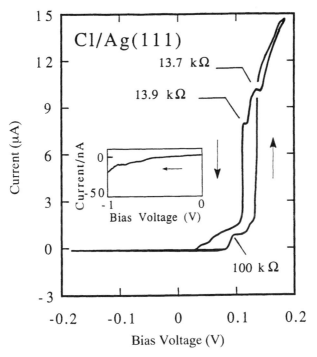

Figure 18. Typical $I-V$ curve obtained in air on a chlorine-covered Ag(111) facet. The curve in the inset shows the $I-V$ response at negative bias potentials. Note the difference in current scales.

resolved STM images of the hydrogen terminated n-Si(111) surfaces have also been reported by Itaya et al.[79]

In preliminary studies, we have observed an unusual $I-V$ response in spectroscopic measurements of Ag halide monolayers (see Figure 6) that is ascribed to a form of resonance tunneling.[80] Figure 18 shows the $I-V$ response of a Cl-covered Ag(111) response, measured in air with the STM tip held at a constant height above the surface. The $I-V$ curve for this system displays highly rectifying behavior, with the tunneling current increasing sharply above a positive sample bias threshold (~ 0.1 V). At more positive sample biases, the $I-V$ curve displays regions of the negative tunneling resistance, as indicated by the peaks in the $I-V$ response. This behavior closely resembles electronic resonances observed in two-dimensional and single-atom quantum well devices, where tuneling occurs across a double barrier system. In such systems, the transmission probability of the tunneling electron can approach unity if the energy of the tunneling electron is equal to the energy of states localized in the junction. When energy matching occurs, the minimum resistance has been

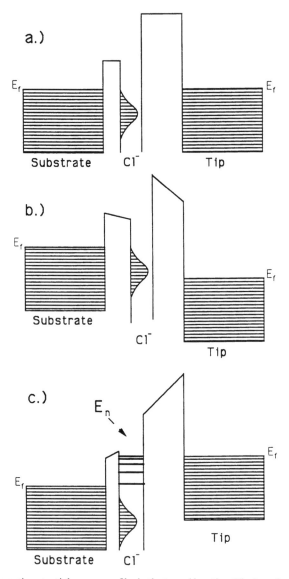

Figure 19. Schematic potential energy profiles in the tunnel junction. The broadened Cl^- $3p$ level is indicated by the bell-shaped distribution in the potential well between the tip and sample. (**a**) Unbiased tunnel junction. (**b**) Negative sample bias. (**c**) Positive sample bias.

shown to be equal to $\pi\hbar/e^2$ ($\sim$ 12.6 kΩ).[81,82] This value is in good agreement with experimental values of the tunnel junction resistance as shown in Figure 18. Similar I–V behavior is observed for I, Br, and F monolayers on Au(111) surfaces. However, the threshold potential at which the resonance occurs is a function of the identity of the adlayer, and it correlates reasonably well with the atom electron affinity. For comparison, I–V curves measured for bare metal surfaces (e.g., Au) or an oxide-coated surface (Ag$_x$O) are essentially ohmic, as is expected for electron tunneling occurring in a single potential barrier. The data in Figure 18 indicate that it is possible to extract information from the spectroscopic measurements concerning the electronic structure of adsorbates. For example, Figure 19 shows an energy-level diagram that is consistent with the I–V data for the halogen-covered Ag sample. At positive biases, electrons may tunnel from filled tip states into empty Ag states, via virtual bound state levels in the Cl potential well. Whenever the energy of the incident electrons matches the energy of such bound states, peaks in the I–V spectrum should occur, in qualitative agreement with the I–V curves in Figure 18. Further studies are necessary in order to quantitate various aspects of this work, but the results clearly suggest the possibility of extending the use of STM in electrochemistry beyond topographic imaging.

IV. CONCLUSIONS

Scanning tunneling and atomic force microscopy will become increasingly important tools for characterization of electrochemical interfaces, providing atomic- and molecular-level understanding of electrode processes. Further work is needed to fully understand the relationship between STM and AFM data and actual physical structure of the interface, a development that will be assisted by the continually growing list of examples. The STM is uniquely poised to address a key issue in electrode characterization—the role of the defects in electron transfer. Tunneling spectroscopy will likely provide substantial insight into the local electronic structure associated with these defects. The relationship between electrode structure (e.g., parent surfaces and their reconstructed forms) and electrochemically measured electron transfer rates also needs to be addressed.

New advances in scanning probe microscopies that build on the capabilities of STM and AFM have appeared recently that are likely to open new opportunities for electrode characterization. The friction force microscope, in which the lateral force exerted upon a tip by a surface is measured, can provide near-atomic-level detail of relative friction coefficients on surfaces. This characterization mode may be valuable for examination of electrode tribology. A particularly interesting new development is the magnetic force microscope, which is capable of spatially resolving different magnetic states on a surface. It is reasonable to speculate that this imaging mode will be useful for characterizing processes involving transformations between paramagnetic and diamagnetic redox states associated with adsorbed species as well as transition metal oxides. Clearly, scanning probe microscopies will be a cornerstone of characterization for electrode interfaces in the foreseeable future.

ACKNOWLEDGMENTS

The authors gratefully acknowledge Joachim Hossick Schott and Andrew Hillier for their contributions to projects related to STM and AFM of electrode interfaces. The authors are also grateful for the financial support provided by the Office of Naval Research and the National Science Foundation.

REFERENCES

1. A. J. Bard and L. R. Faulkner, *Electrochemical Methods*, Jonh Wiley & Sons, New York (1980).
2. D. A. Buttry and M. D. Ward, *Chem. Rev.*, **92**, 1355 (1992).
3. P. N. Ross and F. T. Wagner, in *Advances in Electrochemistry and Electrochemical Engineering*, Vol. XIII, H. Gerischer and C. W. Tobias, eds., Wiley-Interscience, New York (1984).
4. A. T. Hubbard, *Acc. Chem. Res.*, **13**, 177 (1985).
5. A. T. Hubbard, *Chem. Rev.*, **88**, 633 (1988).
6. P. M. A. Sherwood, *Chem. Soc. Rev.*, **14**, 1 (1985).
7. D. M. Kolb, *Ber. Bunsenges. Phys. Chem.*, **92**, 1175 (1988).
8. E. Yeager, A. Homa, B. D. Cahan, and D. Scherson, *J. Vac. Sci. Technol.*, **20**, 628 (1982).
9. M. P. Soriaga, *Prog. Surf. Sci.*, **39**, 325 (1992).
10. J. Augustynski and L. Balsenc, in *Modern Aspects of Electrochemistry*, Vol. XIV, J. O.'M. Bockris and B. E. Conway, eds., Plenum, New York (1979).
11. H. D. Abruña, J. H. White, M. J. Albarelli, G. M. Bommarito, M. J. Bedzyk, and M. McMillan, *J. Phys. Chem.*, **92**, 7045 (1988).
12. G. Binnig, H. Rohrer, Ch. Gerber, and E. Weibel, *Phys. Rev. Lett.*, **49**, 57 (1982).
13. G. Binnig, C. F. Quate, and Ch. Gerber, *Phys. Rev. Lett.*, **56**, 930 (1986).
14. R. B. Morris, D. J. Franta, and H. S. White, *J. Phys. Chem.*, **91**, 3559 (1987).
15. R. M. Penner, M. J. Heben, T. L. Login, and N. S. Lewis, *Science*, **250**, 1118 (1990).
16. T. R. I. Cataldi, I. G. Blackham, G. A. D. Briggs, J. B. Pethica, and H. A. O. Hill, *J. Electroanal. Chem.*, **290**, 1 (1990).

17. S. Sonnenfeld, J. Schneir, and P. K. Hansma in *Modern Aspects of Electrochemistry*, Vol. 21, R. E. White, J. O.'M. Bockris, and B. E. Conway, eds., Plenum, New York (1990), p. 1.

18. S. R. Snyder and H. S. White, *Anal. Chem.*, **64**, 116R (1992).

19. D. A. Bonnel, ed., *Scanning Tunneling Microscopy and Spectroscopy*, VCH, New York (1993).

20. I. Giaever, *Phys. Rev. Lett.*, **5**, 464 (1960).

21. (a) J. Tersoff and D. R. Hamann, *Phys. Rev. B*, **3**, 805–813 (1985). (b) J. Tersoff, *Phys. Rev. Lett.* **57**, 440 (1986).

22. (a) E. Tekman and S. Cirai, *Phys. Rev. B*, **40**, 10286 (1989). (b) R. M. Tromp, *J. Phys. Cond. Matter*, **1**, 10211 (1989).

23. A. J. Bard and F.-R. Fan, in *Scanning Tunneling Microscopy and Spectroscopy*, D. A. Bonnel, ed., VCH, New York (1993).

24. (a) A. L. Weisenhorn, P. K. Hansma, T. R. Albrecht, and C. F. Quate, *Appl. Phys. Lett.* **54**, 2651 (1989). (b) J. Israelachvilli, *Intermolecular and Surface Forces*, 2nd edition, Academic Press, New York (1991). (c) U. Hartmann, *Phys. Rev. B*, **43**, 2404 (1991).

25. (a) J. E. Dzyaloshinskii, E. M. Lifshitz, and L. B. Pitaevskii, *Adv. Phys.*, **10**, 165 (1961). (b) A. L. Weisenhorn, P. Maivald, H.-J. Butt, and P. K. Hansma, *Phys. Rev. B*, **45**, 11226 (1992).

26. J. Hossick Schott and H. S. White, *Langmuir*, **9**, 3471 (1993).

27. X. Gao, A. Hamelin, and M. J. Weaver, *J. Chem. Phys.*, **95**, 6993 (1991).

28. J. Hossick Schott and H. S. White, *Langmuir*, **8**, 1955 (1992).

29. J. B. Pethica, *Phys. Rev. Lett.*, **57**, 3235 (1986).

30. U. Landman, W. D. Luedtke, and M. W. Ribarsky, *J. Vac. Sci. Technol.*, **A7**, 2829 (1989).

31. N. A. Burham, R. J. Colton, and H. M. Pollock, *Nanotechnology*, **4**, 64 (1993).

32. F. Ohnesorge and G. Binnig, *Science*, **260**, 1451 (1993).

33. H. D. Abruña, J. Hossick Schott, J. E. Hudson, S. R. Snyder, and H. S. White, *Comments on Inorganic Chemistry*, **15**, 213 (1993).

34. R. Traomp, R. Hamers, and J. Demuth, *Phys. Rev. B*, **34**, 1388 (1986).

35. A. Hamelin, *J. Electroanal. Chem.*, **255**, 299 (1982).

36. X. Gao, A. Hamelin, and M. J. Weaver, *Phys. Rev. Lett.*, **67**, 618 (1991).

37. A. Hamelin, X. Gao, and M. J. Weaver, *J. Electroanal. Chem.*, **323**, 361 (1991).

38. A. Friedrich, B. Pettinger, D. M. Kolb, G. Lupke, R. Steinhoff, and G. Morowsky, *Chem. Phys. Lett.*, **163**, 123 (1989).

39. D. M. Kolb and J. Schneider, *Electrochim. Acta*, **31**, 929 (1986).

40. Ch. Woll, S. Chiang, R. J. Wilson and P. H. Lippel, *Phys. Rev. B*, **42**, 9307 (1990).

41. J. V. Barth, R. J. Behm, and G. Ertl, *Surface Sci.*, **302**, L319 (1994).

42. N. J. Tao and S. M. Lindsey, *Surf. Sci.*, **274**, L546 (1992).

43. D. D. Chambliss, R. T. Wilson, and S. Chiang, *J. Vac. Sci. Technol. B*, **9**, 2933 (1991).

44. J. Hossick Schott, C. R. Arana, H. D. Abruna, H. Hurrell Petach, C. M. Elliot, and H. S. White, *J. Phys. Chem.*, **96**, 5222 (1992).

45. T. Hachiya, H. Honbo, and K. Itaya, *J. Electroanal. Chem.*, **315**, 275 (1991).

46. J. Hossick Schott and H. S. White, *J. Phys. Chem.*, **98**, 297 (1994).

47. (a) C. P. Smith and H. S. White, *Anal. Chem.*, **64**, 2398 (1992). (b) C. P. Smith and H. S. White, *Langmuir*, **9**, 1 (1993).

48. H. P. Bonzel, *Surf. Sci. Rep.*, **8**, 43 (1987).

49. J. Hossick Schott and H. S. White, *Langmuir*, **10**, 486 (1994).

50. (a) *CRC Handbook of Chemistry and Physics*, 72nd edition, D. R. Lide, ed., 1991–1992. (b) L. Pauling, *The Nature of the Chemical Bond*, 3rd edition, Cornell University Press, Ithaca, NY (1960).

51. N. W. Ashcroft and N. D. Mermin, *Solid State Physics*, WB Saunders Philadelphia (1976).

52. C. Chen and A. A. Gewirth, *J. Am. Chem. Soc.*, **114**, 5439 (1992).

53. (a) J. J. Kulys, *Biosensors*, **2**, 3 (1986). (b) W. J. Albery and P. N. Bartlett, *J. Chem. Soc. Chem. Commun.*, 234 (1984). (c) W. J. Albery, P. N. Bartlett, A. E. G. Cass, D. H. Cranston, and B. G. D. Hagget, *J. Chem. Soc. Farad. Trans. 1*, **82**, 1033 (1986). (d) W. J. Albery, P. N. Bartlett, A. E. G. Cass, D. H. Cranston, and K. W. Sim, *J. Electroanal. Chem.*, **218**, 127 (1987). (e) S. Zhao, U. Korell, L. Cuccia, and R. B. Lennox, *J. Phys. Chem.*, **96**, 5641 (1992). (f) S. Zhao and R. B. Lennox, *J. Electroanal. Chem.*, **346**, 161 (1993).

54. J. L. Kawagoe, D. E. Nichaus, and R. M. Wightman, *Anal. Chem.*, **63**, 2961 (1991).

55. M. D. Ward, *Electroanalytical Chemistry*, Vol. 16, A. J. Bard, ed., Marcel Dekker, New York (1989).

56. S. Pan, A. L. Delozanne, and R. Fainchtein, *J. Vac. Sci. Technol. B*, **9**, 1017 (1991).

57. T. Sleator and R. Tycko, *Phys. Rev. Lett.*, **60**, 1418 (1988).

58. S. N. Magonov, J. Schuchhardt, S. Kempf, E. Keller, and H.-J. Cantow, *Synth. Met.*, **40**, 59 (1991).

59. S. Li, H. S. White, and Ward, M. D. *Chem. Mat.*, **4**, 1082 (1992).

60. R. Fainchtein and J. C. Murphy, *J. Vac. Sci. Technol. B*, **9**, 1013 (1991).

61. S. Li, H. S. White, and M. D. Ward, *J. Phys. Chem.*, **96**, 9014 (1992).

62. S. N. Maganov, G. Bar, E. Keller, E. B. Yagubskii, E. E. Laukhina, and H.-J. Cantow, *Ultramicroscopy*, **42–44**, 1009 (1992).

63. H. Bando, S. Kashiwaya, H. Tokumoto, H. Anzai, N. Kinoshita, and H. Kajimura, *J. Vac. Sci. Technol.*, *A*, **8**, 479 (1990).

64. M. Yoshimura, H. Shigekawa, H. Nejoh, G. Saito, Y. Saito, and A. Kawazu, *Phys. Rev. B*, **43**, 13590 (1991).

65. S. N. Magonow, J. Schuchhardt, S. Kempf, E. Keller, and H.-J. Cantow, *Synth. Met.*, **40**, 59 (1991).

66. S. N. Magonov, S. Kempf, H. Rotter, and H.-J. Cantow, *Synth. Met.*, **40**, 73 (1991).

67. C. Bai, C. Dai, C. Zhu, Z. Chen, G. Huang, X. Wu, D. Zhu, and J. D. Baldeschwieler, *J. Vac. Sci. Technol. A*, **8**, 484 (1991).

68. M. Yoshimura, N. Ara, M. Kageshima, R. Shiota, A. Kawazu, H. Shigekawa, Y. Saito, M. Oshima, H. Mori, H. Yamochi, and G. Saito, *Surf. Sci.*, **242**, 18 (1991).

69. A. C. Hillier and M. D. Ward, *Science*, **263**, 1261 (1994).

70. Phillip W. Carter, A. C. Hillier, and M. D. Ward, *J. Am. Chem. Soc.*, **116**, 944 (1994).

71. R. Feenstra, J. Stroscio, and A. Fein, *Surf. Sci.*, **181**, 295 (1987).

72. F.-R. Fan and A. J. Bard, *J. Phys. Chem.*, **94**, 3761 (1990).

73. E. Tomita, N. Matsuda, and K. Itaya, *J. Vac. Sci. Technol.*, **A8**, 534 (1990).

74. S. Gilbert and J. H. Kennedy, *Langmuir*, **5**, 1969 (1991).

75. F.-R. Fan and A. J. Bard, *J. Phys. Chem.*, **95**, 1969 (1990).

76. N. Casillas, S. R. Snyder, W. H. Smyrl, and H. S. White, *J. Phys. Chem.*, **95**, 7002 (1991).

77. N. Casillas, S. Charlebois, W. H. Smyrl, and H. S. White, *J. Electrochem. Soc.*, **140**, L142 (1993).

78. E. Tomita, N. Matsuda, and K. Itaya, *J. Vac. Sci. Technol.*, *A*, **8**, 534 (1989).

79. K. Itaya, R. Sugawara, Y. Morita, and H. Tokumoto, *Appl. Phys. Lett.*, **60**, 2534 (1992).

80. J. Hossick Schott and H. S. White, *J. Phys. Chem.*, **98**, 291 (1994).

81. V. Kalmeyer and R. B. Laughin, *Phys. Rev. B*, **35**, 9805 (1987).

82. N. D. Lang, *Phys. Rev. B.*, **36**, 8173 (1987).

CHAPTER

4

STRIPPING ANALYSIS

HOWARD D. DEWALD

Department of Chemistry
Clippinger Laboratories
Ohio University
Athens, Ohio 45701

Modern Techniques in Electroanalysis, Edited by Petr Vanýsek, Chemical Analysis Series, Vol. 139.
ISBN 0-471-55514-2 © 1996 John Wiley & Sons, Inc.

Since World War II, demands for faster analysis time and lower detection limits on smaller samples have been made in analytical chemistry. Trace analysis at the parts-per-billion (ppb, or $1:10^9$) level in complex matrices (e.g., hazardous waste and blood) is routinely accomplished today. Still, legislative and regulatory mandates continually push the thresholds of chemical reagents to ever lower levels. One of the most successful electroanalytical techniques used to meet these demands has been stripping analysis.

Stripping analysis is a combination of electrochemical procedures performed in two main steps. During the first step, the analyte is concentrated at the surface of a microelectrode under controlled conditions (e.g., potential, stirring, and time). This step serves to "extract" the analyte from the sample solution and deposit it on the electrode. Thus the concentration of the analyte at the surface of the electrode is far greater than it is in the bulk solution. The deposition is performed either by physical adsorption on the electrode surface, by electroplating onto a solid electrode, or by electrolytic deposition into a mercury electrode, forming an amalgam or insoluble mercury salt. The concentrated analyte is then electrolytically or chemically redissolved (stripped) from the electrode, and the analysis is performed by either potentiometric or voltammetric methods. A peak appears in the voltammetric stripping curve (current–potential curve), and a plateau occurs in the potentiometric stripping curve (potential–time curve) with the value of the potential a qualitative indication of the analyte. Since stripping curves for various analytes occur at characteristic potentials, several species can often be determined simultaneously. The voltammetric stripping peak current or the potentiometric stripping time forms the basis for quantitative determination in stripping analysis. Detection limits on the order of 10^{-10} M with excellent precision have been attained for many species.

A variety of series chapters,[1–4] monographs,[5–9] reviews,[10–22] technical notes,[23,24] and manuals[25,26] have been written in the past 30 years that give in-depth coverage of the history and theory of stripping analysis. The basic principles, experimental methodology, and procedures, along with applications of stripping techniques, will be presented in this chapter. Additionally, hybrid combinations of stripping analysis with flow injection analysis, modified electrodes, immunoassays, and chromatography will also be presented.

I. METHODS

A. Anodic Stripping Voltammetry

Anodic stripping voltammetry (ASV) is used to determine the concentration of trace metals and is the most common stripping technique.[11,12,15] In ASV, metal ions in solution are concentrated onto an inert solid or into a mercury electrode by reduction to the metallic state. The deposition potential is more negative than the half-wave potential of the metal ions to be determined. During deposition the metals dissolve in the mercury and form an amalgam. A positive-going potential scan is applied to oxidize (strip) the metal from the amalgam back into solution to its original state (Figure 1). The resulting anodic peak currents are proportional to the concentrations of the metal ions in the sample.

The principle of the anodic stripping experiment for the determination of lead (Pb^{2+}) is illustrated in Figure 2. A deposition potential of -1.0 V (versus the saturated calomel electrode, SCE) is applied for several minutes to concentrate the Pb^{2+} from solution into the mercury drop electrode as Pb(Hg):

$$Pb^{2+} + Hg + 2e^- \rightarrow Pb(Hg)$$

In order to maximize the amount of Pb^{2+} reduced at the electrode, the deposition is usually accompanied by convection by electrode rotation, solution flow, or stirring. Since the electrolysis is not exhaustive, all operating conditions (e.g., convection rate, deposition time) must remain the same

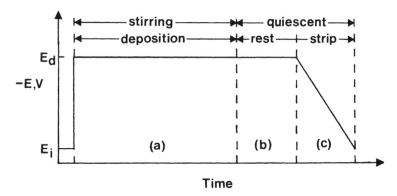

Figure 1. Potential–time waveform used in anodic stripping voltammetry. (**a**) Deposition at E_d, stirred solution. (**b**) Rest period, stirrer off. (**c**) Stripping, positive potential scan to E_i.

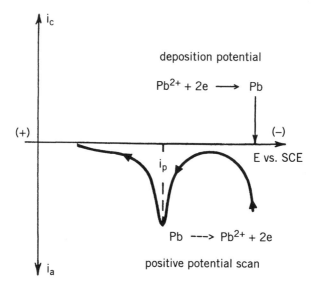

Figure 2. Anodic stripping voltammogram for Pb^{2+}.

throughout the entire stripping experiment to obtain high precision and accuracy in the sample and standard solution data. At the end of the deposition period the forced convection is stopped and a short rest period on the order of several seconds (depending on the type of mercury electrode) is observed. During the rest period the deposition current drops to near zero, the concentration of metal in the amalgam becomes more uniform, and the solution is allowed to become quiescent. Deposition by diffusional mass transport continues during the rest period. During the stripping step an excitation waveform is applied which causes the amalgamated lead to be oxidized back into the solution:

$$Pb(Hg) \rightarrow Pb^{2+} + Hg + 2e^-$$

The excitation waveform consists of a positive-going potential scan using one of the common voltammetric methods. Differential-pulse stripping discriminates against background charging current better than linear-scan stripping voltammetry, but the slower scan rates used with differential-pulse stripping increase the analysis time. Square-wave stripping techniques result in similar sensitivity to differential-pulse stripping and allow a faster scan rate. The lead is stripped from the amalgam at a potential close to the polarographic half-wave potential and results in a current peak that is proportional to the concentration of Pb^{2+} in the sample.

Table 1. Metals That Have Been Determined by ASV

Antimony	Germanium	Rhodium
Arsenic	Gold	Selenium
Bismuth	Indium	Silver
Cadmium	Lead	Thallium
Copper	Manganese	Tin
Gallium	Mercury	Vanadium
	Nickel	Zinc

ASV is commonly used for the determination of metal ions at the trace level. The determination of lead and cadmium are perhaps the most widespread applications. ASV has been used extensively for the analysis of clinical samples (such as lead in blood[16]) and environmental samples (such as water from numerous sources[15]). An application to forensic chemistry is the determination of antimony and lead in the analysis of firearm discharge residues.[27,28] ASV is restricted to about 30 metals. A dozen or so are easily detected at mercury electrodes in aqueous solutions: antimony, bismuth, cadmium, copper, gallium, germanium, indium, nickel, lead, thallium, tin, and zinc. The use of solid electrodes, such as glassy carbon or gold, enable arsenic, gold, mercury, platinum, and silver to be determined by ASV. Table 1 lists several examples of metals which have been determined by ASV.

B.   Cathodic Stripping Voltammetry

Cathodic stripping voltammetry $(CSV)^{3,13}$ is used to determine a variety of anions that form insoluble salts with mercury(I) ion. In CSV the mercury electrode is not inert. A relatively positive potential is applied to a mercury electrode where mercury is oxidized to mercury(I):

$$2Hg \rightarrow Hg_2^{2+} + 2e^-$$

Then in the presence of an anion X^- an insoluble film forms on the electrode surface following the general reaction:

$$Hg_2^{2+} + 2X^- \rightarrow Hg_2X_2$$

Stripping in CSV consists of a negative-going potential scan which reduces the mercury salt to Hg and X^-:

$$Hg_2X_2 + 2e^- \rightarrow 2Hg + 2X^-$$

Table 2. Substances That Have Been Determined by CSV

Arsenic	Molybdate	Sulfide
Chloride	Oxalate	Thiocyanate
Chromate	Selenium	Thiols
Bromide	Succinate	Tungstate
Iodide	Sulfate	Vanadate

The stripping peak position and height are characteristic of the type of anion and its concentration in solution. CSV procedures have been developed for determination of chloride, bromide, iodide, sulfide, selenide, chromate, molybdate, oxalate, succinate, sulfate, thiocyanate, tungstate, vanadate, arsenic, selenium, and thiols. A few metal cations [e.g., cerium(III), ion(II), manganese(II), and thallium(I)] can be concentrated as insoluble hydroxides by oxidation on a carbon electrode. The hydroxides are then stripped from the electrode by reduction with a negative-going potential scan. Table 2 lists examples of several substances which have been determined by CSV.

C. Adsorptive Stripping Voltammetry

In adsorptive stripping voltammetry (AdSV) a spontaneous adsorption process is utilized to concentrate analytes at the surface of an electrode.[18-21] Most commonly a hanging mercury drop electrode is immersed into a stirred solution of the analyte for several minutes. Accumulation of the analyte occurs by physical adsorption rather than by electrolytic deposition. Depending on the oxidation–reduction properties of the accumulated analyte, it is determined by scanning the potential in the appropriate negative or positive direction. Nonelectroactive species are analyzed from the tensammetric (adsorption/desorption) peaks produced during the stripping step. Many organic molecules have a strong tendency to adsorb from aqueous solution onto a mercury surface. For neutral compounds a potential close to the potential of zero charge (ca. -0.5 V versus SCE) is recommended. Other solution properties (e.g., pH) can be adjusted to improve adsorption. Table 3 illustrates some adsorptive stripping measurements of organic compounds.

Additionally, many inorganic cations have been determined by AdSV. In these applications the cations are generally complexed with surface-active complexing agents, L, (e.g., catechol, dimethylglyoxime, bipyridine):

$$M^{n+} + xL \rightarrow ML_x^{n+}$$

$$ML_x^{n+} + Hg \rightarrow ML_{x,ads}^{n+}(Hg)$$

Table 3. Adsorptive Stripping Measurements of Organic Compounds

Adriamycin	Digoxin	Progesterone
Bilirubin	Dopamine	Riboflavin
Butylated	DNA	Testosterone
hydroxyanisole (BHA)	Heme	Thiourea
Chlorpromazine	Monensin	Trichlorobiphenyl
Cimetidine	Phenanthrenequinone	
Cocaine	Polyethylene	
Codeine	glycols (PEGs)	

The cation is then released from the complex by reduction:

$$ML_{x,ads}^{n+}(Hg) + me^- \rightarrow M^{(n-m)+} + xL + Hg$$

Most of the cations measured by AdSV cannot be measured conveniently by ASV as a result of their low solubility in mercury, extreme redox potentials, or irreversiblity. Paneli and Voulgaropoulos[21] have reviewed the applications of AdSV in the determination of metal ions.

D. Potentiometric Stripping Analysis

Potentiometric stripping analysis (PSA) is an alternate method to ASV.[14,26,29,30] The first step is performed in a manner similar to that of ASV. By means of a negative potential, metal ions are reduced and dissolved in a thin mercury film on a glassy carbon electrode. Again, metal ions which cannot be determined with amalgamation can be deposited on a solid electrode, such as gold. In the stripping step, the potential applied to the electrode is removed and the metals are oxidized (in order of their oxidation potentials) from the film chemically as a result of the reaction

$$M(Hg) + \text{oxidant} \rightarrow M^{n+} + ne^-$$

As the concentration of the metal in the amalgam is exhausted, the potential drops toward more positive values. The duration of each plateau (at the standard potential of the metal–metal ion couple) represents the time required for oxidation of the metal, which is proportional to its concentration (Figure 3). Either dissolved oxygen or mercury(II) ions present in the solution are used typically as the oxidants. The oxidation reaction with mercury(II) is

$$M(Hg) + n/2Hg^{2+} \rightarrow M^{n+} + n/2Hg$$

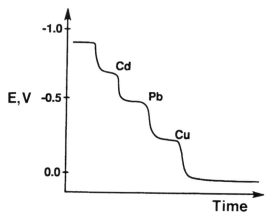

Figure 3. Potentiometric stripping analysis potential–time curve for Cd, Pb, and Cu.

and with dissolved oxygen the reaction is

$$M(Hg) + n/4O_2 + nH^+ \rightarrow M^{n+} + n/2H_2O$$

Alternatively, a small constant current $(5–20\,\mu A)$ can be applied to strip the metals from the amalgam. The current density and not the concentration of oxidant determines the sensitivity. Constant current stripping analysis (CCSA) has been used in an AdSV determination of nickel ion complexed with dimethylglyoxime, using a reducing current to strip the nickel from the mercury electrode.[31] Applications of PSA and CCSA are highlighted in Table 4. It is also possible to perform anodic preconcentration with reductive potentiometric stripping.[32]

Table 4. Applications of PSA and CCSA

PSA		CCSA	
Bismuth	Indium	Antimony	Mercury
Cadmium	Lead	Arsenic	Molybdenum
Copper	Thallium	Cobalt	Nickel
Gallium	Tin	Gold	Selenium
	Zinc	Iron	Silver
		Manganese	

II. APPARATUS

A. Electrochemical Cell

A typical cell for stripping analysis is shown in Figure 4. The three-electrode system consists of the working electrode, a reference electrode (e.g., Ag/AgCl or the SCE), and the auxiliary electrode which is frequently a platinum wire or foil. In many cases, tubing for deoxygenation and gas blanketing of the solution is required. Sample solution volumes range between 5 and 50 ml, but smaller volumes as low as a few microliters can be used with microcells.[33] Many stripping analyzers have a standard cell station. The cell container can be an ordinary lab glass beaker, sample vial, or weighing bottle. For ultratrace analysis, acid-washed quartz, Teflon (DuPont), or disposable plastic beakers are preferable to reduce contamination from leaching or ion loss through surface adsorption. As with all analytical methods that are capable of analysis at the trace level, contamination of the sample must be avoided. Stringent cleaning and rinsing of all cell components is necessary. To avoid leaching metals from the cell, it is wise to adopt a standard practice

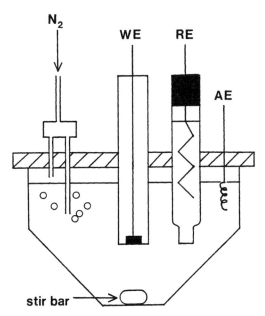

Figure 4. Electrochemical cell for stripping analysis. WE, working electrode; RE, reference electrode; AE, auxiliary electrode.

of storing 6 M nitric acid in the cell for an hour (or overnight, when possible) prior to an analysis. Only high-quality water should be used to rinse the cell.

Contamination from other species in the solution or from the reference electrode solution or a salt bridge can present problems. The reference and auxiliary electrodes can be isolated with porous Vycor or porcelain or fiber wicks that generally have low leakage rates. Reference electrodes should be checked for drift (as a result of low Cl$^-$ levels in Ag/AgCl and SCE electrodes) and clogging. Reference electrodes should be properly stored in a recommended soaking solution to prevent chloride depletion and to help prevent clogging of the electrode tip.[34] A Teflon-covered stirring bar may be used with a magnetic stirrer. Erratic stir bar behavior affects results. Better control and precision are often obtained with arrangements that employ a rotating working electrode, a flow-through cell, or a propeller stirrer. Often, the solution convection is initiated several seconds before the deposition is started so that uniform convection currents are established in the cell. On–off control is recommended over a variable speed control for more reproducible convection rates between samples. Also, the cell bottom should not be in direct contact with the surface of a magnetic stirrer to prevent heating of the solution. A styrofoam or cork insulator is frequently placed between the cell and magnetic stirrer.

B. Working Electrodes

1. Solid Electrodes

A variety of materials are used for working electrodes in stripping analysis. The ideal electrode has a reproducible surface, a reproducible area, and a low residual current. Two groups of materials are the most practical as electrodes in stripping analysis: mercury and inert solids such as gold, platinum, and carbon. Indeed, solid electrodes must be used for the determination of metals with oxidation potentials more positive than mercury (e.g., Ag, Au, and Hg itself) or for metals that are insoluble in mercury (e.g., As). Generally, poorer deposition and stripping results are obtained with solid electrodes than with mercury.

Carbon has the advantages of a wide potential range (± 1 V), relative chemical inertness, and low cost. Glassy carbon,[35] pyrolytic carbon,[36] and wax-impregnated graphite[37] have been used extensively. The carbon paste electrode,[38] a mixture of graphite and a viscous organic binder, has a low background current and an easily renewable surface. The carbon paste is packed into a shallow well containing an electrical contact. The surface is then smoothed on a flat surface such as an index card.

Gold and platinum are the most common metallic solid electrodes. Each has a hydrogen overpotential lower than that of mercury, but they are not as easily oxidized. Noble-metal electrodes adsorb hydrogen and are subject to formation of oxide films. Some of the noble metals have a tendency to form intermetallic compounds (e.g., AuHg). Carbon-based gold films have been prepared to circumvent these experimental difficulties.[39] All solid electrodes require specific cleaning, polishing, and pretreatment steps which are dependent on the material in use. On solid surfaces, deposited metals are not interdiffusible. When several metals are codeposited, a homogeneous layer does not form and overlapping, multiple, and split peaks may arise during stripping. The underlying material may be prevented from stripping until the overlying deposit has been oxidized. Consequently, although good stripping results can be obtained at solid electrodes, a mercury electrode is preferred. Two types of mercury electrodes are popular: the hanging mercury drop electrode (HMDE) and the mercury film electrode (MFE).

2.   Hanging Mercury Drop Electrode

The HMDE is shown in Figure 5. The drop may be suspended from the tip of a platinum wire embedded in glass or plastic[40] or it may be suspended from a mercury thread at the tip of a glassy capillary or microsyringe.[41] The entire stripping experiment is performed on one mercury drop. The drop is then dislodged and a new drop is formed. A typical drop diameter is 0.5 mm. Reproducibility of the drop radius is important for precise quantitative analysis. A homemade HMDE is constructed from a platinum wire embedded in glass or plastic. The wire is then made flush with the embedding surface,

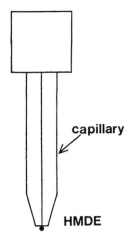

Figure 5. Hanging mercury drop electrode (HMDE).

sanded flat, and polished. The platinum contact is then plated with mercury to make it receptive to the mercury drop. A few drops of mercury from a dropping mercury electrode (DME) are collected in a glass or plastic spoon and transferred to the PtHg surface where the drop attaches. Commercial HMDEs are available which extrude a mercury drop through a glass capillary consisting of a microsyringe with a micrometer for control of drop size. The static mercury drop electrode (SMDE) has been developed which pumps a reproducible drop through a fine-bore glass capillary.[42]

Special attention should be paid to the proper operation and maintenance of the HMDE.[23] The HMDE should have a clean, dry siliconized capillary. The capillary is cleaned by aspirating first 3 M HNO_3 followed by high-quality water. After air-drying, a 5% solution of dichlorodimethylsilane in CCl_4 is drawn through the capillary. Finally, the capillary is air-dried and filled with mercury. The siliconizing procedure enhances the hydrophobic nature of the capillary and helps prevent solution creep between the mercury and capillary glass wall. Entrapped air in the HMDE can lead to nonreproducible drop sizes and premature drop dislodgment. Entrapped air is evidenced by retraction of the mercury thread when the drop is dislodged. The capillary filling procedure specified by the manufacturer should be followed.

The HMDE has the advantage of a low residual current of pure mercury and a wide potential range limited on the positive end by mercury oxidation (ca. 0 V versus Ag/AgCl) and by hydrogen gas evolution on the negative side (-1.8 V versus Ag/AgCl depending on pH). HMDEs are generally restricted

Table 5. Solubility and Diffusion Coefficients of Metals in Mercury

Metal	Solubility[a] (% w/w)	Diffusion Coefficient[b] (10^{-5} cm^2/sec)
Ag	0.035	1.5
Bi	1.1	0.87
Cd	5.3	0.80
Cu	0.003	0.68
In	57	
Pb	1.1	0.92
Sn	0.6	
Tl	42.8	1.88
Zn	1.99	

[a]Temperature, 20°C. Adapted from reference 7.
[b]Temperature, 25°C. T. Pavlopoulos and J. D. H. Strickland, *J. Electrochem. Soc.*, **104**, 116–123 (1957).

to rather slow stirring rates. Rapid stirring is impractical because of distortion or dislodgment of the drop at rates above a few hundred revolutions per minute (rpm). The HMDE has a relatively low surface area-to-volume ratio. The small area reduces the deposition efficiency while the large volume yields a low concentration of metal in the mercury. These factors result in long deposition times. If very long deposition times are used, the metals may diffuse into the mercury column in the capillary. Additionally, the large volume leads to broader stripping peaks and limits the ability to resolve adjacent peaks. This is a result of the finite time required for the metals to diffuse from the drop interior. An advantage of the large volume of the HMDE, as compared to the MFE (*vide infra*), is less susceptibility to problems associated with intermetallic compound formation and with exceeding the solubility of metals in mercury as shown in Table 5.

3. Mercury Film Electrode

The MFE is prepared by depositing a coating (1–1000 nm) of mercury onto an inert, electrically conductive solid support. The MFE is prepared by electrochemical reduction of Hg(II) to Hg onto a conductive support; procedures for both *in situ* and *ex situ* deposited films have been developed. The preformed (*ex situ*) film is typically prepared by placing the conductive support in a well-stirred, acidic, deoxygenated solution of $5–50 \times 10^{-6}$ M reagent-grade mercury(II) nitrate at -0.4 to -1.0 V versus Ag/AgCl for several minutes. Removal of oxygen from the solution prevents oxidation of the film. The film thickness is controlled by the Hg(II) concentration and deposition time. The electrode is then removed from the solution, rinsed with supporting electrolyte, and used for sample analysis. More commonly, the MFE is prepared *in situ* by adding the micromolar Hg(II) plating solution directly to the sample solution and simultaneously depositing the mercury and analyte metals.[43] The *in situ*-prepared film can be prepared fresh with each analysis. The film is removed following completion of the experiment either mechanically by wiping it off with a paper tissue or electrolytically by stripping it into a solution that is absent of anions that may cause precipitation of the soluble mercury species.

The ideal substrate for the MFE must be electrically conductive, chemically inert to the mercury and analyte solution, and electrochemically inert in the potential range of interest. Additionally, the substrate condition is a major consideration, because the same electrode surface is used for repetitive analysis. Various metals and forms of carbon have been used as substrates for MFEs.[44] Metallic substrates (e.g., Ag, Au, Ni, and Pt), although highly conductive, have the disadvantages of low hydrogen overpotential, solubility in mercury, adherent oxide films, and intermetallic compound formation.

These factors result in irreproducible mercury films and stripping peaks. Thus metal substrates are not generally used for the MFE.

Carbon is well-suited as a substrate for a MFE. Carbon has a good electrical conducitivity and a reasonably high hydrogen overpotential, is insoluble in mercury, and is chemically inert. Glassy carbon and wax-impregnated graphite (WIG) have been most widely used. Wax impregnation of porous graphites is important for preventing solution creep. A variety of graphites and waxes have been evaluated. In general, the wax should have a low degree of crystallinity (e.g., Sonneborn) to minimize cracking and separation from the graphite. Several preparation procedures have been reported for the WIG–MFE.[45–47] Nonuniform mercury deposition can occur if steps are not taken to keep the WIG electrode surface free of wax films. An MFE prepared on a WIG is relatively stable in neutral or weakly acidic media (3–4 weeks), but it deteriorates within a few hours in highly acidic solution (pH 1–2) as indicated by rapid loss of hydrogen overpotential from the formation of carboxyl groups on the graphite surface.[48] Several attempts have been made to extend the longevity in acidic media by impregnation with epoxy[49] and using gamma radiation[50] to cross-link styrene monomer to the graphite substrate.

Glassy carbon is very hard and has a high density and small pore size. The properties and electroanalytical use of glassy carbon have been evaluated.[35] Deposition of mercury on glassy carbon and WIG substrates results in a coating of mercury drops rather than a continuous liquid film[51] (a true film has recently been obtained on iridium[52]). The size and distribution of these drops depend on the deposition potential. The average thickness, l, of a MFE may be calculated from the equation

$$l = 2.43it/r^2$$

where i is the deposition current, t is the deposition time, and r is the electrode disc radius.

Several commercial manufacturers supply carbon and metal electrodes that can be used for stripping analysis. In most cases, a disc-shaped electrode is used. The electrode (1–5 mm in diameter) is sealed flush with the surface of a glass or plastic sleeve. Steps must be taken to ensure that the surface of the MFE substrate is reproducible. Most commercial glassy carbon and metal electrodes have been polished to a mirror-like finish by standard metallographic procedures. Prior to mercury film formation, aqueous suspensions of diamond paste and alumina are used with a polishing cloth. Often other "conditioning" steps are used to "electrolytically clean" the electrode surface. A conditioning potential or cycling the potential between empirically determined values is applied for a controlled time. Conditioning is not required with an HMDE because a new mercury drop is used for each determination.

The MFE has a larger surface area-to-volume ratio than the HMDE. Thus more sample is concentrated into the mercury for a specific deposition time, and greater sensitivity is achieved for the MFE as compared to the HMDE. Shorter deposition times are required with the MFE in order to obtain the same sensitivity with the HMDE. Unlike the HMDE, the MFE is compatible with convection. The ability to stir the solution or rotate the electrode increases mass transfer during the concentration step and decreases the time required to deposit the sample. However, the MFE becomes more susceptible to problems associated with intermetallic compound formation and to exceeding the solubility of metals in mercury.

The thinness of the MFE as compared to the HMDE results in sharp stripping peaks, which minimizes the overlap of peaks for multielement analysis and reduces uncertainties in the baseline estimation for the calculation of the peak current. In the MFE the metals deposited are constrained near the mercury–solution interphase, while in the HMDE the metals deposited diffuse towards the center of the drop. Thus the distance that the metals must diffuse for oxidation during stripping is much greater for the HMDE than for the MFE.

Precision is less with the MFE than with the HMDE as a result of the care required to form the film. However, increasing needs for sensitive measurements at the parts-per-trillion (pptr) level and for simultaneous multielement determination of metals have resulted in a wider application of the MFE as compared with the HMDE for ASV and PSA. The HMDE is most common in AdSV measurements. The HMDE and MFE have been directly compared.[53]

Only triple-distilled (Bethlehem Apparatus Co.) mercury should be used for HMDE operation. A mercury purity test involves creating a suspension in water. The mercury is pure if it does not immediately separate from the suspension. For very low-level work it may be desirable to further treat the triple-distilled mercury.[23,54]

C. Hydrodynamic Systems

Stripping analysis is easily adapted to hydrodynamic (convective) designs (i.e., solution stirring, rotating electrodes, and flow-through systems). Stripping sensitivity and precision are dependent on the effectiveness of the hydrodynamics that control the amount of analyte concentrated during the deposition step.

When the deposition potential is applied, the concentration of the analyte at the electrode surface is diffusion-limited and becomes effectively zero. By employing hydrodynamics, the concentration of the analyte is maintained at the uniform bulk level up to the distance δ, the diffusion layer thickness,[55] from the electrode surface. Thus both diffusion and convection contribute to the deposition current, which can be described by the equation

$$i_l = nFADC/\delta$$

where i_l is the limiting current for deposition, n is the number of electrons transferred, F is the Faraday constant (96, 487 C/equiv), A is the electrode area, D is the diffusion constant (cm^2/sec), and C is the analyte concentration. Empirically, δ is related to the convection rate which is dependent on the electrode geometry and flow parameters. By reducing the value of δ, the deposition current increases. Thus precise control of convection rates is essential for obtaining reproducible deposition. A stir bar and magnetic stirrer are adequate to achieve sufficient convection rates, but better control is obtained with rotating working electrodes or flow-through cells.

1. Rotating Electrodes

The rotating disc electrode (RDE) and rotating ring-disc electrode (RRDE) provide well-defined hydrodynamics.[56] Generally, for stripping analysis the RDE consists of a carbon disc embedded in a rod of insulating Teflon or Kel-F. Thus a mercury film can be easily formed. The electrode is attached to a motor and rotated at a specified frequency.[57] At the RDE the hydrodynamic flow pattern results from centrifugal forces that move the solution horizontally out and away from the center of the disc while fresh solution continually replaces it by a flow normal to the electrode surface.

The RRDE consists of the RDE that is surrounded by a band of insulating material and a ring of electrode material (e.g., gold, platinum, or glassy carbon). Dual potentiostats are required to control the RRDE operation. Deposition and stripping are performed at the disc. The materials stripped from the disc are then "collected"[58] at the ring as the radial flow patterns sweep the metal ions from the disc to the ring.

2. Flow-Through Cells

The adaptation of stripping analysis to flow-through systems is straightforward and offers several advantages.[17] Higher sample throughput (on-line monitoring) and improved precision will result from automation[59] that a flow system provides. Minimal sample handling in a flow system leads to a lower risk of contamination, especially for trace-level analysis. In a flow-through system, medium exchange (*vide infra*) can be used where the stripping step is performed in an electrolyte solution that is different from the solution used during the deposition step.[60] Thus greater control over selectivity and sensitivity is possible because the medium exchange solution can be chosen to complex an interfering species, to separate the peak potentials of overlapping substances, or to reduce background solvent effects. Precision can be improved in PSA because medium exchange allows better control on the oxidant concentration.

The combination of flow injection analysis (FIA) and stripping analysis offers a simple means for processing microliter samples.[61] The concepts of FIA are based on the principles of sample injection, controlled dispersion, and reproducible timing.[62] In FIA the sample is introduced as a discrete bolus which disperses as it moves along the flow stream to the detector. Contamination between consecutive samples (carry-over) is eliminated, and sensitivity can be improved on the basis of the design of the FIA manifold.

One drawback to flow-through systems is that oxygen removal from the solution (as required in ASV) is often cumbersome and inefficient. Approaches to overcoming this problem include using square-wave stripping[63] and background subtraction[64] procedures.

On-line monitoring and FIA concepts have led to the development of a number of flow-through cells. A variety of cell configurations are available: thin-layer (channel), tubular, and the "wall-jet" (flow normal to the electrode surface).[65] In all cases the convection rate is determined by the solution flow rate. The thin-layer design is more widely used (as a hold-over from popular HPLC systems), but the wall-jet design has been shown to be superior for fast response and minimum carry-over.[66]

D. Instrumentation

The electronic requirements for stripping analysis consist of a circuit that applies the potential to the cell, a circuit that measures the current or potential, and a recording device for display. Various "stripping analyzers" are available commercially at relatively modest prices. Most instruments have the capability of applying several potential waveforms for the ultimate in resolution, sensitivity, and speed. Many instruments are microprocessor-controlled, for which the stripping procedure and data reduction is automated.

The heart of the instrumentation is a three-electrode potentiostat. In the potentiostatic circuitry a reference electrode of fixed potential (e.g., Ag/AgCl or SCE) is placed in close proximity to the working electrode and connected through a circuit that draws no current from it. The potential between the working electrode and the reference electrode is measured and compared with the desired controlled potential. The difference is adjusted to zero by a negative feedback circuit consisting of a high-gain operational amplifier, resistors, and the working and auxiliary electrodes. The cell current is passed between the working electrode and the auxiliary electrode. In PSA the reference electrode is connected to a high-input impedance voltmeter (above $10^{12}\,\Omega$) so that oxidation of the amalgamated metals through the reference electrode is avoided.[14] Basic potentiostats are easily constructed from simple operational amplifier circuitry.[7]

The range of potentials should extend from $+1$ V to -2 V versus Ag/AgCl or SCE. The current range should extend from microamperes to nanoamperes (or even picoamperes) for ultratrace analyses. Typically, several potential scan rates extending from mV/sec to several V/sec are available.

III. TECHNICAL ASPECTS AND TECHNIQUES

A. Standard Solutions and Supporting Electrolytes

Usually, quantitative analysis in stripping analysis is achieved by the method of standard additions because the analysis and quantitation can be done in a single electrochemical cell. Standard solutions should be stored no more dilute than 10^{-3} M. Frequently, commercial solutions prepared for atomic absorption spectroscopic (AAS) analysis are used. Dilute solutions are then prepared by serial dilution on a daily basis (since nonspecific adsorption of trace components can occur upon standing). If possible, the same flask should be retained for these dilute solutions. Water of high purity should always be used.

Prior to the stripping analysis, a supporting electrolyte is added to the sample in order to increase the conductivity of the solution to eliminate mass transfer by migration during the deposition step and to minimize peak distortion by IR drop during the stripping step.[67] When a potential is applied to the electrochemical cell, the concentration of the metal ions on the electrode surface is practically zero. As the electrolysis proceeds, the concentration of the ions in the sample is constant up to the thickness of the diffusion layer from the electrode. As a result of the difference in concentration, the metal ions move towards the electrode by diffusion. The ions are influenced by a potential gradient caused by the electrical field around the electrode, namely, migration. In order to minimize the effect of the potential field, an inert supporting electrolyte is added to the sample (thus the ionic strength and conductivity of the solution are increased). Common electrolytes for use in stripping analysis include inorganic salts (e.g., KCl or KNO_3), acids (e.g., HCl), and buffer solutions (e.g., acetate or ammoniacal) when pH control is needed. Many supporting electrolytes have a tendency to form complexes with metal ions. Improved selectivity is possible as a result of shifts in peak potentials along the potential axis from the complex formation.

The reagents used for the preparation of the standards and supporting electrolyte should be of the highest purity possible. Even when reagent-grade chemicals are used for the preparation of supporting electrolytes, low-level heavy metal contamination (which originates in the reagent) can become a significant factor. Purification of supporting electrolytes can be accomplished by ion-exchange or controlled potential electrolysis (for most inorganic salts) or isothermal distillation (for HCl and NH_4OH).[23,68]

In the controlled potential electrolytic treatment, the supporting electrolyte is placed in a large electrochemical cell ($\sim$ 1-liter capacity) and is deaerated. A mercury pool cathode covers the cell bottom. The solution is stirred and an electrolysis potential (ca. -1.5 V versus Ag/AgCl or SCE) is applied to the mercury for approximately 24 hours. The electroactive metals are removed by reduction into the mercury cathode. The purified electrolyte is removed with the potential still applied to the cell (to prevent the metals deposited into the mercury pool from stripping back into the solution).

B. Sample Pretreatment

A sample pretreatment step may be required to convert the metal ions in the sample matrix into a form which can be deposited at the electrode. In many biological samples the metal ions are often bound to proteins. Water samples can be polluted with organic materials. In order to release the metal ions, such samples are treated with acid digestion or dry-ash techniques. During these techniques the matrix constituents are converted to carbon dioxide and water, leaving nonvolatile metals as residue. This residue should be diluted before analysis. Acid concentrations stronger than 1 M can oxidize an MFE. If the matrix contains volatile metals (e.g., As and Se), digestion may need to be performed under reflux.[23]

Polyethylene is preferred for sample containers. Such containers have fewer adsorption problems and do not remove or release metal ions to the sample.[69–71] Containers for samples or reagents should be cleaned with dilute nitric acid and rinsed with high-purity water.

The temperature of all samples and standards should be at room temperature before the analysis is performed.

C. Dissolved Oxygen

Once the sample with supporting electrolyte has been added to the cell, the solution is deoxygenated (except in PSA). Oxygen is capable of dissolving in aqueous solutions at millimolar levels at room temperature and atmospheric pressure. Depending upon the solution pH, dissolved oxygen undergoes reduction in two steps:

$$O_2 + 2H^+ + 2e^- \rightarrow H_2O_2$$
$$H_2O_2 + 2H^+ + 2e^- \rightarrow 2H_2O$$

The potentials of these steps are approximately -0.05 V and -0.9 V (versus SCE), respectively. These reductions result in an increased background current that obscures the stripping peaks. Additionally, oxygen may oxidize the amalgamated metals (which is desirable in PSA). In neutral to slightly basic media, hydroxyl ions form during the reduction of oxygen and can precipitate metals ions at the electrode–solution interface.

Deoxygenation (deaeration) is performed by vigorous bubbling with nitrogen for 5–15 min (depending upon solution volume, bubbling rate, and bubble size) using a purge tube (Pasteur pipet, plastic tubing) that produces small bubbles of gas. The nitrogen should be saturated with the supporting electrolyte prior to purging to prevent evaporative losses from the sample solution. The nitrogen stream is then diverted to pass over the solution, maintaining a nitrogen blanket over the solution and a positive pressure of nitrogen in the cell. Specialized purge tubes and automated solenoid switches are commercially available.

For the most efficient purging of dissolved oxygen, a nitrogen stream passed through an aqueous solution of vanadous chloride (purple-colored) should be used. A blue color indicates that there is insufficient amalgamated zinc in the oxygen scrubber solution.[72]

Standard procedures for deoxygenation of volatile supporting electrolytes require the use of a scrubbing tower filled with the supporting electrolyte. This allows the gas to become saturated with the volatile components in the electrolyte and eliminates the possibility of changes in the pH and concentration of the supporting electrolyte.[23]

D.   Interference

1.   Overlapping Peaks

Overlapping stripping signals may arise as a result of similar redox potentials of the species determined. Overlapping peaks may occur during simultaneous measurement of the following: Cd and In, Pb and Tl, Pb and Sn, Sb and Bi, or Cu and Bi. Resolution in stripping analysis is determined by the peak potentials and the peak width at half peak height.[53,73]

One approach to solving the problem of overlapping peaks is in the selection of the deposition potential. In this instrumental approach a more positive deposition potential is applied until the metal ion with the more negative stripping peak is not deposited. For example, antimony and bismuth will both deposit at an MFE at a potential of -0.5 V versus SCE. If a deposition potential of -0.2 V versus SCE is used, then very little antimony is deposited and only the bismuth stripping peak is observed.[74] When this approach is used, the ion with the more negative peak potential cannot be determined directly.

Several chemical approaches have been developed in the solution of unresolved stripping peaks. Prior to the stripping measurement, a separation step can be introduced based on coprecipitation,[75] solvent extraction,[76,77] and chromatographic processes[78] coupled with on-line stripping procedures. Many of these processes are time-consuming and can result in contamination of the sample.

Another chemical approach is to use a supporting electrolyte consisting of a suitable complexing agent. The peak potentials will shift according to the composition and stability of the complexes formed and the concentration of the complexing reagent. Additionally, sample acidity influences the peak potentials of many metal ions.

Masking an undesirable metal ion by using surface-active agents[79,80] or by chelation[81] has been used in some situations to improve resolution. The technique of medium exchange[82-84] can be used in some situations. In this approach, the deposition is carried out in the sample solution, but during the rest period before the stripping step the supporting electrolyte solution is exchanged for one that is more suitable for stripping. Medium exchange works best in flow-through systems[60,85,86] that permit rapid solution exchange (via a multiport valve), and thus the exchange can be made without breaking the electrical circuit or exposing the electrode to the atmosphere. Frequently, the medium exchange solution is a buffer or a complexing medium.

2. Intermetallic Compounds

The second major interference in stripping analysis is the formation of intermetallic compounds between the metals deposited in the mercury electrode. Table 6 shows intermetallic compounds that have been reported.[2] Intermetallic compounds often oxidize at potentials different from the constituent metals and quantitative results are usually enhanced for one metal and depressed for the other. Intermetallic effects are least severe with the HMDE. The higher concentration of the metals in the MFE increases its susceptibility to the formation of intermetallic compounds. The stoichiometry of intermetallic compounds is not a simple ratio.[87]

The most common intermetallic compound formed in the MFE is the compound between copper and zinc.[88] In the determination of Cu^{2+} in the

Table 6. Intermetallic Compounds[a]

	Ag	Au	Cd	Co	Cu	Fe	Ga	In	Mn	Ni	Pt	Sb	Sn	Tl	Zn
Ag			+		+										+
Au			+				+		+				+		+
Cd	+	+			+										
Co															+
Cu	+		+			+	+	+	+	+		+	+	+	+
Fe									+						
Ga		+			+					+					
In						+									
Mn		+			+	+				+					
Ni					+		+		+			+	+		+
Pt												+	+		+
Sb					+					+	+				
Sn		+			+					+	+				
Tl					+										
Zn	+	+		+	+					+	+				

[a]Adapted from reference 26.

presence of Zn^{2+}, the stripping signal from copper will be enhanced because the Cu–Zn compound strips at a potential close to copper. A proportional decrease can be seen in the magnitude of the zinc response.

The zinc interference in copper analysis can be eliminated by controlling the deposition potential so that only copper is deposited into the electrode. This approach is analogous to that used to avoid overlapping peaks.

The copper interference in zinc analysis can be avoided by the preferential formation of another intermetallic compound. This approach is based on the addition of a "third" element that forms a more stable intermetallic compound with one component of the binary system.[89] The addition of Ga^{3+} to the sample results in the formation of the more stable Ga–Cu intermetallic compound, and the zinc can be determined without interference from copper.[90]

Another approach to minimizing intermetallic compound interference includes the use of dual electrodes to separate the metals at the different electrodes and thus prevent the formation of the intermetallic compound.[91–93] Also, by decreasing the amount of metals concentrated into the mercury through shorter deposition times and the use of pulse voltammetric techniques, errors caused by intermetallic compound formation can be reduced. Titration procedures[94] and suitable choice of supporting electrolyte[95] can also be used to minimize intermetallic interferences.

3. Surface Reactions

The presence of organic compounds in samples, particularly surface-active substances, can complicate the direct determination of metal ions by ASV, but it serves as the basis of analysis in AdSV. Surface-active compounds tend to adsorb on the electrode and inhibit the stripping and/or deposition processes, depending on the stripping technique used. In most cases a decrease in the stripping peak current is observed as a result of hindered mass transfer of metal from the electrode surface due to the adsorbed organic film. The peak potential may shift to more positive values as a result of the increased irreversibility of the metal oxidation.[96] In PSA, the presence of surface-active compounds has no influence on the stripping process because no current passes through the electrode. These substances only affect the double-layer capacity of the working electrode and, therefore, the potential versus time background signal.[97] Several studies have been made on the effects of organic compounds on the response in ASV and PSA.[96,98] However, the potential shift and peak depression effects of surface-active agents can be used to improve resolution in stripping measurements.

Ultraviolet (UV) irradiation[99] and ozone oxidation[100] methods are often used to either remove or destroy organic matter prior to the analysis of organic-rich samples (e.g., blood, urine, and waste water). Alternatively, the electrode surface can be covered with a porous membrane material [e.g., cellulose acetate,[101–105] Nafion[106–108] (DuPont)] that blocks the adsorption of the organic matter but permits the metal ions to be pass through and be reduced.

Inorganic surface reactions can also be a source of interference. For example, in the simultaneous determination of bromide and chloride by CSV,[109,110] the reaction of deposited Hg_2Cl_2 with Br^- in solution to form the more stable Hg_2Br_2 decreases the stripping peak for Cl^- and increases the Br^- peak.

E. Speciation

The chemical form in which a metal ion exists is of considerable importance in understanding its bioavailability, reactivity, and toxicity. Measurement of the total concentration of a metal (as obtained by atomic spectroscopic methods) gives no direct information about its reactivity. Stripping analysis is characterized by speciation capabilities. Speciation studies have been one of the most important applications of stripping analysis in water chemistry.[111]

Trace metal speciation studies must be designed to avoid any step that may alter the equilibria between the various oxidation states and chemical species. Measurements can be performed directly[112,113] on a sample, through titrimetric[114,115] stripping procedures and by coupling stripping analysis with chromatographic[116,117] techniques. Typically, by altering the treatment of the sample before analysis (e.g., pH control, complexation, digestion, and UV or microwave irradiation) the stripping method can determine either the total, naturally complexed, adsorbed, or precipitated metal.

F. Voltammetric Techniques

The stripping voltammogram is a plot of peak current as a function of the applied potential waveform. Several different potential–time waveforms are used to strip the deposited analyte from the electrode and obtain both qualitatitve and quantitative information from the peak potential (E_p) and peak current (i_p), respectively.

1. Linear Sweep Voltammetry

Linear sweep voltammetry is a simple technique for stripping the electrode. The potential–time waveform is a linear ramp (Figure 6a). The potential scan is initiated at the deposition potential in either a positive direction for ASV or a negative direction for CSV, and the resulting peak current is measured.

At the HMDE the peak current (i_p, expressed in μA) is described by the Randles–Sevcik equation[118–121] for linear sweep voltammetry under conditions of planar, semi-infinite diffusion:

$$i_p = 2.72 \times 10^5 n^{3/2} A D^{1/2} C v^{1/2}$$

where A is the surface area (cm^2) of the HMDE, C is the concentration (mM) of the analyte metal in the mercury drop, D is the diffusion coefficient (cm^2/sec) of the metal in the mercury (Table 5), n is the number of electrons transferred, and v is the potential scan rate (V/sec). The expression for the peak potential is

$$E_p = E_{1/2} - 1.1RT/nF$$

where $E_{1/2}$ is the polarographic half-wave potential, F is the Faraday constant (96,487 C/equiv), R is the gas constant (8.3144 J/(mol·K)), and T is the absolute temperature (K).

The peak current for a thin MFE (4–100 μm) as derived by deVries and van Dalen[122] is

$$i_p = 1.1157 \times 10^6 n^2 A C l v$$

where l is the film thickness (cm). Roe and Toni[123] derived the following

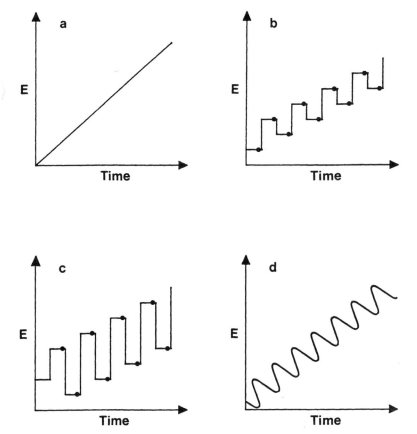

Figure 6. Potential–time waveforms used in stripping analysis. (**a**) Linear sweep. (**b**) Differential pulse. (**c**) Square-wave. (**d**) Alternating current.

equation for the stripping potential at a thin MFE ($< 10\,\mu$m):

$$E_p = E^{\circ\prime} + 2.3RT/nF \log nF\delta lv/RTD$$

where $E^{\circ\prime}$ is the formal electrode potential and δ is the thickness of the diffusion layer.

As both equations for peak current show, the parameter for enhancing the stripping current is the scan rate. The equations derived are generally valid for a range of scan rates. In typical stripping experiments, scan rates of 20 mV/sec to 100 mV/sec are used.

Unfortunately, the tradeoff for higher peak currents with increasing scan rate is an increase in the residual nonfaradaic capacitive (or charging) current which can be approximated by the equation:

$$i_c = AC_d v$$

where i_c is the capacitive current and C_d is the capacitance of the double layer. Ultimately, the capacitive current limits the detection limit of linear sweep voltammetry.

2. Differential Pulse Voltammetry

The differential pulse voltammetric waveform (Figure 6b) consists of a slow linear potential ramp (5–10 mV/sec) upon which small fixed-height potential pulses (5–100 mV) are superimposed every 0.5 to 5 sec. A current sample lasting 16.7 msec is taken just prior to the pulse application. The pulse lasts about 60 msec. During the first 40 msec, the nonfaradaic capacitive current induced by the pulse application decays and the last 16.7 msec of the pulse are used to measure the faradaic current. The current sampling times are synchronized to eliminate the effect of 60-Hz (in North America) line interference. The first current sample is subtracted from the second, and the peak-shaped output is displayed versus the potential ramp. The scan is slow so that the potential does not change significantly during the pulse application, and thus the capacitive current from the ramp increase is small.

The capacitive current decays at approximately $e^{-t/RC}d$, whereas the faradaic current is governed by the Cottrell equation and decreases at approximately $t^{-1/2}$, where t is the time (sec), R is the solution resistance (Ω), and C_d is the capacitance (F). As a result of this discrimination against the capacitive current, detection limits are improved.[10,124]

A second advantage of the differential pulse technique is the redeposition of metal ion analyte during the rest period between pulses. As the potential is pulsed through that at which the metal is stripped, the oxidized metal does not have time to diffuse from the electrode surface. At the end of the potential pulse the stripped metal is redeposited into the electrode to be stripped again during the next pulse cycle (as long as the potential ramp is negative of the reduction potential for the metal ion). Short waiting periods between pulses result in decreased peak currents from increasing pulsing frequencies and decreased redeposition.

With newer computer-software-driven instruments, the timing of the pulse duration can be shortened. The effect of a short pulse is that less metal ion diffuses to the bulk of the solution during stripping and redeposition is more effective.[125]

The equation for the differential pulse stripping peak current (for small pulse amplitudes) at the HMDE is[126]

$$\Delta i_p = kn^2 r \, \Delta E v^{1/2} t_d C$$

where ΔE is the pulse amplitude (V) and t_d is the deposition time (sec). The peak potential at the HMDE is given by[53]

$$E_p = E_{1/2} - 1.1RT/nF - \Delta E/2$$

While the best sensitivity is obtained at the largest pulse amplitude, the peak resolution is improved when small amplitudes are used. Typically, pulse amplitudes of either 25 or 50 mV are used as a compromise.

Osteryoung and Christie[127] have shown that for large pulse amplitudes (> 50–100 mV) the peak current is related to the amount of material deposited into the MFE:

$$i_p = 0.138 Q_m/t$$

where Q_m is the charge required to oxidize the metal in the film (C) and t is the pulse width (sec). Thus the peak current can be enhanced with short pulse widths, but extremely short pulse widths are not advised because they do not allow sufficient time for the capacitive current to decay.

Overall, differential pulse stripping voltammetry is more sensitive than linear sweep stripping voltammetry. The tradeoff is a slower analysis. For routine analysis at relatively high concentrations (> 100 ppb), linear sweep voltammetry may be more advantageous.

3. Square-Wave Voltammetry

The square-wave potential–time waveform was introduced by Barker in the late 1950s.[128] However, it was not until the mid-1980s that commercial (computer-based) instrumentation became available containing the square-wave waveform. It offers the same ability to discriminate against the capacitive current as differential pulse voltammetry and a fast scan rate.

Square-wave voltammetry (Figure 6c) utilizes a square-wave potential pulse superimposed upon a staircase ramp. In the waveform introduced by Osteryoung and Osteryoung[129] the optimal conditions for square-wave stripping utilize a 50-mV pulse amplitude with a step height of 10 mV applied every 16.7 msec. The current is sampled each time the square wave changes polarity. A net current is obtained as the difference between the forward and reverse pulses of a square-wave period and is plotted versus the potential.

Under coulometric conditions for the deposition, the expression for the differential current at an MFE for $\Lambda < 0.1$ is[130]

$$\Delta i_p = 0.298 nit/\tau$$

where $\Lambda = 1/(D\tau)^{1/2}$ and τ is the square-wave period.

The high frequency of the square wave requires a high concentration supporting electrolyte solution to allow the capacitive current to decay rapidly. If the pulse amplitude is larger than 10 mV, redeposition of metal occurs during the reverse pulse. With square-wave techniques the stripping step can be performed in a matter of seconds as compared to 2–3 min for differential pulse voltammetry. Detection limits of square-wave stripping are comparable to differential pulse stripping.

4. Alternating Current Voltammetry

The waveform for alternating current (ac) voltammetry is a linear potential ramp with a small-amplitude (5–20 mV) sine wave (10–1000 Hz) superimposed on it as shown in Figure 6d. With ac voltammetry the faradaic current is separated from the capacitive current by judicious choice of phase relationships. The ac faradaic current from a reversible reaction (cf. Warburg impedance) precedes the applied ac potential by a 45° phase angle (1/8 cycle), while the maximum capacitive current is 90° out of phase. Use of a phase-sensitive detection system (lock-in amplifier) can separate the faradaic current from the nonfaradaic current components.[131]

The equation for the ac stripping peak current at the HMDE is given by Underkofler and Shain[132]:

$$i_p = n^2 F^2 A C \omega^{1/2} D^{1/2} \Delta E/4RT$$

where ω is the sine wave frequency (Hz).

G. Procedures

Once the electrochemical cell has been prepared, a general procedure for analysis by stripping voltammetry is followed as outlined below[3,13,133]:

1. Deoxygenate supporting electrolyte/sample by vigorous bubbling with solvent-saturated nitrogen gas for 5–15 min.
2. Divert nitrogen gas stream over the solution to blanket the solution and to maintain a positive pressure of nitrogen in the cell.
3. Activate stirring and apply the deposition potential for a precisely measured time interval. (Deposition usually varies from 1 to 30 min

depending on the concentration level, electrode type, and stripping technique.)

4. Switch off stirring (at the precribed time) and allow convection to cease during a rest period for 10–60 sec (with the potential still applied).

5. Activate potential scan for stripping and record the voltammogram.

The procedure is first performed on the supporting electrolyte solution to make a background measurement of residual currents (i.e., impurities). The procedure is repeated identically for all samples and standards. It is important to reproduce parameters such as deposition time, stirring rate, rest period, scan rate, and electrode area for a series of measurements.

Quantitative analysis is based on the measurement of the peak current, i_p. The background residual current is subtracted from the total peak current to obtain the net i_p. The concentration is determined by a calibration curve or by the method of standard additions. The latter method has the advantage of minimizing matrix effects.

The standard addition method is performed by adding a small aliquot of a standard solution to the sample solution, and the analysis procedure is repeated. The concentration of the original solution is calculated as follows:

$$C_u = i_1 v C_s / [i_2 v + (i_2 - i_1) V]$$

where i_1 is the original sample net peak current, i_2 is the net peak current of the sample plus the standard added, v is the volume of the standard solution added, V is the original sample volume C_s is the concentration of the standard solution, and C_u is the original concentration. Experimental precision can be improved by making two or more additions to the same solution and preparing a graph of i_p (y-axis) versus concentration (x-axis) of the standard in the solution. The dilution factor must be included when calculating the concentration added.

The possible hazard of the metals analyzed in the experiment requires caution in the handling of solutions. Rubber bulbs must be used when pipetting metal ion stock solutions. Eye and skin contact with the metal solutions and acids should be treated by flushing with water. Additionally, disposal of metal ion and mercury-based solutions must be performed in an approved and safe manner.

IV. APPLICATIONS

Stripping techniques have been applied to the determination of many species, especially trace metals, in a variety of sample types. Space does not permit a comprehensive discussion of the literature; therefore, selected publications will be used to indicate the utility of these techniques.

Stripping analysis has been performed on many water sources (e.g., seawater, municipal waters, rainwater, surface waters, spring water, and mine water)[3,8,15] and other environmental samples.[15] An advantage of stripping analysis to aquatic chemistry has been its ability to determine various oxidation states of a metal ion and to distinguish between labile and bound metal ions.

Electroanalytical methods based on stripping techniques have become widely used for analysis of biological[16] and clinical fluids[134] with urine and blood (and blood products) receiving the greatest attention.[135,136] A nonisotopic alternative to radioimmunoassay for human serum albumin based on ASV of indium has been developed.[137] Many applications of CSV and AdSV exist for measurements of drug levels in urine[18,138,139]. Normally, organic matter is destroyed prior to the stripping measurement.

Also, before a food sample can be analyzed with stripping techniques, organic matter is destroyed first.[140] Typical examples of foods that have been subject to stripping analysis of heavy metal content include fruit juices,[141] milk,[142] and wine.[86] As mentioned previously, ASV has been used as a procedure for the identification of gun powder residues in forensic chemistry.[27,28] Many other applications of stripping analysis to the oil and gas, steel, paint, pharmaceutical, and semiconductor industries can be found in the literature.

The capabilities of stripping analysis compare favorably with other analytical techniques for trace metal analysis. ASV has been compared directly with atomic absorption spectroscopy, which is probably the most widely used spectroscopic technique for metal analysis.[12] Detection limits are typically below 1 ng/ml. Stripping analysis has proved to be precise, sensitive, capable of simultaneous multielement determination, and easy to implement. Automation of stripping procedures continues to receive attention.[143] Utilization of stripping analysis continues to expand with new applications[144] and new techniques.[145-148]

REFERENCES

1. I. Shain, Stripping analysis, in *Treatise on Analytical Chemistry*, Vol. 4, Part I, I. M. Kolthoff and P. J. Elving, eds., Wiley, New York (1963), pp. 2533–2568.

2. E. Barendrecht, Stripping voltammetry, in *Electroanalytical Chemistry*, Vol. 2, A. J. Bard, ed., Marcel Dekker, New York (1967), pp. 53–110.

3. W. R. Heineman, *Pollut. Eng. Technol.*, **18**, 125–150 (1981).

4. J. Wang, Voltammetry following nonelectrolytic preconcentration, in *Electroanalytical Chemistry*, Vol. 16, A. J. Bard, ed., Marcel Dekker, New York (1989), pp. 1–88.

5. R. Neeb, *Inverse Polarographie und Voltammetrie*, Verlag Chemie, Weinheim (1969).

6. Kh. Z. Brainina, *Stripping Voltammetry in Chemical Analysis*, Halsted Press–Wiley, New York (1974).

7. F. Vydra, K. Štulík, and E. Juláková, *Electrochemical Stripping Analysis*, Halsted Press–Wiley, New York (1976).

8. J. Wang, *Stripping Analysis*, VCH, Deerfield Beach, FL (1985).

9. Kh. Brainina and E. Neyman, *Electroanalytical Stripping Methods*, Wiley, New York (1993).

10. H. Siegerman and G. O'Dom, *Am. Lab.*, **4**(6), 59–68 (1972).

11. W. D. Ellis, *J. Chem. Educ.*, **50**, A131–A147 (1973).

12. T. R. Copeland and R. K. Skogerboe, *Anal. Chem.*, **46**, 1257A–1268A (1974).

13. W. M. Peterson and R. V. Wong, *Am. Lab.*, **13**(11), 116–128 (1981).

14. D. Jagner, *Analyst*, **107**, 593–599 (1982).

15. J. Wang, *Environ. Sci. Technol.*, **16**, 104A–109A (1982).

16. J. Wang, *J. Electroanal. Chem.*, **139**, 225–232 (1982).

17. J. Wang, *Am. Lab.*, **15**(7), 14–23 (1983).

18. J. Wang, *Am. Lab.*, **17**(5), 41–50 (1985).

19. C. M. G. van den Berg, *Analyst*, **114**, 1527–1530 (1989).

20. R. Kalvoda and M. Kopanica, *Pure Appl. Chem.*, **61**, 97–112 (1989).

21. M. G. Paneli and A. Voulgaropoulos, *Electroanalysis*, **5**, 355–373 (1993).

22. A. W. Bott, *Curr. Sep.*, **12**, 141–147 (1993).

23. *Stripping Voltammetry: Some Helpful Hints*, Technical Note 109A, Princeton Applied Research, Princeton, NJ (1974).

24. *Anodic Stripping Voltammetry of Pb & Cd*, EC Capsule 150, Bioanalytical Systems, West Lafayette, IN, 1984.

25. *Operation/Instruction Manual for the CV-27 Cyclic Voltammograph*, Bioanalytical Systems, West Lafayette, IN (1984), Chapter 9.

26. *Stripping Potentiometry*, BU1102, Radiometr Analytical A/S, Denkark (1990).

27. N. K. Konanur and G. W. vanLoon, *Talanta*, **24**, 184–187 (1977).

28. S. Chouchoiy and R. C. Briner, *Curr. Sep.*, **4**(2), 20–22 (1982).

29. D. Jagner and A. Graneli, *Anal. Chim. Acta*, **83**, 19–26 (1976).

30. A. Hussam and J. Coetzee, *Anal. Chem.*, **57**, 581–585 (1985).

31. Y. Zie and C. O. Huber, *Anal. Chim. Acta*, **263**, 63–70 (1992).

32. J. K. Christensen, K. Keiding, L. Kryger, J. Rasmussen, and H. J. Shov, *Anal. Chem.*, **53**, 1847–1851 (1981).

33. T. P. DeAngelis and W. R. Heineman, *Anal. Chem.*, **48**, 2262–2263 (1976).

34. C. Duda, *Curr. Sep.*, **11**, 32–35 (1992).

35. W. E. Van Der Linden and J. W. Dieker, *Anal. Chim. Acta*, **119**, 1–24 (1980).

36. D. L. Manning and G. Mamantov, *J. Electroanal. Chem.*, **7**, 102–108 (1964).

37. V. F. Gaylor, A. L. Conrad, and J. H. Landerl, *Anal. Chem.*, **29**, 224–228 (1957).

38. C. Olson and R. N. Adams, *Anal. Chim. Acta*, **22**, 582–589 (1960).

39. R. W. Andrews, *Anal. Chim. Acta*, **119**, 47–54 (1980).

40. W. L. Underkofler and I. Shain, *Anal. Chem.*, **33**, 1966–1967 (1961).

41. W. Kemula and K. Kublik, *Anal. Chim. Acta*, **18**, 104–111 (1958).

42. W. M. Peterson, *Am. Lab.*, **11**(12), 69–78 (1979).

43. T. M. Florence, *J. Electroanal. Chem.*, **27**, 273–281 (1970).

44. Kh. Z. Brainina, *Talanta*, **18**, 513–539 (1971).

45. W. R. Matson, D. K. Roe, and D. E. Carritt, *Anal. Chem.*, **37**, 1594–1595 (1965)

46. R. G. Clem, G. Litton, and L. D. Ornelas, *Anal. Chem.*, **45**, 1306–1317 (1973).

47. T. R. Copeland, J. H. Christie, R. A. Osteryoung, and R. K. Skogerboe, *Anal. Chem.*, **45**, 2171–2174 (1973).

48. R. G. Clem, *Anal. Chem.*, **47**, 1778–1784 (1975).

49. J. E. Anderson and D. E. Tallman, *Anal. Chem.*, **48**, 209–212 (1976).

50. R. G. Clem and A. F. Sciamanna, *Anal. Chem.*, **47**, 276–280 (1975).

51. M. Stulikova, *J. Electroanal. Chem.*, **48**, 33–45 (1973).

52. S. P. Kounaves, *Anal. Chem.*, **64**, 2998–3003 (1992).

53. G. E. Batley and T. M. Florence, *J. Electroanal. Chem.*, **55**, 23–43 (1974).

54. G. C. Whitnack and R. Sasselli, *Anal. Chem. Acta*, **47**, 267–274 (1969).

55. C. N. Reilley and R. W. Murray, Introduction to electrochemical techniques, in *Treatise on Analytical Chemistry*, Vol. 4, Part I, I. M. Kolthoff and P. J. Elving, eds., Wiley, New York (1963), pp. 2164–2232.

56. W. J. Albery and M. L. Hitchman, *Ring-Disc Electrodes*, Clarendon, Oxford (1971).

57. R. N. Adams, *Electrochemistry at Solid Electrodes*, Marcel Dekker, New York (1969), Chapter 4.

58. D. C. Johnson and R. E. Allen, *Talanta*, **20**, 305–313 (1973).

59. A. Zirino, S. H. Lieberman, and C. Clavell, *Environ. Sci. Technol.*, **12**, 73–79 (1978).

60. E. B. Buchanan, Jr., and D. D. Soleta, *Talanta*, **29**, 207–211 (1982).

61. J. Wang, H. D. Dewald, and B. Greene, *Anal. Chim. Acta*, **146**, 45–50 (1983).

62. J. Ruzicka and E. H. Hansen, *Flow Injection Analysis*, 2nd ed., Wiley, New York (1988).

63. M. Wojciechowski and J. Balcerzak, *Anal. Chem.*, **62**, 1325–1331 (1990).

64. J. Wang and H. D. Dewald, *Anal. Chem.*, **56**, 156–159 (1984).

65. S. G. Weber and W. C. Purdy, *Ind. Eng. Chem. Prod. Res. Dev.*, **20**, 593–598 (1981).

66. H. Gunasingham, *TrAC, Trends Anal. Chem.*, **7**, 217–221 (1988).

67. T. R. Copeland, J. H. Christie, R. K. Skogerboe, and R. A. Osteryoung, *Anal. Chem.*, **45**, 995–996 (1973).

68. R. P. Maas and S. A. Dressing, *Anal. Chem.*, **55**, 808–809 (1983).

69. D. E. Robertson, *Anal. Chim. Acta*, **42**, 533–536 (1968).

70. G. Scarponi, G. Capodaglio, P. Cescon, B. Cosma, and R. Frache, *Anal. Chim. Acta*, **135**, 263–276 (1982).

71. L. M. Petrie and R. W. Baier, *Anal. Chim. Acta*, **82**, 255–264 (1976).

72. *Deaeration... Why and How*, Application Note D-2, Princeton Applied Research, Princeton, NJ (1980).

73. Z. Stojek and Z. Kublik, *J. Electroanal. Chem.*, **105**, 247–259 (1979).

74. T. M. Florence, *J. Electroanal. Chem.*, **49**, 255–264 (1974).

75. T. M. Florence and Y. J. Farrar, *J. Electroanal. Chem.*, **51**, 191–200 (1974).

76. T. V. Nghi and F. Vydra, *J. Electroanal. Chem.*, **71**, 325–332 (1976).

77. T. V. Nghi and F. Vydra, *J. Electroanal. Chem.*, **71**, 333–340 (1976).

78. R. W. Andrews and D. C. Johnson, *Anal. Chem.*, **48**, 1056–1060 (1976).

79. S. Glodowski and Z. Kublik, *Anal. Chim. Acta*, **115**, 51–60 (1980).

80. J. H. Mendez, R. C. Martinez, and M. E. G. Lopez, *Anal. Chim. Acta*, **138**, 47–54 (1982).

81. J. E. Bonelli, H. E. Taylor, and R. K. Skogerboe, *Anal. Chim. Acta*, **118**, 243–256 (1980).

82. S. L. Phillips and I. Shain, *Anal. Chem.*, **34**, 262–265 (1962).

83. M. Ariel, U. Eisner, and S. Gottesfeld, *J. Electroanal. Chem.*, **7**, 307–314 (1964).

84. E. Desimoni, F. Palmisano, and L. Sabbatini, *Anal. Chem.*, **52** 1889–1892 (1980).

85. G. Koster and M. Ariel, *J. Electroanal. Chem.*, **33**, 339–349 (1971).

86. L. Anderson, D. Jagner, and M. Josefson, *Anal. Chem.*, **54**, 1371–1376 (1982); Correction: *Anal. Chem.*, **55**, 416 (1983).

87. M. S. Shuman and G. P. Woodward, Jr., *Anal. Chem.*, **48**, 1979–1983 (1976).

88. W. Kemula, Z. Galus, and Z. Kublik, *Nature*, **182**, 1228–1229 (1958).

89. E. Ya Nieman, L. G. Petrova, V. I. Ignatov, and G. M. Dolgopolova, *Anal. Chim. Acta*, **113**, 277–285 (1980).

90. T. R. Copeland, R. A. Osteryoung, and R. K. Skogerboe, *Anal. Chem.*, **46**, 2093–2097 (1974).

91. T. P. DeAngelis, R. E. Bond, E. E. Brooks, and W. R. Heineman, *Anal. Chem.*, **49**, 1792–1797 (1977).

92. D. A. Roston, E. E. Brooks, and W. R. Heineman, *Anal. Chem.*, **51**, 1728–1732 (1979).

93. J. Wang and H. D. Dewald, *Anal. Chem.*, **55**, 933–936 (1983).

94. D. Jagner and L. Kryger, *Anal. Chim. Acta*, **80**, 255–266 (1975).

95. R. W. Gerlach and B. R. Kowalski, *Anal. Chim. Acta*, **134**, 119–127 (1982).

96. P. L. Brezonik, P. A. Brauner, and W. Stumm, *Water Res.*, **10**, 605–612 (1976).

97. J. Mortensen and D. Britz, *Anal. Chim. Acta*, **131**, 159–165 (1981).

98. P. Sagberg and W. Lund, *Talanta*, **29**, 457–460 (1982).

99. G. E. Batley and Y. J. Farrar, *Anal. Chim. Acta*, **99**, 283–292 (1978).

100. R. G. Clem and A. T. Hodgson, *Anal. Chem.*, **50**, 102–110 (1978).

101. J. Wang and L. D. Hutchins, *Anal. Chem.*, **57**, 1536–1541 (1985).

102. J. Wang and L. D. Hutchins-Kumar, *Anal. Chem.*, **58**, 402–407 (1986).

103. E. E. Stewart and R. B. Smart, *Anal. Chem.*, **56**, 1131–1135 (1984).

104. R. B. Smart and E. E. Stewart, *Environ. Sci. Technol.*, **19**, 137–140 (1985).

105. J. H. Aldstadt and H. D. Dewald, *Anal. Chem.*, **65**, 922–926 (1993).

106. B. Hoyer, T. M. Florence, and G. E. Batley, *Anal. Chem.*, **59**, 1608–1614 (1987).

107. B. Hoyer and T. M. Florence, *Anal. Chem.*, **59**, 2839–2842 (1987).

108. H. Huilang, D. Jagner, and L. Renman, *Anal. Chim. Acta*, **207**, 17–26 (1988).

109. G. Colovos, G. S. Wilson, and J. L. Moyers, *Anal. Chem.*, **46**, 1045–1050 (1974).

110. G. Colovos, G. S. Wilson, and J. L. Moyers, *Anal. Chem.*, **46**, 1051–1054 (1974).

111. T. M. Florence, *Analyst*, **111**, 489–505 (1986).

112. T. M. Florence, *Anal. Chim. Acta*, **141**, 73–94 (1982).

113. H. W. Nurnberg, *Anal. Chim. Acta*, **164**, 1–21 (1984).

114. M. S. Shuman and G. P. Woodward, Jr., *Anal. Chem.*, **45**, 2032–2035 (1973).

115. D. Jagner and K. Aren, *Anal. Chim. Acta*, **134**, 201–209 (1982).

116. P. Figura and B. MuDuffie, *Anal. Chem.*, **52**, 1433–1439 (1980).

117. G. E. Batley and T. M. Florence, *Anal. Chem.*, **52**, 1962–1963 (1980).

118. J. E. B. Randles, *Trans. Faraday Soc.* **44**, 327–338 (1948).

119. A. Sevcik, *Collect. Czech. Chem. Commun.* **13**, 349–377 (1948).

120. W. H. Reinmuth, *Anal. Chem.*, **33**, 185–187 (1961).

121. R. S. Nicholson and I. Shain, *Anal. Chem.*, **36**, 706–723 (1964).

122. W. T. deVries and E. van Dalen, *J. Electroanal. Chem.* **14**, 315–327 (1967).

123. D. K. Roe and J. E. A. Toni, *Anal. Chem.* **37**, 1503–1506 (1965).

124. G. D. Christian, *J. Electroanal. Chem.*, **23**, 1–7 (1969).

125. P. Valenta, L. Mart, and H. Rutzel, *J. Electroanal. Chem.*, **82**, 327–343 (1977).

126. W. Lund and D. Onshus, *Anal. Chim. Acta* **86**, 109–122 (1976).

127. R. A. Osteryoung and J. H. Christie, *Anal. Chem.* **46**, 351–355 (1974).

128. G. C. Barker, *Anal. Chim. Acta*, **18**, 118–131 (1958).

129. J. G. Osteryoung and R. A. Osteryoung, *Anal. Chem.*, **57**, 101A–110A (1985).

130. S. P. Kounaves, J. J. O'Dea, P. Chandrasekhar, and J. Osteryoung, *Anal. Chem.*, **59**, 386–389 (1987); Errata: *Anal. Chem.*, **59**, 1888 (1987).

131. D. J. Curran, *Am. Lab.*, **13**(6), 27–40 (1981).

132. W. L. Underkofler and I. Shain, *Anal. Chem.* **37**, 218–222 (1965).

133. J. Wang, *J. Chem. Educ.*, **60**, 1074–1075 (1983).

134. A. Jensen, J. M. Christensen, and P. Persson, *Met. Ions Biol. Syst.*, **16**, 185–199 (1983).

135. G. Morrell and G. Giridhar, *Clin. Chem.*, **22**, 221–223 (1976).

136. P. Ostapczuk, *Clin. Chem.*, **38**, 1995–2001 (1992).

137. M. J. Doyle, H. B. Halsall, and W. R. Heineman, *Anal. Chem.*, **54**, 2318–2322 (1982).

138. I. E. Davidson and W. F. Smyth, *Anal. Chem.*, **51**, 2127–2133 (1979).

139. E. N. Chaney, Jr., and R. P. Baldwin, *Anal. Chem.*, **54**, 2556–2560 (1982).

140. W. Holak, *J. Assoc. Off. Anal. Chem.*, **58**, 777–780 (1975).

141. S. Mannino, *Analyst*, **108**, 1257–1260 (1983).

142. L. Almestrand, D. Jagner, and L. Renman, *Talanta*, **33**, 991–995 (1986).

143. A. Cladera, J. Estela, and V. Cedra, *Talanta*, **37**, 689–693 (1990).

144. J. M. Estela, C. Tomas, A. Cladera, and V. Cerda, *Crit. Rev. Anal. Chem.*, **25**, 91–141 (1995).

145. F. Scholz and B. Lange, *TrAC, Trends Anal. Chem.*, **11**, 359–367 (1992).

146. J. H. Aldstadt and H. D. Dewald, *Anal. Chem.*, **64**, 3176–3179 (1992).

147. T. Horiuchi, O. Niwa, M. Morita, and H. Tabei, *Anal. Chem.*, **64**, 3206–3208 (1992).

148. N. A. Madigan, T. J. Murphy, J. M. Fortune, C. R. S. Hagan, and L. A. Coury, Jr., *Anal. Chem.*, **67**, 2781–2786 (1995).

CHAPTER

5

ELECTROCHEMICAL DETECTORS IN AUTOMATED ANALYTICAL SYSTEMS

MAREK TROJANOWICZ

Laboratory for Flow Analysis and Chromatography
Department of Chemistry
University of Warsaw
02-093 Warsaw, Poland

I. INTRODUCTION

Nowadays, chemical analysis has been founded on a solid theoretical basis. It is also subject to a permanent technological progress. The progress is observed in better instrumentation for analytical determinations, in an increased number of the sample pretreatment operations already mechanized, in the full automation of many analytical procedures, and in computerization so common at present. The development of instrumental techniques has largely reduced human labor engaged in analytical procedures. However, various aspects of the analytical procedure are not equally susceptible to the development. As a result, it is usually the sample pretreatment prior to the detection stage that has remained the most tedious and time-consuming step in the analytical procedure. Therefore the sample pretreatment and the proper

Modern Techniques in Electroanalysis, Edited by Petr Vanýsek, Chemical Analysis Series, Vol. 139.
ISBN 0-471-55514-2 © 1996 John Wiley & Sons, Inc.

detection stage focus much concern to "replace, refine, extend, or supplement human effort" engaged in it. The quotation comes from the IUPAC definition of mechanization of measuring methods in analytical chemistry.[1] The progress in detection methods provides new possibilities for the existing concepts of mechanization to be applied to whole analytical procedures with a given detection. It also yields developments of entirely new and unique concepts.

Electrochemical methods remain of great importance in modern chemical analysis. Potentiometric determination of pH is the most widespread of all chemical analytical measurements in various areas of science, technology, and everyday life. Measurements of pH and conductivity are common in process analysis, environmental protection, biotechnology, and clinical analysis. Potentiometric measurements with membrane ion-selective electrodes have completely dominated analysis of electrolytes in clinical analysis. High-performance liquid chromatography (HPLC) belongs to the most versatile and powerful techniques of the modern routine analysis. Electrochemical detectors are amongst the most sensitive detectors used in HPLC. Examples illustrating the importance of electrochemical methods in chemical analysis could be further multiplied.

Progress in electroanalytical methods yields more possibilities to improve their routine applications. The progress takes place in many areas and directions. One of the most important trends in electroanalysis is the use of new electrode materials. The materials include new active bulk electrode materials for voltammetric techniques and membrane potentiometry and also electrode active surface modifiers of inorganic, organic, and biological origin.

An even larger impact on modern instrumentation for routine chemical analysis comes from miniaturization of detectors and other devices. It involves combining detectors with signal transducers employing electronics of the large scale of integration. It stimulates development of novel computerized methods for electroanalytical measurements and for processing of the measured data.

It has to be realized that most often it takes quite a substantial delay before the results of the basic research are put into practice in the routinely used analytical instrumentation. The process of transformation of scientific achievements to the form of manufactured, commercially available devices encounters numerous technological barriers. A new device faces the competition from numerous already existing developments in other fields of analytical instrumentation widely used in analytical laboratories.

II.   GENERAL ASPECTS OF MECHANIZATION AND AUTOMATION OF ANALYTICAL PROCEDURES

In the past 20 years, increasing attention has been paid to mechanization and automation of analytical measurements. This is evidenced in the published monographs, in reviews in analytical journals,[2-11] and especially in the ever-increasing number of original research papers in the analytical literature. Technological progress influences all areas of human activity and also stimulates mechanization and automation of chemical analysis in particular for the following reasons:

- The increased demand for analytical determinations in various fields of contemporary life, science, and technology.
- The increasing necessity of environmental and process monitoring independent of direct supervision of a human operator.
- The increased demand for the higher accuracy and precision of analytical determinations.
- The common tendency to reduce the cost of analytical determinations.

The distinction of the terms *mechanization* and *automation* in chemical analysis is not observed consistently in the scientific and technological literature, although the appropriate recommendations of IUPAC have already been accepted for a long time.[1] The meaning of *mechanization* in chemical analysis is the use of mechanical devices to perform some of the operations of the whole analytical procedure and to eliminate the manual performance of these operations. When considering *automation* of chemical analysis, two different levels of devices are distinguished: the *automatic* and *automated* analyzers. In general, mechanized devices require supervision of a human operator unless they are automatic or automated. Automatic and automated devices are equipped with electronic control systems and do not require supervision. Automatic analyzers are equipped with control units able to

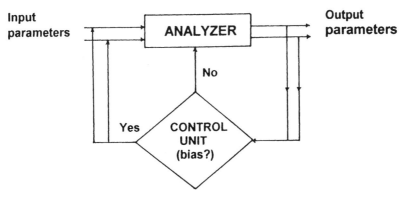

Figure 1. The feedback loop principle in automated chemical analyzers.[8]

make selections of sequences of operations and their working conditions based on the operator-defined input parameters.

In the measuring systems that run with the least operator engagement the control unit interacts with the analyzer on the feedback principle (Figure 1). The object of regulation usually consists of modules performing individual operations—that is, sampling, sample pretreatment, detection, and so on. The set of input parameters to control the analyzer performance may include the following: sample volume, concentration of reagents, incubation time, temperature, time lengths of various unit operations (e.g., separation, preconcentration), and the functioning parameters of the detector. The output usually consists of the determination results and those values of control parameters which are subject to change during the determination. One is often forced to consider the role of *interfering factors*. They make the output parameters deviate from the expected behavior of the system. All variables with undetermined values associated with the analyzed sample may be interfering factors. When the control unit can adjust the input parameters so as to reduce or eliminate interfering factors to perform an optimal determination, the analyti-

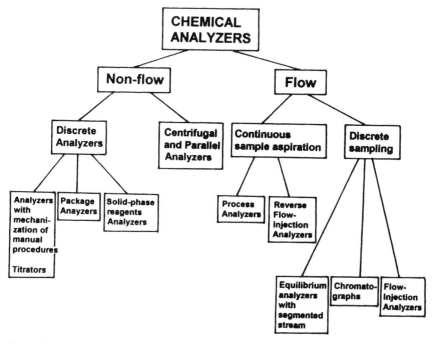

Figure 2. Classification of the chemical analyzers according to the hydrodynamic conditions of detection and the sampling mode.

cal system is considered as *automated*. The titrator is a very simple example of a device exploiting the feedback loop with a simple analog electronic circuit. When the titration approaches the endpoint, it decreases the rate of the titrant delivery to the measuring vessel. Very complex automated systems are often equipped with a control unit with an expert software based on decisive algorithms enabling it to make decisions about working conditions of the operations based on the acquired knowledge.[12,13]

In the above considerations, *chemical analyzer* means an analytical measuring instrument equipped with a detection system and a device able to carry out at least one mechanized operation of the sample pretreatment.[8] However, a different meaning of this term can also be found in the literature. Contemporary chemical analyzers can be classified according to the detection method used, the type of material analyzed, the number of components analyzed in a sample, the aim of their installation, and their design.[3] In the scheme presented in Figure 2, analyzers are classified on the basis of the mode of conducting measurement, the way of sampling, and the sample pretreatment prior to detection. Although this arbitrary classification contains some unavoidable inconsistencies, it should serve as a sufficient guidance to the present review on electrochemical detection in analytical instrumentation, where various concepts of mechanization and automation in chemical analysis are used.

III. ELECTROCHEMICAL METHODS IN DISCRETE ANALYZERS

The difference of the nonflow and flow analyzers is of special importance to electrochemical detection. The electrochemical measurements in a mechanized or automated nonflow instrumentation are not essentially different from those in manual devices. The differences occur in the detector design, in the organization of the solution transport, and in the data processing. The conditions of the continuous flow of the analyte solution through the electrochemical flow detector have a much greater impact on the performance of the analytical determination.

Nonflow analytical determinations can be carried out as a sequence of operations performed individually for each sample to be analyzed. Such systems are commonly referred to as *discrete analyzers*. In other types of analyzers the sequence of operations can be performed practically simultaneously for several samples. The centrifugal and parallel analyzers work this way. The first one is widely used in clinical analysis.[14] It has been applied also to environmental analysis.[15] None of these types employs electrochemical detection.

There are several concepts to organize performance of the whole analytical procedure in discrete analyzers of single samples. The most widely applied design of discrete analyzers in general and also of those employing electrochemical detection is that it follows a conventional manual procedure doing all its steps mechanically. In DuPont's ACA and Abbott's Vision analyzers the whole analytical procedure—and, in particular, various operations of the sample pretreatment—can be conducted in specially designed packages. So far, however, such analyzers have not employed electrochemical detection. Analyzers using solid-phase reagents represent an entirely different approach to the analytical procedure. In these analyzers and in titrators, electrochemical detection has found several applications.

As far as the analyzers following conventional manual procedures in a mechanized way are concerned, electrochemical detection can be found both in dedicated single component instruments and in multicomponent analyzers for simultaneous determination of several analytes in the sample introduced to the analyzer.

As a classical example of a dedicated electrochemical analyzer, let us mention the Beckman clinical analyzer for determinations of several substrates in physiological fluids. Glucose is determined using an amperometric oxygen sensor based on the oxygen consumption in a biocatalytic oxidation of glucose in the presence of the glucose oxidase in solution. Washing the detector cell, exchanging the solution, equilibration, calibration, and all other stages are carried out automatically (Figure 3). A similar analyzer design is that developed for the determination of the blood urea nitrogen using conductometric detection and urease in the solution. In more recent versions of such analyzers offered by various manufacturers, enzymes are usually immobilized on various supports. A mechanized analytical system for pH determinations in soil extracts is reported by Baker.[16]

Coulometric titration after an appropriate sample treatment is employed in the Dohrmann analyzers to determine the total halide content as well as the content of halide bound in organic compounds. Some analyzers involve microcoulometry to determine the total nitrogen content in water.[17] A 10- to 50-μl sample is introduced to the analyzer. The sample is delivered to the pyrolitic reactor with Ni catalyst, where all nitrogen-containing compounds are converted into ammonia, which is then absorbed in an acidic solution. The platinum black electrode and the $Pb/PbSO_4$ electrode serve as an indicating system. Hydrogen ions are generated at the platinum black anode. Coulometry is also employed in CO_2 measurements in determinations of the total carbon content and the organic carbon content in waters.[18] Dedicated titrators with potentiometric detection for determinations of chloride in blood, milk, and water samples of various origin are also very widely used. Amperometric detection is most commonly used in the coulometric titrators for the Karl Fisher determination of water in nonaqueous solvents.[19]

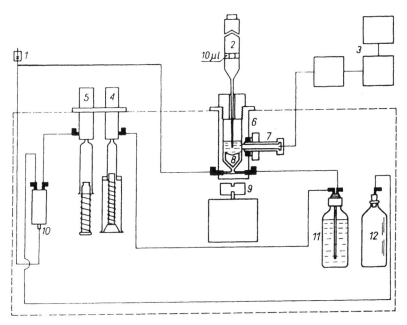

Figure 3. Schematic diagram of the Beckman glucose analyzer with amperometric detection [S. Engel, *Beckman Report*, 1971, p. 29] 1, Magnetic valve; 2, pipette; 3, transducer of the oxygen sensor; 4, filling syringe; 5, emptying syringe; 6, measuring vessel; 7, oxygen sensors; 8, stirring bar; 9, stirrer; 10, three-way valve; 11, reagent; 12, waste.

Numerous applications of amperometric detection can be also found in the mechanized and automatic volumetric titrators for the determination of chloride,[20] sulfate[21] and zirconium.[22] The pH electrode for the acid–base titrations and the silver electrode for the halide precipitation titrations are most often used in the potentiometric detection in titrators. The latter ones can be also carried out with the sulfide-sensitive ion-selective membrane electrode.[13] The automatic titrator Mettler System Titrator SR 10, equipped with the potentiometric and amperometric detection, consists of several independent stations performing weighing, dilution, extraction of solid samples, and the final titration with various detection methods.[24]

In the case of recording of the whole potentiometric titration curve in a mechanized system, it is advantageous to use electronic linearization of the output signal for the quantitative interpretation of the experimental data.[25] In the literature there are many examples of the on-line computer data processing used just for this purpose.

Various types of electrochemical detection have also been applied in multicomponent discrete analyzers. In the clinical analyzer of the acid–base equilibrium in blood, Radiometer ABL 3 analyzer, carbon dioxide, dissolved oxygen, and pH can be electrochemically determined in a 125-μl sample. In recent years, electrolyte and CO_2 potentiometric analyzers based on ion-selective membrane electrodes are most commonly introduced to clinical diagnostics. At present they are available from many manufacturers and they have completely replaced flame photometry instruments. A schematic diagram of the multielectrode potentiometric tubular detector used in the Kone Microlyte-6 analyzer is shown in Figure 4 as an example of this type of devices, and its functional parameters are given in Table 1. In a determination of several species, insufficient selectivity of sensors poses a problem. It requires an additional data processing with the corrections using estimated selectivity coefficients. Modules with potentiometric detectors are also often attached to multicomponent clinical analyzers with spectrophotometric detectors.

Discrete analyzers based on spectrophotometric measurements in the layers of solid-phase reagents obtained, for example, by photographic film technology are much less common. In order to extend clinical applications of this kind of instrument, similar instruments for potentiometric measurements of several electrolytes and CO_2 have also been developed based on the

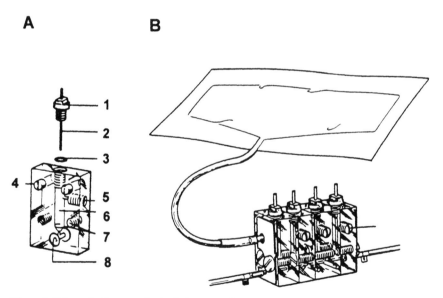

Figure 4. Schematic diagram of a single detector cell (**A**) and the arrangement of a three-electrode detection unit with a common reference electrode (**B**) used in the clinical analyzer Kone Microlyte 6. 1, Internal electrode cap; 2, internal wire reference electrode; 3, O-ring; 4, positioning dowel pins; 5, side port for filling; 6, electrode chamber; 7, holes for screws; 8, measurement channel.

Table 1. Functional Characteristics of Potentiometric Clinical Analyzer KONE Microlyte 6 for Determination of pH and Electrolytes in Physiological Fluids

	Measurement Ranges (mM)		Accuracy (%)	Standard Deviation[b] (mM)
Analyte	Serum, Plasma	Urine[a]		
K^+	1.8–10	10–200	±2	0.04
Na^+	80–200	10–300	±2	1.3
Ca^{2+}	0.2–8.0		±3	0.04
Cl^-	50–150		±2	1.0
pH	6–8.5[c]			0.02[c]
Mg^{2+}	0.2–3.0		±3	0.04

[a] With dilution.
[b] For serum.
[c] In pH units.

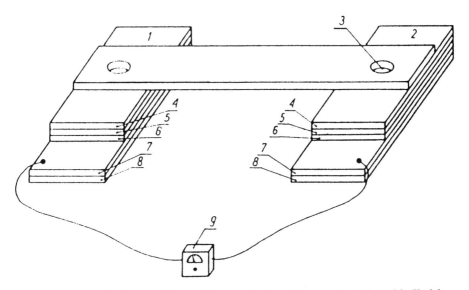

Figure 5. Schematic diagram of the multilayer potentiometric measuring cell used in Kodak Ektachem analyzers. 1, Sample half-cell; 2, half-cell for standard solution; 3, measuring wells for solutions; 4, ion-selective membrane; 5, electrolyte gel layer; 6, silver chloride; 7, silver foil; 8, support; 9, meter.

ion-selective electrode principle.[26] A schematic configuration of such a detector is shown in Figure 5. The multilayer structure of this device is analogous to a membrane half-cell. Its properties depend on the ion-selective membrane composition. The determination is carried out by introducing 10 µl of the sample and 10 µl of the standard solution to each well of the stripe-detector and then measuring the potential difference between the two membrane half-cells.

IV.   FLOW ANALYSIS WITH ELECTROCHEMICAL DETECTION

Conducting sample pretreatment and detection in flow conditions is an effective method to improve performance of analytical procedures. Some instrumentation designed for similar measurements in process analysis had been available for a considerably long time[27] before the laboratory applications of flow measurements started. Such applications with effective on-line operations of sample pretreatment were initiated in the pioneering work by Skeggs,[28] who invented the concept of flow analysis with segmentation of the flowing streams. Many years later a secondary substantial impact on the progress in the field of flow analysis came from the development of flow injection analysis.[29]

A.   Equilibrium Systems with Segmented Stream

In spite of the very late development of the theoretical physicochemical description of the dispersion in flow systems with air segmentation,[30,31] this concept of automation of the analytical procedure was very quickly and widely accepted first in clinical but also in environmental, agricultural, and industrial laboratories. These methods are treated in some monographs,[32,33] and reviews[9,31,34,35] and in numerous original research papers. Initially, spectrophotometric detection was used. Very soon, practically all instrumental methods of analysis, including electrochemical, started to be employed in this attractive methodology.

Potentiometric detection was applied in flow systems with air segmentation relatively early. The main reason is that the membrane electrodes can be highly selective; see Table 2 for a list of applications. The applications are especially numerous for the highly selective fluoride electrode.[36-40] Early studies on the detection with this electrode report many fundamental observations applicable generally to potentiometric detection with membrane electrodes in flow measurements. It is found that the baseline stability and the response rate can be improved by the addition of a certain level of fluoride to the buffering TISAB (total ionic strength adjustment buffer) solution.[37] The baseline insta-

Table 2. Examples of Applications of Potentiometric Detection with Ion-selective Electrodes in Continuous Flow Measurements with Stream Segmentation

Analyte	Conditions	Matrix	Reference
Al	KNO_3, hydroxylamine, acetate buffer	Tap water	50
Ammonium	Tris[a] buffer pH 7.5	Blood, serum	44
Ca	KNO_3, iminodiacetate, acetylacetone, and ammonium buffer	Water	47
Cu(II)	KNO_3, citric acid	Refining electrolyte	48
K	Tris and Met_4NCl	Serum	53
Na	NH_4NO_3	River water	54
	Tris and Met_4NCl	Serum	53
Chloride	KNO_3	Mineral Waters	55
Cyanide	$KAg(CN)_2$ in NaOH	—	43
	NaOH	—	42
Fluoride	TISAB[b]	Water	36
	TISAB	Urine	37
	TISAB	Serum	38
	Acetate buffer	Air	39
	NaCl, citrate, acetate buffer	Plants	40
Nitrate	Ag_2SO_4, phosphate buffer, pH 2.3	Water	47
Sulfate	$NaClO_4$ in methanol (indirect measurement with Pb-ISE)	Water	49
Sulfide	Titration with Hg(II)	—	52

[a] Tris, tris(hydroxymethyl)aminomethane.
[b] TISAB, total ionic strength adjustment buffer.

bility observed for clinical samples is caused by adsorptive accumulation of proteins on the electrode surface and can be avoided by the addition of the proteolytic enzyme rennin to the TISAB solution.[38] The iodide-selective electrode is found to yield almost a steady-state signal with a small carry-over at a sufficient flow rate even for the sampling rate of 360/hr.[41]

The response dynamics in flow measurements is significantly influenced by the design of the flow-through detector. For conventional electrodes the flow-through cap with a flat sensing surface is most often used (Figure 6A) with the reference electrode placed at the outlet. Several other designs have also been employed (Figure 6), including the quasitubular electrodes, where polymeric membrane is built in the wall of the measuring channel (Figure 6D). A frequently reported trouble in the potentiometric flow measurements is the

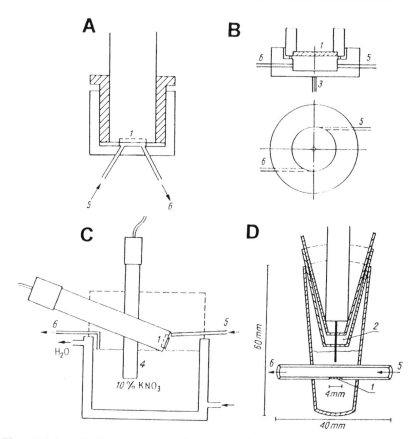

Figure 6. Schematic diagrams of potentiometric flow-through cells used in continuous flow measurements with a segmented stream. (**A**) Thin-layer cap for the indicator electrode[41] (**B**) flow-through cap for indicator electrode with mixing compartment[42]; (**C**) cascade-type detector[43]; (**D**) tubular arrangement with polymer membrane. 1, Ion-selective membrane; 2, internal reference electrode; 3, liquid junction; 4, outer reference electrode; 5, inlet; 6, outlet.

signal oscillation caused by changes of the streaming potential.[45] The response stability is favored by a short distance between the indicating and the reference electrode, by a large tubing diameter, a large electrolyte concentration in the background solution, and, finally, by a grounding of the solution. Commonly used reference electrodes with the liquid junction at which the diffusion potential difference forms are reported to cause transient but often large potential changes as a result of changes in the sample concentration.[46] Therefore, electrodes with the controlled flow of the internal reference solution are especially recommended.

A significant aspect of the flow system design is to make the chemical conditions of the measurement provide the optimum selectivity and detectability of the analyte. Hence the aspirated sample is usually on-line mixed with an appropriate solution adjusting a constant ionic strength, pH, masking interferences, or decomposing complexes, in which the analyte is bound. Examples of the latter procedure are the following: masking of magnesium with acetylacetone in the potentiometric determination of calcium,[47] masking of iron(III) with citrate in the determination of copper(II),[48] and the release of fluoride from complexes with aluminum or iron(III).[36-38,40] In the determination of analytes convertible into gaseous species, on-line dialyzers are employed in order to improve the determination selectivity.[43,44]

Several indirect flow methods have been developed for the determination of the analytes directly undeterminable with the available potentiometric sensors. The differential determination of sulfate in the flow system shown in Figure 7 is carried out with two lead ion-selective electrodes,[49] whereas the determination of aluminum in a flow cell with two fluoride electrodes.[50] In a flow system with a measuring equilibrium steady-state response, but without segmentation of the stream, phosphate is determined using a lead electrode.[51] Also gradient titrations of sulfide with mercury(II)[52] and a complexometric gradient titration of calcium[47] are reported in flow systems.

Both polarographic detection and voltammetric detection have been widely applied in equilibrium flow measurements (Table 3). Yet only few reports can be found on the use of both detection methods in a typical system with stream segmentation. The flow polarographic measurements have been applied first in a liquid chromatography detection.[56,57] They turn out advantageous in flow determinations of metals forming amalgams[58] and have been used in a flow system with stream segmentation.[61] For a determination of calcium and magnesium the decrease of the anodic wave of oxidation of EDTA and EGTA has been used.[60] The mercury drop electrode (MDE) has been applied to the determination of ascorbic acid[59] and proteins.[62] The latter method is based on the complex formation of proteins with Ti(IV). Some authors point out that the detection limit can be improved by the use of differential pulse polarography rather than constant current measurements.[61,62] In more recent applications of flow polarography the MDE is replaced either by the static mercury drop electrode or by the mercury film electrode. A marked improvement of detectability can be achieved with the static electrode owing to the strict control of the drop size as well as performing the measurement in the time range when the charging current can be neglected. Such a technique has been employed for a continuous monitoring of cadmium, copper, and zinc in industrial electrolytes.[63] Optimization of polarographic flow detectors and measuring techniques is reported in several papers.[64-66]

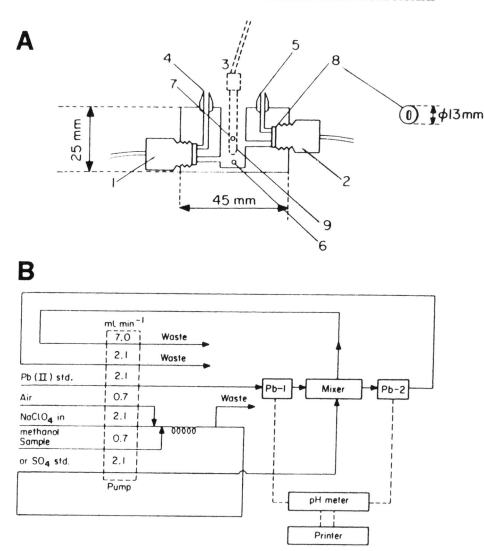

Figure 7. Schematic diagram of a detector cell (**A**) and flow manifold (**B**) for indirect potentiometric determination of sulfate with lead ion-selective electrodes.[49] 1, 2, Lead electrodes; 3, reference electrode; 4, inlet for standard lead(II) solution to the Pb-1 electrode chamber; 5, outlet for lead(II) solution after Pb-2 electrode; 6, inlet for solution containing sulfate; 7, outlet for the excess of solution in mixing chamber; 8, spacer; 9, mixing chamber.

Table 3. Examples of the Application of Polarographic and Voltammetric Detection to the Continuous Flow Measurements

Analyte	Working Electrode[a]	Measuring Technique[b]	Reference
Ascorbic acid	MD	dc	59
	GC rotating	dc	83
	C porous, rotating	dc	101
Bi, Cd, Cu, Pb, Co(NH$_3$)$_6^{3+}$	(Pt)Hg	dc	61
Ca, Mg (indirect)	MD, HMD, Pt, (Pt)Hg	dc	60
Cd, Pb, Zn	MD	dc	59
	SMD	dp	63
Chlorite	Ag porous	dp	97
Cu, Fe(III)	Pt	dc, ac	100
Cyanide	Ag	dc	98
	Au	dc	99
Dopamine	C porous, rotating	dc	101
Hexacyanoferrate	Pt	dc	102
	C porous, rotating	dc	83
H$_2$O$_2$	Ag porous	dc	97
NADH	C porous, rotating	dc	83
Nitrozepam	C(Hg)	dc	81
p-Nitroso-N,N-diethylaniline	GC	dc, np	81
Proteins (indirect)	MD	dp	62
2-Thiobarbituric acid	(C)Hg	dc	81

[a] Electrodes: GC, glassy carbon; HMD, hanging mercury drop; MD, mercury drop; SMD, static mercury drop.
[b] Techniques: ac, alternating current; dc, direct current; dp, differential pulse; np, normal pulse; rp, reverse pulse.

Much larger attention is focused on voltammetric detection using working electrode materials other than mercury in flow measurements. The replacement of mercury with solid electrodes results in a limited range of available cathodic potentials. However, it greatly expands the range of anodic polarization for such materials as various forms of carbon, platinum, and gold. The easy renewal of the electrode surface characteristic for the mercury drop is lost for solid electrodes. The solid electrodes, however, offer a larger flexibility in the design of flow-through detectors (Figure 8). A detailed discussion on the advantages and drawbacks of various detectors configurations can be found in numerous original research papers,[67-73] in review articles,[74-77] and in the book by Štulík and Pacáková.[78] The thin-layer and wall-jet detectors are

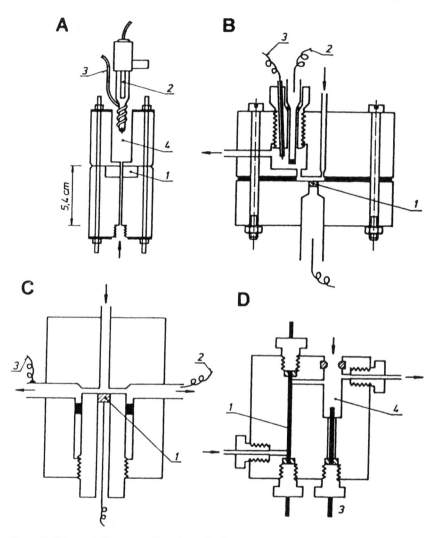

Figure 8. Schematic diagrams of flow-through cells used in voltammetric detection in continuous flow measurements. **(A)** Cell with a tubular working electrode[67]; **(B)** thin-layer cell[67]; **(C)** wall-jet cell[68]; **(D)** arrangement with a wire electrode. 1, Working electrode, 2, reference electrode; 3, auxiliary electrode; 4, reference electrode compartment.

most commonly used. An essential improvement of the detection limit can be obtained in the wall-jet detector with the ring-disk pair of electrodes. With this kind of a detector, copper and zinc have been determined on a mercury film electrode while manganese(II) and nickel have been determined on a platinum electrode.[79]

Not only the choice of a suitable material for the working electrode and the appropriate design of the flow-through detector, but also the use of various measuring techniques, can affect sensitivities of polarographic and voltammetric detections. A decrease of the thickness of the diffusion layer at the working electrode surface results in an improvement of the detection limit. In most of the flow-through detectors, the thinning of the diffusion layer can be obtained by the increase of the solution flow rate and by rotation or vibration of the working electrode. The replacement of a constant polarization by pulse techniques where short pulses of polarizing voltage are applied to the working electrode leads to an increase of the signal to noise ratio. The increase is caused by the different rates of change of two currents: the capacitance current and the faradaic current. An evident increase of the current with the decrease of the pulse width is observed in the determination of cyanide on a platinum supported mercury film electrode.[80] The potential pulse is so selected that at the initial potential value the electrode process of the analyte does not occur. If the analyte or products of its electrode reaction are strongly adsorbed on the electrode surface, a properly selected value of the initial potential should ensure desorption of these species. A frequently emphasized advantage of the pulse techniques is the current being independent of the flow rate and the reversibility of the electrode process.[81] A limitation of the residual current contribution in amperometric flow measurements can be obtained by a temporary flow stop[82] or by the electrode rotation switching off and on.[83]

Selectivity of a voltammetric measurement depends on the polarization of the working electrode, chemical conditions of the determination, and the measuring technique applied. It can be additionally changed by a chemical modification of the working electrode surface and a suitable processing of the output signal. Either in the measurements at a constant potential or using the normal pulse technique, determination of one analyte in a sample is associated with the reduction of all the other sample components, the reduction potentials of which are less negative than the analyte reduction potential. The differential pulse measurement, where the measured signal is a difference of current intensities at the end and at the beginning of a pulse, allows for the selection of signals from different depolarizers. It has been used in flow determinations with the (AulHg film electrode[84] and the graphite paste electrode.[85] Much attention is focused also on the use of chemically modified electrodes in the flow conditions.[86,87] Most often the role of such a modification is a selective catalysis of the electrode process of the analyte or differentiation of the transport of different depolarizers to the working electrode surface. Platinum electrodes covered with a layer of conducting polymers (polypyrrole, polyaniline) have been used for the amperometric detection of electrochemically inactive anions.[88,89] Covering of the electrode surface with cellulose acetate suppresses interference from large organic molecules[90] which are size excluded, whereas a Nafion layer with cation-exchange properties excludes anions and some electroneutral molecules.[91] The rapid-scan technique within a suitable potential interval yields a series of voltammograms for the passage of the sample zone through the detector.[92,93] It is suitable for a selective amperometric determination. Data processing allows us subtract the background and filter the noise.

In recent years there has been an increasing interest in the use of micro- and ultramicroelectrodes in flow measurements.[94,95] These electrodes attract attention with their low noise level and the independence of the limiting current of the flow rate. The microelectrode array in a thin-layer cell has increased the signal-to-noise ratio as compared to the glassy carbon macroelectrode.[96]

Polarographic and voltammetric detection methods, especially those employing fast potential scanning in a wide range, can be suitably applied in multicomponent determinations. Lead, cadmium, and zinc have been simultaneously determined using the hanging mercury drop electrode and the square-wave technique. A large number of such determinations carried out by the flow-injection method have been reported.

B. Flow-Injection Analysis

A steady-state equilibrium response is not necessary for a successful analysis in flow conditions either in a segmented stream system or in a continuous flow system without segmentation. It had been demonstrated several years before the invention of flow-injection analysis (FIA).[103,104] As mentioned above, not earlier than in the mid-1970s, the series of works by Ruzicka and Hansen[105-107] provided fundamentals for this attractive methodology of analytical measurements. The concept of analytical procedure mechanization in FIA is based on the following principles: (i) The volume of the liquid sample introduced into the measuring system is small and insufficient to give the equilibrium response of the detector, (ii) the sample dispersion from the injection point to the detector is strictly controlled, and (iii) residence time of the successive samples in the measuring system is maintained constant. Almost 20 years of the development of FIA has brought its broad acceptance in research laboratories. The number of FIA applications in routine analytical laboratories has increased recently owing to the fact that many FIA instruments have become commercially available.

Spectrophotometric methods of detection predominate also in this area of analytical instrumentation. Nevertheless, numerous applications of electrochemical detection, in FIA, particularly of potentiometry and voltammetry, are also reported.

There are several advantages of potentiometric detection in FIA, but also some difficulties. The indicator half-cell, often a membrane electrode, is fairly

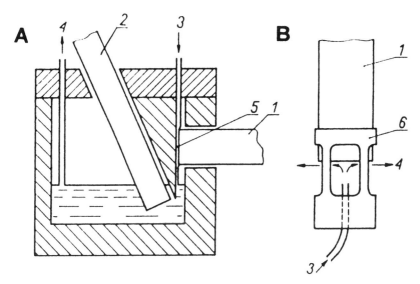

Figure 9. Schematic diagrams of detector arrangements used in potentiometric detection with ion-selective electrodes in flow-injection analysis: (**A**) Thin-layer cell[108]; (**B**) exchangable cap on indicator electrode for a large volume wall-jet arrangement.[109] 1, Indicator electrode; 2, outer reference electrode; 3, inlet for measuring solution; 4, outlet; 5, layer of measured solution; 6, electrode cap.

slow in its response to changes in the analyte concentration. It is considerably slower than the spectrophotometric detector responding to the solution optical property changes induced by the analyte. Hence, a success of this detection in FIA will substantially depend on the appropriate design of the flow-through detector. Basic types of flow-through detectors shown in Figure 6 have been employed also in FIA. Satisfactory applications are reported for the thin-layer detector shown in Figure 9. It supports various conventional electrodes with a flat sensing surface (Figure 9A). With the cap on the electrode a simple wall-jet configuration is obtained (Figure 9B). It works immersed into the solution with a reference electrode.

The above-mentioned tubular electrode design can be used not only for the polymeric membranes, but even more readily with solid-state membrane materials.[110-112] In flow-injection measurements, also a flow-through cell with a chloride-sensitive coated-wire electrode has been employed.[113] In FIA systems using ion-selective field effect transistors the thin-layer design of a flow cell is mainly employed.[114-116]

In potentiometric detection the indicator electrode potential depends on the activity of the main ion sensed. It is considered as a principal advantage of this detection also in flow measurements that the dependence is semilogarithmic in a wide analyte activity range according to the Nikolsky–Eisenman equation. At the low analyte concentrations the validity of the semilogarithmic linear relationship is limited by the solubility of the membrane material in the measured solution. It is limited by contaminations of the measured solutions in the case of solid-state electrodes with an extremely insoluble membrane material. In the range of sub-Nernstian concentrations of the analyte and small potential changes, the dependence of the potential on the activity (or the concentration at a constant ionic strength) of the analyte is usually linear and can be used for analytical purposes in FIA.[117]

As already mentioned, potentiometric detectors with membrane electrodes are rather slow. Their dynamic properties depend on the range of the main ion concentration, on the sign of the measured concentration change, and on the state of the membrane surface at the interface with the measured solution.[118] Stabilization of the potential value at the low concentrations is slower than at the high ones. It may be responsible for the super-Nernstian sensitivities observed in flow-injection measurements with ion-selective membrane electrodes.[119] As in the equilibrium measurements, a better dynamic behavior of the detector with a higher sampling rate can be achieved by the introduction of a small concentration of the main ion to the carrier solution. This has been demonstrated in the measurements with the fluoride,[119,120] sulfide,[121] and halide[120] electrodes. Concentrations of the analyte lower than that of the carrier solution after the addition of the main ion can be determined in the injected samples.[122]

The most favorable feature of the potentiometric detection in fast flow-injection measurements is that the kinetic discrimination of interferences can be applied. The effect is generally attributable to the dynamic selectivity of membrane electrodes which depends on the substrate diffusion at the phase boundary.[123] In FIA measurements, this has already been observed for chloride-,[117,124] bromide- and iodide-sensitive solid-state electrodes[112] as well as for the nitrate electrode with a polymeric membrane.[125] For the ions that strongly interfere with the electrode function (e.g., bromide and iodide for the chloride-sensitive electrode), the apparent selectivity coefficients in flow-injection conditions can be even a few orders of magnitude lower than those in nonflow conditions.

Some examples of potentiometric detection with membrane electrodes in FIA are shown in Table 4. Owing to easy miniaturization of the potentiometric flow-through detectors, several designs of the multicomponent detectors are reported.[140–142] A schematic diagram of a measuring system and a detector with a tubular polymeric membrane for the determination of calcium and nitrate in water is shown in Figure 10. When detectors are insufficiently

Table 4. Examples of the Application of Potentiometric Detection with Ion-Selective Electrodes in Flow-Injection Analysis

Analyte	Membrane Type	Analyzed Sample	References
Ammonia	Polymeric	Blood serum	126
	gas sensing	Culture media	145
Ca	Polymeric	Blood serum	127
		Water, wastes	127, 128, 140
Cu(II)	Solid	Refining electrolytes	129
K	Polymeric	Blood serum	130, 131
		Soil extracts	130
		Fertilizers	132
Li	Polymeric	Blood serum	133
Na	Glass	Blood serum, soil extracts	130
	polymeric	Blood serum	134
pH	Glass	Water	120
	polymeric	Soil extracts	139
Bromide	Solid	Water	120
Chloride	Solid	Water	117
		Blood serum	135
Chlorine residual	Solid (iodide)	Water	136
Fluoride	Solid	Water	120, 122
		Soft drinks, urine	122
Nitrate	Polymeric	Blood serum	130
		Waters	137, 140
		Soil extracts	130, 137
		Fertilizers	132, 137
Sulfide	Solid	Wastes	121
		Diisopropanolamine	138

selective for a multicomponent determination, then sophisticated computer data processing is necessary to help the situation.[141]

Several indirect methods of potentiometric detection have been developed in FIA. Let us mention the following examples: the determination of residual chlorine with the iodide electrode,[136] the determination of cyanide with the sulfide electrode,[143] and the determination of sulfite with the solid-state electrode based on a mixture of HgS and Hg_2Cl_2 as an active membrane material.[144]

Polarographic and especially voltammetric detection have found wide-spread applications in FIA. However, the field of the most extensive flow voltammetry applications is detection in high-performance liquid chromatog-

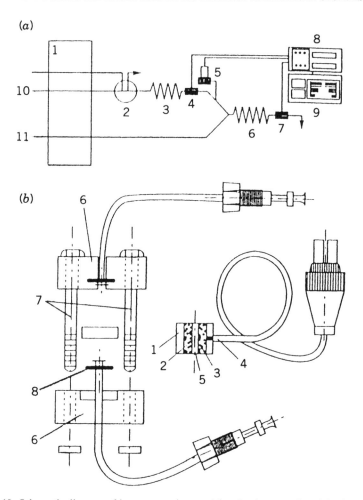

Figure 10. Schematic diagram of instrumentation used for simultaneous flow-injection poten-tiometry of calcium and nitrate in water.[140] (**A**) Manifold. 1, Peristaltic pump; 2, injection valve; 3, 6, mixing coils; 4, calcium tubular sensor; 5, common reference electrode; 7, nitrate tubular sensor; 8, potentiometer; 9, two-channel recorder. (**B**) Arrangement of tubular detector. 1, Epoxy body; 2, conducting epoxy cylinder; 3, PVC membrane; 4, cable; 5, channel for solution; 6, perspex supports; 7, screws; 8, O-rings.

raphy. The wall-jet detector with a vertically positioned capillary is most often employed in flow-injection polarography because of the essential role of the sample dispersion in flow-injection measurements.[146–148] However, the de-tector with a horizontally placed capillary is also reported to give satisfactory results.[65] Constant potential pulse polarography allows determination of cathodically active metals without deoxygenation.[148] The polarographic FIA determinations of penicillic acid[146] and thiourea[65] are based on the anodic current measurement.

The variety of materials used for the working electrode in flow-injection votammetry is about as large as in the discussed earlier equilibrium flow measurements. Numerous applications are reported for the platinum electrode designs: tubular,[149,150] wire,[151-153] and disk[154] in thin-layer and wall-jet flow cells. In the theoretical considerations on the flow cell with a tubular electrode, it is found that the peak height depends both on the flow rate and on the flow system dispersion. In other words the peak height is a function of the flow rate, but the function depends on the dispersion in the flow system.[155] For a wire electrode the vibration of the electrode increases the turbulent convection and results in a marked improvement of the detection limit.[151]

Voltammetric detection with the platinum electrode and the three-step pulse polarization have found interesting applications in the determinations of carbohydrates[156] and both organic and inorganic sulfur compounds.[157] The three-step pulse measurement prevents adsorption of products of the electrode process on the electrode surface. It is also found that gold used as an electrode material improves the detection limit several times as compared to platinum.[154] The gold electrode has also been employed in the flow-injection amperometry of free chlorine, hypochlorite, and monochloramines.[158]

Electrochemically inactive compounds and ions may affect the rate of formation of oxides on the platinum electrode surface during the anodic phase of a three-step pulse measurement. The phenomenon can be used for a quantitative determination of such analytes for example, chloride and cyanide.[159] The detection based on $Pt(OH)$ formation on the electrode surface in one of the phases of a three-step pulse measurement can be used for determinations of much concentrated strong acids or bases.[160]

Many different applications of indirect biamperometric detection have been developed in FIA most often with the use of a flow-through detector with two platinum wire electrodes[153,161-163] also for a simultaneous determination of nitrate and nitrite in natural waters (Figure 11).[164] The triiodide–iodide couple is the most frequently used indicator system. An amperometric flow-injection determination of nitrite has also been developed using a three-electrode detection system and the platinum working electrode modified with adsorbed iodine or a layer of poly(4-vinylpyridine).[165]

Less noble materials have also been employed for the working electrode in several determinations for example, silver in anodic determinations of cyanide[166] or the copper amalgam film electrode in determinations of metal ions.[167] Glassy carbon is a relatively often used material (Table 5), for determinations of, for example, nitrite,[168] nitrate[169] and sulfur dioxide.[170]

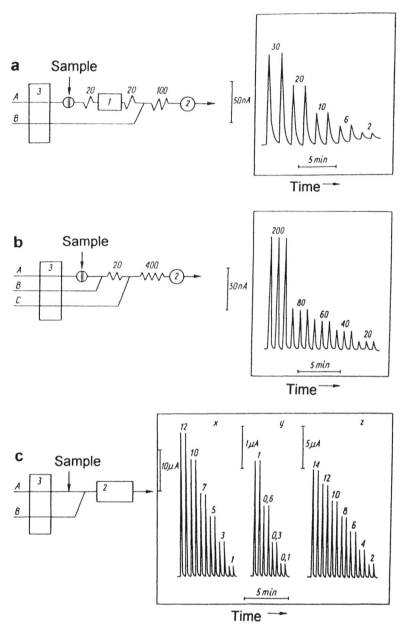

Figure 11. Schematic diagrams of flow-injection systems and signals recorded using biamperometric detection for indirect determination of (**a**) nitrate[162] [A, distilled water; B, potassium iodide in sulfuric acid (concentrations in micromoles)], (**b**) molybdenum using catalytic procedure[163] [A, distilled water; B, hydrogen peroxide in sulfuric acid; C, potassium iodide (concentrations shown in micrograms per liter)], and (**c**) acids using single-point titration technique [A, distilled water; B, solution of potassium iodide and sodium iodate (concentration of sulfuric acid (x, y) and acetic acid (z) shown in micromoles); 1, microcolumn with copperized cadmium; 2, detector; 3, peristaltic pump; length of mixing coils in centimeters.]

Table 5. Examples of the Application of Polarographic and Voltammetric Detection in Flow-Injection Analysis[a]

Analyte	Working Electrode	Measuring Technique	Reference
As(III)	Pt	dc	174
Cd, Cu, Pb	MD	?	148
Cd, Cu, Zn	(Cu)Hg	dc	167
Cu	MD	ac	65
Fe(II), Fe(III)	GC	dc	176
Cyanide	Ag	dc	166
Hexacyanoferrate	GC	rp	175
H_2O_2	GC	dc	177
Iodide	Pt	dc	174
Phosphate	GC	dc	179
Adrenaline	Pt	dc	149
Amino acids	Cu	dc	173
Ascorbic acid	C(Kel-F)	dc	171
Benzoquinone	GC	rp	175
Catechol	Pt	dc	149
Isosorbide dinitrate	MD	dc	147
Ferrocene	C(Kel-F)	dc	171
NADH	GC	rp	175
Nitrobenzene	MD	np	178
Pencillic acid	MD	dc	146
Phenol	GC	rp	175
	C (Kel-F)	dc	171
Thiourea	MD	ac	65

[a] C(Kel-F) is a working electrode made of mixture of graphite and Kel-F polymer; other abbreviations are defined in Table 3.

A mixture of graphite and Kel-F as a working electrode material has been applied to determinations of many analytes.[171] Carbon electrodes modified with hexacyanorutenium(II) have been used for a determination of arsenic(III)[165] and modified with a layer of poly(4-vinylpyridine) containing complexes $IrCl_6^{4-}$ and $IrCl_6^{3-}$ for a determination of nitrite.[172]

Voltammetric multicomponent detection with fast scanning of the working electrode potential is more difficult in flow-injection measurements because there is a short time period available for the measurement and a precise synchronization of the detection with the sample injection is necessary. A number of metals have been determined on the mercury film electrode using differential pulse voltammetry, whereas organic reducing analytes were determined on the carbon paste electrode.[180] Square-voltammetry on the mercury film electrode yields poor resolution of Tl(I) and Pb(II), but the appropriate computer data processing enables a simultaneous determination of these species.[181]

The flow-injection methodology can also be successfully applied to stripping analysis. The main advantage of such a procedure is a very simple exchange of the media: The one used for the electrolytic preconcentration is exchanged for the one more suitable for the dissolution step. The preconcentration step is often carried out at the stopped flow regime.[182,183] In a stripping flow-injection measurement the analytical signal is not obtained directly during the sample passage through the detector. Because of this the height of the dissolution peak is not influenced by the dispersion in the flow system.[184] An appropriate design of the FIA system with anodic stripping voltammetric detection enables measurements with signals corrected for the background electrolyte (Figure 12).[185] It eliminates the effects of (i) the charging current, (ii) the oxidation of mercury, (iii) the reduction of oxygen, and (iv) the faradaic currents corresponding to contamination of carrier solution. It has been applied in trace determination of zinc and copper in vitamin preparations[185] and of lead in blood.[186]

In a similar fashion the flow-injection procedure is used in potentiometric stripping determinations. Changes in the working electrode potential are measured as a function of time after the electrolytic preconcentration in the mercury film. The changes result from the chemical oxidation of amalgams with the oxidant present in the solution. Schematic diagrams of typical detectors used for this purpose are shown in Figure 13. The support most often employed for a deposition of the mercury film is glassy carbon[187–189] Also other supports are used—for example, a porous graphite electrode[190] or a gold electrode for the determination of mercury(II) by oxidation with an acidic Cr(VI) solution containing bromide.[191] In trace determinations where the analyte concentration is very low the electrolytic preconcentration stage is carried out at a very low flow rate or at the stopped flow. Also in this case the flow-injection methodology is convenient for the exchange of matrices between the preconcentration and the dissolution phases. It often improves the determination selectivity.[192,193]

In recent years several different concepts have evolved on the analysis involving manipulation of a small sample volume and a nonequilibrium measurement during the sample passage through a flow-through detector when the sample is in contact with the detector for a short period of time. In these cases some possibilities to use electrochemical detectors have already been demonstrated and more such possibilities certainly do exist.

The principle of a *sequential injection* technique[194,195] is shown in Figure 14. The plunger of the piston pump moves in discrete steps to aspirate the

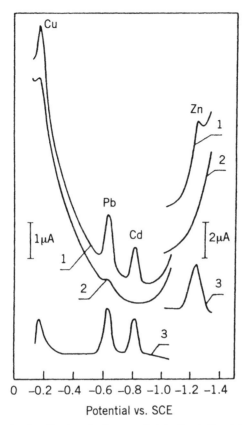

Figure 12. An example of application of voltammetric detection in flow-injection measurements for the determination of metal ions using the differential pulse anodic stripping technique.[185] Recorded curves for (1) solution containing 0.5 μM Cu, 0.25 μM Cd, and 0.25 μM Pb without background correction, (2) background electrolyte, (3) the same solution as 1 with background correction.

wash solution, the sample solution, and the reagent solution into a channel subdivided into three sections: the holding coil, the reactor, and the detector. The selector valve provides sequential injections of the solutions into the channel in programmed steps synchronized with the piston movement. The readout shown is a typical double peak, the trough of which is located at the time when the flow has been reversed. So far this technique has been used in conjunction with spectrophotometric detection. However, it could be used also in conjunction with electrochemical measurements.

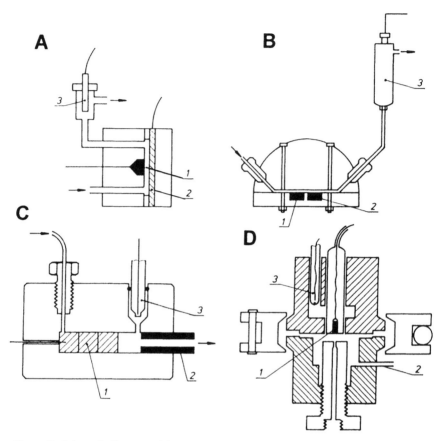

Figure 13. Schematic diagrams of detector cells used in flow-injection potentiometric stripping measurements. (**A**) Thin-layer cell[187]; (**B**) thin-layer cell[188]; (**C**) cell with a porous working electrode[190]; (**D**) wall-jet cell[189]. 1, Working electrode; 2, auxiliary electrode; 3, reference electrode.

A miniaturized system has been designed for liquid handling in FIA with ion-sensitive field effect transitors.[196] With micromachined silicon structures such as the micropumps and the flow manifolds for liquid handling, an important reduction in the overall size of the system has been achieved. The micropumps are piezoelectrically driven. The elements are arranged in a three-dimensional flow system where they are stacked directly on top of each other (Figure 15). The complete system consists of two micropumps for washing and sampling, one detector cell placed at the sample input, and another detector

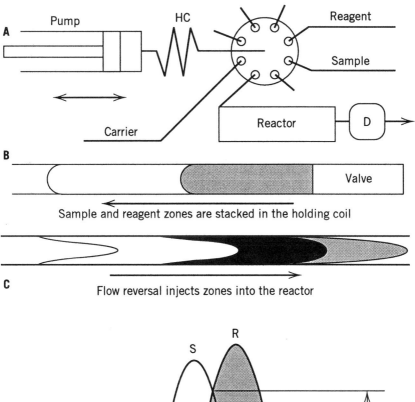

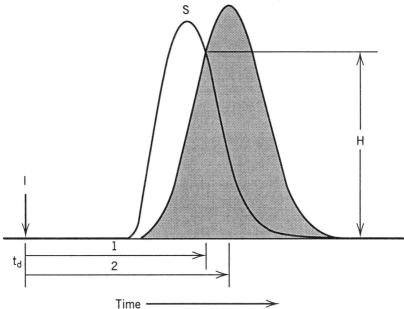

Figure 14. Schematic diagrams showing the principle of sequential injection technique.[195] **(A)** The flow scheme, **(B)** the structure of stacked and injected zones, **(C)** profiles of sample (S) and reagents (R) as seen by the detector. HC, holding coil; D, detector; H, peak height; I, point of injections.

calibration solution

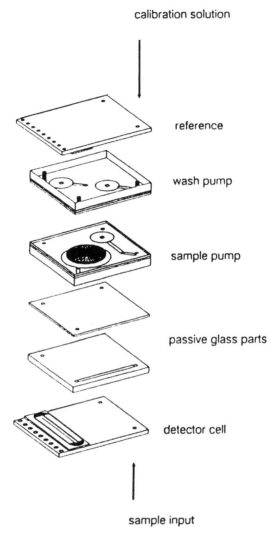

reference

wash pump

sample pump

passive glass parts

detector cell

sample input

Figure 15. Arrangement of the modules of the microsystem for flow-injection analysis with ISFET sensors.[196]

cell serving as a separate reference for the sensing ISFETs. The system is a valveless injection device with a dead volume of the detector of about 5 μl.

Batch injection analysis is a concept of mechanization of analytical measurements in which a very small sample volume can be injected and the transient signal recorded.[197] The technique involves injecting the small

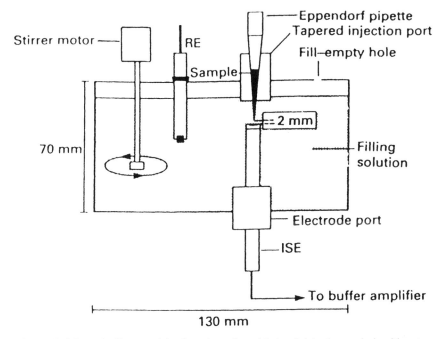

Figure 16. Schematic diagram of the detection cell used in batch injection analysis with potentiometric detection.[199]

sample volume onto a flat sensor surface in a dilution cell and monitoring the transient response to the arrival, passage, and dispersion of the sample zone over the sensor surface. A schematic diagram of the cell used in potentiometric measurements is shown in Figure 16. This technique has been used either with amperometric detection[197] or with the potentiometric detection employing indicating electrodes with solid-state[198] or polymeric[199] membranes. The results reported so far indicate that a wider use of this technique in conjunction with electrochemical measurements is possible. But in contrast to FIA the technique does not offer the on-line operations of the sample treatment.

C. Biosensors in Flow Analysis

The field of electrochemical biosensors i.e., measuring systems, in which electrochemical detection is preceded by biocatalytic transformation of the analyte, belongs to the most rapidly growing fields of the contemporary electroanalysis. An electrochemical biosensor can be described as a device with a spatially and conceptually integrated design, in which the biocatalytic reaction step takes place in the direct vicinity of the active surface of the electrochemical sensor. Electrochemical biosensors, as well as their design, properties, and applications, are discussed in several monographs,[200-205] numerous reviews, and original research papers in the analytical, biochemical, medical, and technological journals. Due to the variety of their potential applications ranging from the use of *in vivo* in living organisms to laboratory, environmental, and industrial applications, many different designs of electrochemical biosensors and measuring systems based on them are reported and manufactured. Considering their contribution to mechanization and automation of analytical procedures, most significant is their application to flow systems, namely, equilibrium and FIA. Some examples of discrete analyzers employing biocatalytic systems are mentioned in the preceding sections.

Internal potentiometric and amperometric detectors are equally common in electrochemical biosensors designed and applied to various analytical problems. The detection method selection depends on the isolated and purified enzyme available, simplicity of the electrochemical determination of the enzymatic reaction product, and the degree of consumption of the other components involved quantitatively in the catalyzed reaction. The abundance and availability of various oxidases results in widespread applications of the anodic detection of hydrogen peroxide and of the oxygen consumption in the reaction medium. The large availability of numerous enzymes catalyzing degradation of nitrogen-containing substrates to ammonia is a reason for numerous applications of detectors for the ammonia nitrogen in biosensors. It is very rare that a biocatalyst is used in flow systems in a dissolved form in the solution, because the cost of such measurements is high.[206-209] Commonly, either in flow systems or in integrated biosensors, enzymes immobilized on various supports are employed. The immobilization can be achieved either by chemical methods of enzyme binding or by an adsorption attachment or by a physical entrapment.[200,201] In addition to the enzyme immobilization on membranes, the direct immobilization on the working electrode surface[210-213] or in the paste of the paste working electrodes[216-218] is used especially in amperometric detectors. In certain cases a stable enzyme immobilization on the surface of a polymer membrane is possible also on the membrane surface of a potentiometric membrane electrode.[214-215]

In flow biocatalytic measurements, especially in the flow-injection mode, much more careful optimization of the dynamic response is required as compared to nonflow enzymatic measuring systems. However, the advantages of the increased sampling rate and of the improved precision of determination outweigh the difficulties, and electrochemical biosensors are applied to a growing number of procedures performed in flow conditions. Examples of flow systems with potentiometric and amperometric detection are shown in Tables 6 and 7.

Table 6. Examples of the Application of Biocatalytic Potentiometric Detection in Flow Analysis

Analyte	Potentiometric Sensor	Placement of Biocatalyst	Kind of Flow Measurement[a]	References
Amino acids	NH_3, semiconductor	Reactor	FI	227
Antibodies	Sulfide	Reactor	CF	225
Creatinine	NH_3, gas sensing	Membrane	CF	221
	NH_3, semiconductor	Reactor	FI	222, 227
	NH_4^+, polymeric	Reactor	FI	223
Glucose	SO_2, gas sensing	Solution	CF	209
	Iodide	Membrane	CF	228
	Fluoride	Membrane, reactor	FI	229
Glutamate	CO_2, gas sensing	Membrane	CF	224
Glutamine	NH_4^+, polymeric	Membrane, reactor	FI	217, 219, 220
IgG	NH_4^+, polymeric	Reactor	FI	226
Mercury compounds	NH_3, gas sensing	Reactor	CF	225
Penicillin	pH	Reactor	FI	233, 234
		Membrane	FI	235
Urea	NH_3, gas sensing	Solution	CF	206
	pH	Solution	FI	208
	NH_4^+, polymeric	Membrane	FI	230
		Reactor	FI	217, 232
	NH_3, semiconductor	Reactor	FI	231

[a]CF, continuous flow; FI, flow injection.

219

Table 7. Examples of the application of amperometric detection in biocatalytic flow measurements.

Analyte	Directly Measured Species	Biocatalyst	Placement of Biocatalyst	Kind of Flow Measurement[a]	References
Alcohols	H_2O_2	Alcohol oxidase, peroxidase	Electrode paste	FI	215
Atrazine	Catechol Violet	Tyrosinase	Electrode surface	FI	213
Ethanol	Oxygen	Alcohol oxidase	Reactor	CF	236
Fructose	H_2O_2	Fructose dehydrogenase	Reactor	FI	237
Glucose	Oxygen	Glucose oxidase, catalase	Solution	FI	207
Glucose	H_2O_2	Glucose oxidase	Membrane	FI	238, 239, 253
Glucose	H_2O_2	Glucose oxidase	Reactor	FI	240–242
Glucose	H_2O_2	Glucose oxidase	Electrode surface	FI	211
Glucose	H_2O_2	Glucose oxidase	Electrode paste	FI	214
Glutamate	H_2O_2	Glutamate oxidase	Reactor	FI	243
Lactate	Oxygen	Lactate oxidase	Membrane	CF	244
Lactate	H_2O_2	Lactate oxidase	Membrane	FI	245
Lactate	H_2O_2	Lactate oxidase	Reactor	FI	246
Lactose	H_2O_2	Lactase, mutarotase, glucose oxidase	?	?	247
Phenolic compounds	Quinones	Tyrosinase	Electrical surface	FI	212
Polyamines, hypoxanthine	H_2O_2	Putrescine oxidase, xathine oxidase	Reactor	FI	248
Starch	H_2O_2	Amyloglucosidase, mutarotase, glucose oxidase	?	FI	249
Succrose	H_2O_2	Invertase, mutarotase, glucose oxidase	Reactor	FI	241
Sulfite	H_2O_2	Sulfite oxidase	Reactor	FI	250
Urea	Ammonia	Urease	Reactor	FI	251

[a] FI, flow injection; CF, continuous flow.

The dynamic characteristics of a biocatalytic flow system is influenced by the response rate of the internal electrochemical sensor and the rate of the biocatalyzed process. It depends on the biocatalyst used and the rate of transport of substrates and products through the membrane or through the flow-through reactor. The effect of the latter factors significantly depends on the biocatalyst immobilization method. A disadvantage of the potentiometric internal detector is its slow response which is usually much slower than that of the amperometric detector. It can be easier, however, to determine some products of enzymatic reactions (e.g., NH_3, CO_2, or hydrogen ions) electrochemically using membrane potentiometric sensors. Unfortunately, selectivity of the potentiometric detection with membrane electrodes is limited by several factors. In the determination of urea and creatinine in physiological fluids and glutamine in biotechnological samples, there is serious interference of the potentiometric detection with a polymeric nonactin-based ammonium-selective electrode appearing in the presence of the alkali metal ions. It can be effectively eliminated by the use of a suitable ion-exchange membrane covering the biosensor[217] or on-line ion-exchange dialyzers.[145,219,220] An elimination of the interference caused by the presence of endogeneous ammonia in the determination of creatinine in blood serum can be achieved using an additional enzyme glutamate dehydrogenase to catalyze binding of the ammonium ions with α-ketoglutarate to produce glutamate.[221,222] An example of an instrumental solution in such a system is shown in Figure 17. Another method of elimination of the interference from ammonium and alkali metal ions can be obtained by the appropriate design of the FIA system with potentiometric detection, where the determination is based on the difference of the two peaks recorded for each injected sample (Figure 18).[223] In such a system a very careful optimization of the measuring conditions is needed.

In biosensors the membrane of the internal potentiometric detector may contain also microorganisms producing enzymes rather than the isolated enzymes. An example of a microbial sensor successfully employed in flow-measurements is the glutamic acid sensor with *Escherichia coli* on the membrane.[224]

The determination of antibodies to benzoate[225] and the immunochemical determination of immunoglobulin G[226] have been carried out in flow measurements with potentiometric detection using the membrane ion-selective electrode sensitive to sulfide or ammonium, respectively.

In numerous enzymatic reactions, oxygen is the acceptor of electrons, whereas hydrogen peroxide is one of the reaction products. Both these species are electrochemically active and very often used for an amperometric detection of biocatalytic processes, in spite of the fact that the indirect detection of oxygen consumption is slower and less sensitive than the determination of H_2O_2.[252] The presence of H_2O_2 formed during the biocatalytic reaction can

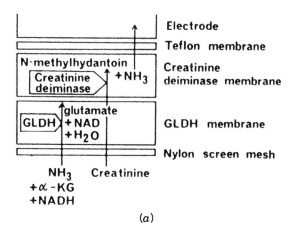

(a)

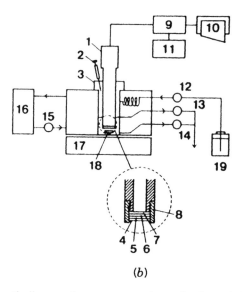

(b)

Figure 17. Schematic diagram of processes occurring at the electrode (**A**) and of the measuring system for enzymatic flow determination of creatinine in the presense of endogeneous ammonia (**B**).[221] 1, Ammonia electrode; 2, microsyringe; 3, cell; 4, nylon screen mesh; 5, GIDH/PVC membrane; 6, creatinine deiminase/PVC membrane; 7, gas permeable Teflon membrane; 8, bottom cap; 9, digital mV meter; 10, recorder; 11, printer, 12–15, peristaltic pumps; 16, water bath; 17, magnetic strirrer; 18, stirer; 19, buffer solution.

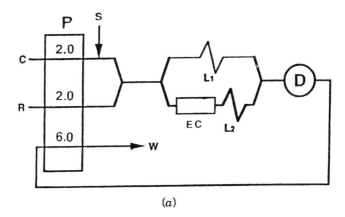

(a)

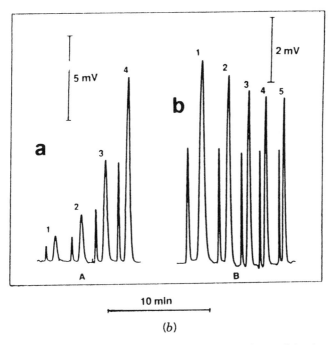

(b)

Figure 18. (A) Schematic diagram of the flow-injection system used for creatinine determination. C, distilled water; R, phosphate buffer; S, sample injection point; ER, enzyme reactor; L_1, L_2, Teflon coils; P, pump; D, detector; W, waste. (B) Flow-injection response recorded for (a) different injection volumes (1) 5, (2) 10, (3) 20, and (4) 50 μl and (b) different flow rates (1) 0.5, (2) 1.0, (3) 1.5, (4) 2.0, and (5) 2.5 ml/min in each line. Injection volume in b is 20 μl. All measurements were performed for $L_2 = 200$ cm and injection of standard solution containing 10 mM ammonium chloride and 10 mM creatinine [223].

interfere with cathodic detection of oxygen in the measurement cell. In a closed-loop system with circulation of the glucose oxidase solution, this effect has been eliminated by the addition of a catalase to decompose hydrogen peroxide.[207]

Biocatalytic systems developed are becoming more complex. Polyphenol oxidase (tyrosinase) catalyzes the oxidation of phenolic compounds with molecular oxygen to catechol and also catalyzes its subsequent dehydrogenation to o-quinone, which readily polymerizes in water to polyaromatic compounds. An electrode with immobilized tyrosinase can be applied to a determination of phenolic compounds based on the electrochemical reduction of quinones.[212] A schematic presentation of the reactions occurring and the design of a flow cell with a biosensor where the enzyme is covalently immobilized on the surface of the carboimide-activated graphite electrode are shown in Figure 19. Another biosensor with the same enzyme cross-linked on a polypyrrole-coated gold electrode surface can be used for a determination of the herbicide atrazine, which inhibits the enzyme activity.[213]

Determinations in real samples of a complex matrix require very often a better selectivity than that provided by a simple biosensor with an immobilized enzyme. In the anodic detection of hydrogen peroxide, all the reducing species, which can be oxidized at the oxidation potential of H_2O_2, interfere in the determination of the substrate. If they are ionic species, their effect can be suppressed by covering the electrode surface with a suitable ion-exchange membrane.[253] Another possibility is to design a more complex flow system splitting the sample into two segments to determine the total amount of the reducing species before and after the enzymatic conversion. Such an FIA system is proposed for the determination of glucose in the presence of ascorbic acid.[242]

Another way to reduce the influence of interfering species is to decrease the potential required for the detection. One method is facilitating electron transfer in the enzymatic reaction using soluble redox mediators. Another method is such chemical modification of the working electrode surface, which makes the enzyme orientation favorable for an electron transfer. The mediator, which is an electron acceptor, rapidly oxidizes the reduced form of oxidases. The reduced form of the mediator is oxidized to the initial form very fast and at a lower potential than the oxidase.[254,255] Ferrocene, its derivatives, and hexacyanoferrate(II)/(III) are most often used as mediators for oxidases. Their use allows for a measurement independent of the oxygen content in the reaction medium.

Nicotinamide adenine dinucleotide (NAD), the indispensable cofactor in the reactions catalyzed by dehydrogenases, is most often used as a mediator. Since about 300 dehydrogenases are known, there are numerous possibilities to design various biosensors. NAD is used either dissolved in the analyzed

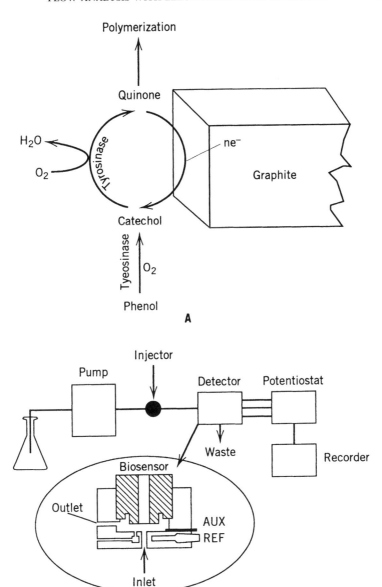

Figure 19. Reaction detection scheme (**A**) and flow-injection manifold (**B**) for the detection of phenol with a tyrosinase graphite electrode.[212]

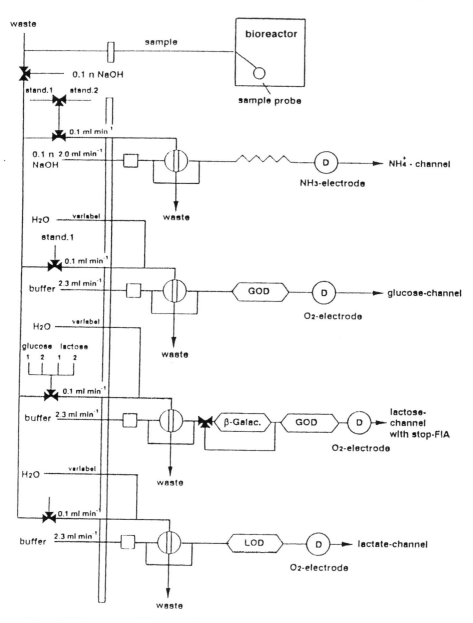

Figure 20. Four-channel FIA system for on-line monitoring and control of alkaline protease production with *Bacillus licheniformis*.[273]

solution[239,256-258] or immobilized on the working electrode surface.[254,259] A marked decrease of the oxidation potential of NADH can be achieved by modification of the working carbon electrode surface with Meldola Blue.[257,258] Flow systems with the NAD cofactor have been designed for determinations of ethanol,[239,254,259] glucose,[257,258] lactate,[254,259] and malate.[259] Using two flow-through reactors with the enzymes amylglucosidase and glucose dehydrogenase, a flow-injection determination of starch has been performed.[256]

Much attention has recently been devoted to the use of conducting organic polymers for the design of biosensors.[260] Electropolymerization as a way to obtain a conducting polymer provides also a convenient method to immobilize enzymes and often to eliminate the effect of some types of interference. However, the role of the polymer in the electron transfer is not clear. Let us recall the applications of a conducting polymer to the determination of atrazine[213] and urea[251] already mentioned above. As a further example, let us mention electrodes with glucose oxidase immobilized in conducting polymers applied to flow-injection determination of glucose.[211,261]

Flow analysis with biocatalytic electrochemical detection has already found numerous applications in routine chemical analysis. For continuous monitoring of insecticides, inhibition of activity of cholinesterases can be utilized.[262] For the control of glucose and glutamine content in cell culture media a flow-injection system with amperometric and potentiometric detectors has been developed.[220] A flow-injection system has also been designed for determination of meat freshness, based on simultaneous determinations of polyamines and hypoxanthine using amperometric detection.[248]

V. ELECTROANALYTICAL METHODS IN PROCESS ANALYSIS

Process analysis is a very important field of contemporary chemical analysis. It is directly associated with the chemical control of technological processes. Quality of the substrates, the semiproducts, and the final product can be controlled to optimize the technological process. The aim of process analysis is also the safety control of the installation and of the environmental pollution. The pH measurements with glass electrodes[263] introduced to the industry in the 1940s are the early example of process analysis, which nowadays is commonly used with all the variety of techniques and methods of modern instrumental analysis.[4,5,8,263-266] Generally, process analysis includes analysis performed off-line in laboratories located apart from the technological installation. However, the on-line analytical process control with automatic or automated analyzers situated at or in the process installation is usually more efficient. A majority of the on-line instruments have their earlier versions as pieces of the off-line laboratory equipment. Their modification to the on-line devices poses several specific problems owing to the usually very severe and chemically aggressive conditions in which they have to be used. The specific difficulties include the necessity to control the signal drift, appropriate validation of calibration, and frequent cleaning and maintenance. There are estimates that 5–10 years have to elapse before a new method or technique widely accepted in the laboratory scale finds its application in process analysis.[268] The selection of a proper sampling procedure is also a very important problem.[267] The economical and construction aspects of process analyzers are thoroughly discussed in the above-mentioned monographs.

Numerous process analyzers are mechanized and automated setups employing electrochemical measurements. Their extensive presentation is not within the scope of this chapter. Nevertheless, some examples should be mentioned in view of the importance of this area of application of electroanalysis in current analytical instrumentation.

In contrast to laboratory instrumentation, conductivity detection is very widely employed in process analyzers. The conductivity probes are equipped with two or four electrodes. Two electrode detectors are used for measurements in solutions of small and moderate conductivity, especially in monitoring of water pollution and in water control for power plants and the electronic industry. Four electrode probes are used for measurements in solutions of large conductivity, where the difficulties associated with the polarization of the electrodes may occur, for example, in wastes, in the food industry, and in electrotechnical processes. In the analyzers of oxygen dissolved in very clean waters the reation of metallic thalium with oxygen is used. Thalium hydroxide is very soluble in water and increases conductivity of the solution. In conductivity analyzers designed for other gases, the gases are absorbed in a suitable solution changing conductivity of the solution. Analyzers of this kind are available for determinations of SO_2, H_2S, HCl, NH_3, chlorine, CO, and methane. The latter two gases are first oxidized to CO_2.

Potentiometric process analyzers employing measurements of pH and redox potential are most common. The pH measurements are usually carried out using suitably protected and continuously cleaned glass electrodes. However, antimony electrodes are also used. The measurements of redox potential are applied to control the treatment of wastes containing chromium and cyanide and to monitor the composition of electroplating baths and the stages of production of synthetic dyes. They are also used in the analyzers of residual chlorine. Platinum and gold electrodes are used as the working electrodes and in some cases also silver and nickel electrodes.

Measurements of pH and potentiometric detection with other indicating electrodes are often applied in process titrators for example, in redox titrations with a tubular platinum electrode,[269] in complexometric titrations of cations of water hardness with the amalgamated platinum electrode,[270] or in titrations of sulfide with the metallic silver electrode. Membrane ion-selective electrodes for direct determination of numerous inorganic species such as fluoride, chloride, iodide, sulfide, cyanide, ammonia, calcium, potassium, copper, and silver have found wide applications in process analyzers. Several manufacturers provide potentiometric analyzers for the low sodium concentration, where the sodium-sensitive glass electrodes are applied. In the hydrogen sulfide analyzer for waste control[271] a cell containing a sulfide-selective electrode and a glass pH electrode is used. Potentiometric monitoring of aluminum for the paper industry is based on an indirect measurement using a fluoride-selective electrode.[272]

Potentiometry is also used in oxygen analyzers as a detection method. For this purpose the corrosion potential of a metallic thallium electrode can be used along with potentiometric zirconium sensors, where the measuring cell is made of porous metallic electrodes with a stabilized zirconium oxide. Such analyzers can also be applied to indirect determinations of organic components, and they find many applications in various branches of industry and environmental protection.

Amperometric detection in process analysis is employed first of all in analyzers of dissolved oxygen for energetics and environmental analysis. In a suitable detector cell (e.g., with a gold cathode and a silver anode), oxygen diffusing through the membrane is reduced on the cathode. In the amperometric chlorine monitor there are two platinum electrodes in the detector cell. In the analyzer for a determination of hydrazine in the power plant waters a three-electrode detector cell is employed with a platinum working electrode. In a determination of humidity of gases and liquids, water vapor is adsorbed in a layer of phosphorus pentoxide and electrolysis is carried out. The magnitude of observed current is proportional to the humidity of analyzed sample.

Numerous electrochemical analyzers are designed as coulometric titrators, where free halogens are most often generated. Iodine generation is commonly employed in water analyzers by the Karl Fischer method. Bromine is generated in coulometric analyzers of volatile sulfur compounds in gases and air. In a free chlorine analyzer Fe(II) is generated, which reacts with the analyte.

In recent years several applications of electrochemical biosensors for the control of biotechnological processes are reported. On-line monitoring using flow-injection analysis is developed to control various processes. A four-channel system with one potentiometric and three amperometric detectors has been developed for monitoring and control of medium components during the production of alkaline protease (Figure 20).[273] The application of appropriate enzymes in flow-through reactors allows for monitoring of glucose, maltose, and L-amino acids. Similar systems are developed to monitor frementation processes[274] and to control glucose during a microbial fed-batch process of the gluconic acid production.[275] Possible faults with FIA analyzers with biosensors are indicated, and a necessity of signal validation is stressed for on-line monitoring of medium components in biotechnological processes.[273]

REFERENCES

1. H. M. N. H. Irving, H. Freiser, and T. S. West, *Compendium of Analytical Nomenclature*, IUPAC, Pergamon, London (1978).

2. J. K. Foreman and P. B. Stockwell, eds., *Automatic Chemical Analysis*, Ellis Horwood, Chichester (1975).

3. J. K. Foreman and P. B. Stockwell, eds., *Topics in Automatic Chemical Analysis*, Ellis Horwood, Chichester (1979).

4. J. Vana, *Gas and Liquid Analysers*, Elsevier, Amsterdam (1982).

5. D. P. Manka, ed., *Automated Stream Analysis for Process Control*, Academic Press, New York (1982).

6. P. Barker, *Computers in Analytical Chemistry*, Pergamon, Oxford (1983)

7. M. Valcarcel and M. D. Luque de Castro, *Automatic Methods of Analysis*, Elsevier, Amsterdam (1988).

8. M. Trojanowicz, *Automatyzacja w Analizie Chemicznej*, WNT, Warsaw (1992).

9. J. T. van Gemert, *Talanta*, **20**, 1045 (1973).

10. P. B. Stockwell, *Talanta*, **27**, 835 (1980).

11. J. Fyffe, P. Daga, and M. J. Roddis, *J. Autom. Chem.*, **10**, 43 (1988).

12. W. A. Schlieper, T. L. Isenhour, and J. C. Marshall, *Anal. Chem.*, **60**, 1142 (1988).

13. H. G. Fouda and R. P. Schneider, *Trends Anal. Chem.*, **6**, 139 (1987).

14. C. P. Price, and K. Spencer eds., *Centrifugal Analysers in Clinical Chemistry*, Praeger Publishers, Eastbourne, England (1980).

15. R. L. Coleman, W. D. Shults, M. T. Kelly, and J. A. Dean, *Int. Lab.*, September, 40 (1971).

16. K. F. Baker, *Analyst*, **95**, 885 (1970).

17. Method D. 3589-83, *Annual Book of ASTM Standards*, Vol. 11.02, ASTM, Philadelphia, (1986), p. 206.

18. Method D.4139-82, *Annual Book of ASTM Standards*, Vol. 11.02, ASTM, Philadelphia (1986), p. 110.

19. D. E. Jordan and J. L. Hoyt, *J. Assoc. Off. Anal. Chem.*, **52**, 569 (1969).

20. A. L. Juliard, *Anal. Chem.*, **30**, 137 (1957).

21. S. Myers and W. B. Swann, *Talanta*, **12**, 133 (1965).

22. H. Kubota and J. G. Surak, *Anal. Chem.*, **35**, 1715 (1963).

23. W. Krijgsman, J. F. Mansveld, and B. F. A. Griepink, *Fresenius Z. Anal. Chem.*, **249**, 368 (1970).

24. P. U. Fruk, L. Meier, H. Rutishauser, and O. Siroky, *Anal. Chim. Acta*, **95**, 97 (1977).

25. J. M. van der Meer and J. C. Smit, *Anal. Chim. Acta*, **83**, 367 (1976).

26. H. Curme, R. L. Columbus, G. M. Dappen, T. W. Eder, W. D. Fellows, J. Figueras, C. P. Glover, C. A. Goffle, D. E. Hill, W. H. Lawton, E. J. Muka, J. E. Pinney, R. N. Rand, K. J. Sanford, and T. W. Wu, *Clin. Chem.*, **24**, 1335 (1978).

27. S. Siggia, *Continuous Analysis of Chemical Process Systems*, Wiley, New York (1959).

28. L. T. Skeggs, Jr., *Am. J. Clin. Pathol.*, **28**, 311 (1957).

29. J. Ruzicka and E. H. Hansen, *Anal. Chim. Acta*, **78**, 145 (1975).

30. L. R. Snyder and H. J. Adler, *Anal. Chem.*, **48**, 1017, 1022 (1976).

31. L. Snyder, *Anal. Chim. Acta*, **114**, 1 (1980).

32. W. B. Furman, *Continuous Flow Analysis. Theory and Practice*, Marcel Dekker, New York (1976).

33. W. A. Coakley, *Handbook of Automated Analysis. Continuous Flow Techniques*, Marcel Dekker, New York (1981).

34. M. K. Schwartz, *Anal. Chem.*, **45**, 739A (1972).

35. L. Snyder, J. Levine, R. Stoy, and A. Conetta, *Anal. Chem.*, **48**, 942A (1976).

36. F. J. Philbert, M. N. Smith, and O. El Kei, *Advances in Automated Analysis, Technicon International Congress, 1976*, Vol. 2, Mediad, Tarrytown, NY (1977), p. 43.

37. D. C. Cowell, *Am. Clin. Biochem.*, **14**, 269 (1977).

38. D. C. Cowell, *Am. Clin. Biochem.*, **14**, 275 (1977).

39. R. T. Oliver, G. F. Lenz, and W. P. Frederik, *Advances in Automated Analysis, Technicon International Congress, 1969*, Vol. 2, Mediad, Tarrytown, (1970), p. 309.

40. J. C. Landry, F. Cupelin, and C. Michal, *Analyst*, **106**, 1275 (1981).

41. P. W. Alexander and P. Seegopaul, *Anal. Chem.*, **52**, 2403 (1980).

42. B. Fleet and H. von Storp, *Anal. Chem.*, **43**, 1575 (1971).

43. R. A. Durst, *Anal. Lett.*, **10**, 961 (1977).

44. Y. M. Fraticelli and M. E. Meyerhoff, *Anal. Chem.*, **53**, 992 (1981).

45. P. Van den Winkel, J. Mertens, and D. L. Massart, *Anal. Chem.*, **46**, 1765 (1974).

46. D. P. Brzezinski, *Analyst*, **108**, 425 (1983).

47. A. Hulanicki, M. Maj-Zurawska, and M. Trojanowicz, *Z. Anal. Khim.*, **32**, 767 (1977).

48. A. Hulanicki, T. Krawczynski vel Krawczyk, and M. Trojanowicz, *Chem. Anal. (Warsaw)*, **24**, 435 (1979).

49. M. Trojanowicz, *Anal. Chim. Acta*, **114**, 293 (1980).

50. M. Trojanowicz and A. Hulanicki, *Mikrochim. Acta*, **II**, 17 (1981).

51. H. Hara and S. Kusi, *Anal. Chim. Acta*, **261**, 411 (1992).

52. B. Fleet and A. Y. W. Ho, *Anal. Chem.*, **46**, 9 (1974).

53. K. J. M. Rao, M. H. Pelavin, and S. Morgenstern, *Advances in Automated Analysis*, Technicon International Congress, 1972, Vol. 1, Mediad, Tarrytown (1973), p. 33.

54. P. Van den Winkel, J. Mertens, G. de Baenst, and D. L. Massart, *Anal. Lett.*, **5**, 567 (1972).

55. M. Vandeputte, L. Dryon, and D. L. Massart, *Anal. Chim. Acta*, **91**, 113 (1977).

56. B. Drake, *Acta Chem. Scand.*, **4**, 545 (1950).

57. W. Kemula, *Rocz. Chem. (Warsaw)*, **26**, 281 (1952).

58. L. D. Wilson and R. J. Smith, *Anal. Chem.*, **25**, 218 (1953).

59. H. G. Lento, *Advances in Automated Analysis, Technicon Symposia 1966*, Vol. 1, Mediad, Tarrytown (1967), p. 598.

60. M. D. Booth, B. Fleet, S. Win, and T. S. West, *Anal. Chim. Acta*, **48**, 329 (1969).

61. T. C. Oesterling and C. L. Olson, *Anal. Chem.*, **39**, 1543 (1967).

62. P. W. Alexander and M. H. Shah, *Talanta*, **26**, 97 (1979).

63. A. M. Bond, H. A. Hudson, and P. A. van den Bosch, *Anal. Chim. Acta*, **127**, 121 (1981).

64. J. Wang, E. Ouziel, Ch. Yarnitzky, and M. Ariel, *Anal. Chim. Acta*, **102**, 99 (1978).

65. H. B. Hanekamp, P. Bos, and O. Vittori, *Anal. Chim. Acta*, **131**, 149 (1981).

66. Z. Kowalski and W. Kubiak, *Anal. Chim. Acta*, **159**, 129 (1984).

67. P. L. Meschi and D. J. Johnson, *Anal. Chem.*, **52**, 1304 (1980).

68. K. Štulík, V. Pacáková and B. Stárková, *J. Chromatogr.*, **213**, 41 (1981).

69. J. A. Lown, R. Koile, and D. C. Johnson, *Anal. Chim. Acta*, **116**, 33 (1980).

70. S. G. Weber, *J. Electroanal. Chem.*, **145**, 1 (1983).

71. E. M. Roosendaal and H. Poppe, *Anal. Chim. Acta*, **158**, 323 (1984).

72. Y. Yamada and H. Matsuda, *J. Electroanal. Chem.*, **44**, 189 (1973).

73. H. Gunasingham and B. Fleet, *Anal. Chem.*, **55**, 1409 (1983).

74. K. Štulík and V. Pacáková, *CRC Crit. Rev. Anal. Chem.*, **14**, 297 (1984).

75. M. Trojanowicz, *Chem. Anal. (Warsaw)*, **30**, 171 (1985).

76. H. Gunasingham and B. Fleet, Hydrodynamic voltammetry in continuous-flow analysis, in A. J. Bard, Ed., *Electroanalytical Chemistry, A Series of Advances*, Vol. 16, Marcel Dekker, New York (1989), pp. 89–180.

77. K. Štulík and V. Pacáková, *Sel. Electr. Rev.*, **14**, 87 (1992).

78. K. Štulík and V. Pacáková, *Electrochemical Measurements in Flowing Liquids*, Ellis Horwood, Chichester (1987).

79. C. M. A. Brett and M. M. P. M. Neto, *J. Chem. Soc. Faraday Trans.*, **52**, 1549 (1980).

80. A. M. Bond, I. D. Heritage, G. G. Wallace, and M. J. McCormick, *Anal. Chem.*, **54**, 582 (1982).

81. A. Ivaska and W. F. Smyth, *Anal. Chim. Acta*, **114**, 283 (1980).

82. J. Wang and H. D. Dewald, *Talanta*, **29**, 453 (1982).

83. W. J. Blaedel and J. Wang, *Anal. Chim. Acta*, **116**, 315 (1980).

84. W. A. MacCrehan, *Anal. Chem.*, **53**, 74 (1981).

85. D. G. Swartzfager, *Anal. Chem.*, **48**, 2189 (1976).

86. J. Wang, *Anal. Chim. Acta*, **234**, 41 (1990).

87. E. Wang, H. Ji, and W. Hon, *Electroanalysis*, **3**, 1 (1991).

88. Y. Ikariyama and W. R. Heineman, *Anal. Chem.*, **58**, 1803 (1986).

89. E. Wang and A. Liu, *Anal. Chim. Acta*, **252**, 53 (1991).

90. L. S. Kuhn, S. G. Weber, and K. Z. Ismail, *Anal. Chem.*, **61**, 303 (1989).

91. J. Wang, P. Tuzhi, and T. Golden, *Anal. Chim. Acta*, **194**, 129 (1989).

92. R. Samuelsson, J. O'Dea, and J. Osteryoung, *Anal. Chem.*, **52**, 2215 (1980).

93. L. J. Nagesl, J. M. Kauffman, C. Dewade, and F. Parmentier, *Anal. Chim. Acta*, **234**, 75 (1990).

94. L. A. Knecht, E. J. Guthrie, and J. W. Jorgenson, *Anal. Chem.*, **56**, 479 (1984).

95. K. W. Bixler and A. M. Bond, *Anal. Chem.*, **58**, 2859 (1986).

96. W. L. Caudill, J. O. Howell, and R. M. Wightman, *Anal. Chem.*, **54**, 2532 (1982).

97. B. Fleet, A. Y. W. Ho, and J. Tenygl, *Talanta*, **19**, 317 (1972).

98. B. Pihlar and L. Kosta, *Anal. Chim. Acta*, **114**, 275 (1980).

99. D. B. Easty, W. J. Blaedel, and L. Anderson, *Anal. Chem.*, **43**, 509 (1971).

100. K. Štulík and V. Hora, *J. Electroanal. Chem.*, **70**, 253 (1976).

101. W. J. Blaedel and J. Wang, *Anal. Chem.*, **52**, 1697 (1980).

102. W. J. Blaedel, C. L. Olson, and L. R. Sharma, *Anal. Chem.*, **35**, 2100 (1963).

103. R. H. P. Reid and L. Wise, *Advances in Automated Analysis, Technicon Symposia 1967*, Vol. 2, Mediad, Tarrytown (1968), p. 159.

104. W. J. Blaedel and G. P. Hicks, *Anal. Chem.*, **34**, 388 (1962).

105. J. Ruzicka and E. H. Hansen, *Flow Injection Analysis*, John Wiley & Sons, New York (1988).

106. M. Valcarcel and M. D. L. de Castro, *Flow Injection Analysis. Principles and Applications*, Ellis Horwood, Chichester (1987).

107. J. Möller, *Flow Injection Analysis*, Springer, Berlin (1988).

108. M. Trojanowicz and W. Matuszewski, *Anal. Chim. Acta*, **138**, 71 (1982).

109. M. Trojanowicz and W. Frenzel, *Fresenius Z. Anal. Chem.*, **328**, 653 (1987).

110. W. E. van der Linden and R. Oostervink, *Anal. Chim. Acta*, **101**, 419 (1978).

111. J. F. van Staden, *Anal. Chim. Acta*, **261**, 381 (1992).

112. I. M. P. L. V. O. Ferreira, J. L. F. C Lima and L. S. M. Rocha, *Fresenius J.Anal. Chem.*, **347**, 314 (1993).

113. S. Motomizu and T. Yoden, *Anal. Chim. Acta*, **261**, 461 (1992).

114. S. Shiono, Y. Hanazato, M. Nakako, and M. Maeda, *Anal. Chim. Acta*, **202**, 131 (1987).

115. S. Alegret, J. Alonso, J. Bartroli, M. Del Valle, N. Jafferzic-Renault, and Y. Duvault-Herrera, *Anal. Chim. Acta*, **222**, 373 (1989).

116. J. Alonso-Chamarro, J. Bartroli, and C. Jimenez, *Anal. Chim. Acta*, **261**, 419 (1992).

117. M. Trojanowicz and W. Matuszewski, *Anal. Chim. Acta*, **151**, 77 (1983).

118. W. E. Morf, E. Lindner, and W. Simon, *Anal. Chem.*, **47**, 1596 (1975).

119. L. Ilcheva, M. Trojanowicz, and T. Krawczynski vel Krawczyk, *Fresenius Z. Anal. Chem.*, **328**, 27 (1987).

120. J. Slanina, W. A. Lingerak, and F. Bakker, *Anal. Chim. Acta*, **117**, 91 (1980).

121. M. G. Gleister, G. J. Moody, and J. D. R. Thomas, *Analyst*, **110**, 113 (1985).

122. W. Frenzel and P. Brätter, *Anal. Chim. Acta*, **188**, 151 (1986).

123. A. Lewenstam, A. Hulanicki, and T. Sokalski, *Anal. Chem.*, **59**, 1539 (1987).

124. L. Ilcheva and K. Cammann, *Fresenius Z. Anal. Chem.*, **320**, 664 (1985).

125. S. Alegret, J. Alonso, J. Bartroli, J. L. F. C. Lima, A. A. S. C. Machado, and J. M. Paulis, *Anal. Lett.*, **18**, 2291 (1985).

126. M. E. Meyerhoff and Y. M. Fraticelli, *Anal. Lett.*, **14**, 414 (1981).

127. E. H. Hansen, J. Ruzicka, and A. K. Ghose, *Anal. Chim. Acta*, **100**, 151 (1978).

128. J. Alonso, J. Bartroli, J. L. F.. C. Lima, and A. A. S. C. Machado, *Anal. Chim. Acta*, **179**, 503 (1986).

129. L. Risinger, *Anal. Chim. Acta*, **179**, 509 (1986).

130. J. Ruzicka, E. H. Hansen, and E. A. G. Zagatto, *Anal. Chim. Acta*, **88**, 1 (1977).

131. M. E. Meyerhoff and P. M. Kovach, *J. Chem. Educ.*, **60**, 766 (1983).

132. E. H. Hansen, F. J. Krug, A. K. Ghose, and J. Ruzicka, *Analyst*, **102**, 714 (1977).

133. R. Y. Yie and G. D. Christian, *Anal. Chem.*, **58**, 1806 (1986).

134. U. Fiedler, *Anal. Chim. Acta*, **89**, 101 (1977).

135. H. Müller, *Anal. Chem. Symp. Ser.*, **8**, 279 (1981).

136. M. Trojanowicz, W. Matuszewski, and A. Hulanicki, *Anal. Chim. Acta*, **126**, 85 (1982).

137. E. H. Hansen, A. K. Ghose, and J. Ruzicka, *Analyst*, **102**, 705 (1977).

138. W. E. van der Linden, *Anal. Chim. Acta*, **179**, 91 (1986).

139. C. Hongbo, E. H. Hansen, and J. Ruzicka, *Anal. Chim. Acta*, **169**, 209 (1985).

140. J. Alonso-Chamarro, J. Bartroli, S. Jun, J. L. F. C. Lima, and M. C. B. S. M. Montenegro, *Analyst*, **118**, 1527 (1993).

141. R. Virtanen, *Anal. Chem. Symp. Ser.*, **8**, 375 (1981).

142. J. F. Van Staden, *Analyst*, **111**, 1231 (1986).

143. P. Pěták and Štulík, *Anal. Chim. Acta*, **185**, 171 (1986).

144. G. B. Marshall and D. Midgley, *Analyst*, **108**, 701 (1983).

145. B. O. Palsson, B. Q. Shen, M. E. Meyerhoff, and M. Trojanowicz, *Analyst*, **118**, 1361 (1993).

146. U. Forsman and A. Karlsson, *Anal. Chim. Acta*, **139**, 133 (1982).

147. B. Persson and L. Rosen, *Anal. Chim. Acta*, **123**, 115 (1981).

148. G. G. Neuburger and D. C. Johnson, *Anal. Chim. Acta*, **179**, 381 (1986).

149. J. Matysik, E. Soczewinski, E. Zminkowska-Halliop, and M. Przegalinski, *Chem. Anal. (Warsaw)*, **26**, 463 (1981).

150. D. MacKoul, D. C. Johnson, and K. G. Schick, *Anal. Chem.*, **56**, 436 (1984).

151. K. W. Pratt, Jr., and D. C. Johnson, *Anal. Chim. Acta*, **148**, 87 (1983).

152. J. Kurzawa, *Anal. Chim. Acta*, **173**, 343 (1985).

153. W. Matuszewski, A. Hulanicki, and M. Trojanowicz, *Anal. Chim. Acta*, **194**, 269 (1987).

154. G. G. Neuburger and D. C. Johnson, *Anal. Chim. Acta*, **59**, 203 (1987).

155. P. L. Meschi and D. C. Johnson, *Anal. Chim. Acta*, **124**, 303 (1981).

156. S. Hughes and D. C. Johnson, *Anal. Chim. Acta*, **132**, 11 (1981).

157. G. G. Neuberger and D. C. Johnson, *Anal. Chem.*, **59**, 150 (1987).

158. A. N. Tsaousis and C. O. Huber, *Anal. Chim. Acta*, **178**, 319 (1985).

159. J. A. Polta and D. C. Johnson, *Anal. Chem.*, **57**, 1373 (1985).

160. J. A. Polta, L.-H. Yeo, and D. C. Johnson, *Anal. Chem.*, **57**, 563 (1985).

161. M. Trojanowicz and J. Michalowski, *J. Flow Injection Anal.*, In press.

162. A. Hulanicki, W. Matuszewski, and M. Trojanowicz, *Anal. Chim. Acta*, **194**, 119 (1987).

163. M. Trojanowicz, A. Hulanicki, W. Matuszewski, M. Palys, A. Fuksiewicz, T. Hulanicka-Michalak, S. Raszewski, J. Szyller, and W. Augustyniak, *Anal. Chim. Acta*, **188**, 165 (1986).

164. M. Trojanowicz, W. Matuszewski, B. Szostek, and J. Michalowski, *Anal. Chim. Acta*, **261**, 391 (1992).

165. J. A. Cox and K. R. Kulkarni, *Talanta*, **11**, 911 (1986).

166. B. Pihlar, L. Kosta, and B. Hristovski, *Talanta*, **26**, 805 (1979).

167. P. W. Alexander and U. Akapongkul, *Anal. Chim. Acta*, **148**, 103 (1983).

168. A. G. Fogg and N. K. Bsebsu, *Analyst*, **109**, 19 (1984).

169. A. G. Fogg, A. Y. Chamsi, and M. A. Abdalla, *Analyst*, **108**, 464 (1983).

170. M. Granados, S. Maspoch, and M. Blanco, *Anal. Chim. Acta*, **179**, 445 (1986).

171. D. J. Chesney, J. L. Anderson, D. E. Weishaar, and D. E. Tallman, *Anal. Chim. Acta*, **124**, 321 (1981).

172. J. A. Cox and K. R. Kulkarni, *Analyst*, **111**, 1219 (1986).

173. W. Th. Kok, H. B. Hanekamp, P. Bos, and R. W. Frei, *Anal. Chim. Acta*, **142**, 31 (1982).

174. J. A. Lown, R. Koile, and D. C. Johnson, *Anal. Chim. Acta*, **116**, 33 (1980).

175. J. Wang and H. D. Dewald, *Talanta*, **29**, 901 (1982).

176. J. W. Dieker and W. E. van der Linden, *Anal. Chim. Acta*, **114**, 267 (1980).
177. H. Lundbäck, *Anal. Chim. Acta*, **145**, 189 (1983).
178. H. B. Hanekamp, W. H. Voogt, and P. Bos, *Anal. Chim. Acta*, **118**, 73 (1980).
179. A. G. Fogg and N. K. Bsebsu, *Analyst*, **107**, 566 (1982).
180. J. Wang and H. D. Dewald, *Anal. Chim. Acta*, **153**, 325 (1983).
181. C. A. Scolari and S. D. Brown, *Anal. Chim. Acta*, **178**, 238 (1985).
182. J. Wang, *Anal. Chem.*, **54**, 598 (1982).
183. J. Wang, H. D. Dewald, and B. Greene, *Anal. Chim. Acta*, **146**, 45 (1983).
184. J. Wang and H. D. Dewald, *Anal. Chem.*, **162**, 189 (1984).
185. J. Wang and H. D. Dewald, *Anal. Chem.*, **56**, 156 (1986).
186. E. W. Kristensen, R. L. Wilson, and R. M. Wightman, *Anal. Chem.*, **54**, 986 (1986).
187. L. Anderson, D. Jagner, and M. Josefson, *Anal. Chem.*, **54**, 1371 (1982).
188. S. Mannino, *Analyst*, **109**, 905 (1984).
189. W. Frenzel and P. Brätter, *Anal. Chim. Acta*, **179**, 389 (1986).
190. G. Schulze, M. Husch, and W. Fenzel, *Mikrochim. Acta*, **I**, 191 (1984).
191. D. Jagner, M. Josefson, and K. Aren, *Anal. Chim. Acta*, **141**, 147 (1982).
192. A. Hu, R. E. Dessy, and A. Graneli, *Anal. Chem.*, **55**, 320 (1983).
193. G. Schulze, W. Bönigk, and W. Frenzel, *Fresenius Z. Anal. Chem.*, **322**, 255 (1985).
194. J. Ruzicka and G. D. Marshall, *Anal. Chim. Acta*, **237**, 329 (1990).
195. J. Ruzicka, *Anal. Chim. Acta*, **261**, 3 (1992).
196. B. H. van der Schoot, S. Jeanneret, A. van der Berg, and N. F. de Rooij, *Anal. Methods Instrum.*, **1**, 38 (1993).
197. J. Wang and Z. Taha, *Anal. Chem.*, **63**, 1053 (1991).
198. J. Wang and Z. Taha, *Anal. Chim. Acta*, **252**, 215 (1991).
199. J. Lu, Q. Chen, D. Diamond, and J. Wang, *Analyst*, **118**, 1131 (1983).
200. P. W. Carr and L. D. Bowers, *Immobilized Enzymes in Analytical and Clinical Chemistry*, Wiley–Interscience, New York (1980).
201. G. G. Guilbault, *Analytical Uses of Immobilized Enzymes*, Marcel Dekker, New York (1984)
202. A. P. F. Turner, I. Karube G. S. Wilson, eds., *Biosensors Fundamentals and Applications*, Oxford University Press, Oxford (1987).
203. R. F. Taylor, ed., *Protein Immobilization Fundamentals and Applications*, Marcel Dekker, New York (1991).
204. F. Scheller and F. Schubert, *Biosensors*, Elsevier, Amsterdam (1992).
205. E. A. H. Hall, *Biosensors*, Open University Press, Buckingham (1990).
206. R. A. Llenado and G. A. Rechnitz, *Anal. Chim.*, **46**, 1109 (1987).
207. C. M. Wolff and H. A. Mottola, *Anal Chim.*, **50**, 94 (1978)
208. J. Ruzicka, E. H. Haansen, A. K. Ghose, and H. A. Mottola, *Anal. Chim.*, **51**, 199 (1979).

209. P. W. Alexander and P. Seegopaul, *Anal. Chim. Acta*, **125**, 55 (1981).

210. H. J. Wieck, G. H. Heider, Jr, and A. M. Yacynych, *Anal. Chim.*, **158**, 137 (1984).

211. M. Trojanowicz, W. Matuszewski, and M. Podsiadla, *Biosensors and Bioelectronics*, **5**, 149 (1990).

212. F. Ortega, E. Dominguez, G. Jönsson-Pettersson, and L. Gorton, *J. Biotechnol.*, **31**, 289 (1993).

213. F. A. McArdle and K. C. Persand, *Analyst*, **118**, 419 (1993).

214. W. Matuszewski and Trojanowicz, *Analyst*, **113**, 735 (1988).

215. K. Johansson, G. Jönnson-Pettersson, L. Gorton, G. Marko-Varga, and E. Csöregi, *J. Biotechnol.*, **31**, 301 (1993).

216. G. S. Cha and M. Meyerhoff, *Talanta*, **36**, 271 (1989).

217. S. A. Rosario, G. S. Cha, M. E. Meyerhoff, and M. Trojanowicz, *Anal. Chim. Acta*, **62**, 2418 (1990).

218. H. S. Yim, C. E. Kibbey, S. C. Ma, D. M. Kliza, D. Liu, S. B. Park, C. E. Torre, and M. E. Meyerhoff, *Biosensors and Bioelectronics*, **8**, (1993).

219. S. A. Rosario, M. E. Meyerhoff, and M. Trojanowicz, *Anal. Chim. Acta*, **258**, 281 (1992).

220. M. E. Meyerhoff, M. Trojanowicz, and B. O. Palsson, *Biotech. Bioeng.*, **41**, 964 (1993).

221. K. Kihara and E. Yasukawa, *Anal. Chim. Acta*, **183**, 75 (1986).

222. F. Winquist, I. Lundström, and B. Danielsson, *Anal. Chim.*, **58**, 145 (1986).

223. W. Matuszewski, M. Trojanowicz, M. E. Meyerhoff, A. Moszczynska, and E. Lange-Moroz, *Electroanalysis*, **5**, 113 (1993).

224. M. Hikuma, H. Obana, T. Yasuda, I. Karube, and S. Suzuki, *Anal. Chim. Acta*, **116**, 61 (1980).

225. R. L. Solsky and G. A. Rechnitz, *Anal. Chim. Acta*, **99**, 241 (1978).

226. I. H. Lee and M. E. Meyerhoff, *Mikrochim. Acta*, **III**, 207 (1988).

227. F. Winquist, A. Spetz, A. Armgarth, I. Lundström, and B. Danielsson, *Sensors and Actuators*, **8**, 91 (1986).

228. J. D. R. Thomas, in E. Pungor, ed., *Ion-Selective Electrodes*, Akademiai Kiado, Budapest (1985), pp. 213–229.

229. W. Matuszewski, M. Trojanowicz, and M. Meyerhoff, *Electroanalysis*, **2**, 525 (1990).

230. K. Yasuda, H. Miyagi, Y. Hamada and Y. Takata, *Analyst*, **109**, 61 (1984).

231. F. Winquist, A. Spetz, I. Lundström, and B. Danielsson, *Anal. Chim. Acta*, **163**, 143 (1984).

232. T. K. V. M. Trojanowicz, and A. Lewenstam, *Talanta*, **41**, 1229 (1994).

233. R. Gnanasekaran and H. A. Mottola, *Anal. Chem.*, **57**, 1005 (1985).

234. M. C. Gosnell, R. E. Snelling, and H. A. Mottola, *Anal. Chem.*, **58**, 1585 (1986).

235. H. Meier, F. Lantreibecq, and C. Tranminh, *J. Autom. Chem.*, **14**, 137 (1992).

236. E. L. Gulberg and G. D. Christian, *Anal. Chim. Acta*, **123**, 125 (1981).

237. K. Matsumoto, O. Hamada, H. Ukeda, and Y. Osajima, *Anal. Chem.*, **58**, 2732 (1986).

238. G. J. Moody, G. S. Sanghera, and J. D. R. Thomas, *Analyst*, **111**, 605 (1986).

239. B. Olsson, H. Lundbäck, G. Johansson, F. Scheller, and J. Nentwig, *Anal. Chem.*, **58**, 1046 (1086).

240. M. Masoom and A. Townshend, *Anal. Chim. Acta*, **166**, 111 (1984).

241. M. Masoom and A. Townshend, *Anal. Proc.*, **22**, 6 (1985).

242. W. Matuszewski, M. Trojanowicz, and L. Ilcheva, *Electroanalysis*, **2**, 147 (1990).

243. K. Matsumoto, K. Sakoda, and Y. Osajima, *Anal. Chim. Acta*, **261**, 155 (1992).

244. J. G. Schindler and M. von Gülich, *Fresenius Z. Anal. Chem.*, **308**, 434 (1981).

245. F. Scheller, F. Schubert, B. Olsson, L. Gorton, and G. Johansson, *Anal. Lett.*, **19**, 1691 (1986).

246. M. Trojanowicz, W. Matuszewski, and T. Krawczynski vel Krawczyk, *J. Flow-Injection Anal.*, **10**, 207 (1993).

247. J. Abdul Hamid, G. J. Moody, and J. D. R. Thomas, *Analyst*, **114**, 1587 (1989).

248. T. Yao, M. Satomura, and T. Wasa, *Anal. Chim. Acta*, **261**, 161 (1992).

249. J. Abdul Hamid, G. J. Moody, and J. D. R. Thomas, *Analyst*, **115**, 1289 (1990).

250. M. Masoom and A. Townshend, *Anal. Chim. Acta*, **179**, 399 (1990).

251. M. Trojanowicz, W. Matuszewski, B. Szczepanczyk, and A. Lewenstam, in G. G. Guilbault and M. Mascini, eds., *Uses of Immobilized Biological Compounds*, Kluwer Academic Publishers, Dordrecht (1993), pp. 141–150.

252. L. Asperger, G. Geppert, and C. Krabisch, *Anal. Chim. Acta*, **201**, 281 (1987).

253. W. Matuszewski, M. Trojanowicz, and A. Lewenstam, *Electroanalysis*, **2**, 607 (1990).

254. J. J. Kulys and M. M. Malinauskas, *Zh. Anal. Khim.*, **34**, 876 (1979).

255. M. J. Green and H. A. O. Hill, *J. Chem. Soc., Faraday Trans.*, **82**, 1237 (1986).

256. J. Enneus, R. Appleqvist, G. Marko-Varga, L. Gorton, and G. Johansson, *Anal. Chim. Acta*, **180**, 3 (1986).

257. R. Appleqvist, G. Marko-Varga, A. Torstensson, and G. Johansson, *Anal. Chim. Acta*, **169**, 237 (1985).

258. G. Marko-Varga, R. Applequist, and L. Gorton, *Anal. Chim. Acta*, **179**, 371 (1986).

259. W. J. Blaedel and R. C. Engstrom, *Anal. Chem.*, **52**, 1691 (1980).

260. M. Trojanowicz, M. L. Hitchman, and K. Cammann, in U. J. Krull and D. P. Nikolelis, eds., *Current Topics in Biophysics*, in press.

261. D. Centonze, A. Guerrieri, C. Malitesta, F. Palmisano, and P. G. Zambonin, *Fresenius J. Anal. Chem.*, **342**, 729 (1992).

262. D. N. Gray, M. H. Keyes, and B. Watson, *Anal. Chem.*, **49**, 1067A (1977).

263. D. J. Huskins, *General Handbook of On-Line Process Analyzers*, Ellis Horwood, Chichester (1981).

264. S. Siggia, *Continuous Analysis of Chemical Process Systems*, John Wiley & Sons, New York (1959).

265. D. J. Huskins, *Electrical and Magnetic Methods in On-Line Process Analysis*, Ellis Horwood, Chichester (1983).

266. K. J. Clevett, *Process Analyzer Technology*, Wiley-Interscience, New York (1986).

267. D. C. Cornich, G. Jepson, and M. J. Smurthwaite, *Sampling Systems for Process Analysers*, Butterworths, London (1981).

268. M. S. Frant and R. T. Oliver, *Anal. Chem.*, **52**, 1252A (1980).

269. W. L. Blaedel and R. H. Laessig, *Anal. Chem.*, **36**, 1617 (1964).

270. W. J. Blaedel and R. H. Laessig, *Anal. Chem.*, **38**, 187 (1966).

271. J. Gulens, H. D. Herrington, J. W. Thorpe, G. Mainprize, M. G. Cooke, P. Dal Bello and S. Macdougall, *Anal. Chim. Acta*, **138**, 55 (1982).

272. F. Oehme and H. Rhyn, *Das Papier*, **31**, 284 (1977).

273. K. Schugerl, *J. Biotechnol.*, **31**, 241 (1993).

274. C. P. Parker, M. G. Gardel, and D. DiBiasio, *Am. Biotechnol. Lab.*, 36 (1985).

275. B. Grundig, B. Strehlitz, H. Kotte, and K. Ethner, *J. Biotechnol.*, **31**, 277 (1993).

CHAPTER

6

STEADY-STATE VOLTAMMETRY AT MICROELECTRODES

CYNTHIA G. ZOSKI

Department of Chemistry
University of Rhode Island
Kingston, Rhode Island

Modern Techniques in Electroanalysis, Edited by Petr Vanýsek, Chemical Analysis Series, Vol. 139.
ISBN 0-471-55514-2 © 1996 John Wiley & Sons, Inc.

I. INTRODUCTION

Voltammetry is that branch of electrochemistry in which the electrode potential or faradaic current, or both, change with time. In transient voltammetry, these three variables are interrelated. However, simplifying conditions can lead to a unique relation solely between current and potential, a relation not involving time or frequency. Such a situation is called a *steady state*, and the wave-shaped graph of current versus potential is a steady-state voltammogram.

True steady-state voltammograms can be obtained at four types of electrode configurations. In the first, the *hydrodynamic electrode configuration*, forced convection is the primary means of transporting electroreactants to within a close distance of an electrode surface. A thin, well-defined boundary is established at the surface of the electrode across which electroactive species are transported by diffusion. This reproducible diffusion of electroactive material results in the development of a stable steady state. The rotating-disc electrode and the wall-jet electrode are examples of successful hydrodynamic electrode configurations[1-5] that operate in a steady-state mode. The second type of electrode configuration employs a thin plastic or gelatinous membrane between an electrode surface and a stirred solution. In this *membrane-covered electrode configuration*, the permeation of electroactive species through the membrane is so slow that the small depletion and enrichment that occurs at its outer surface is easily dissipated by convection in the surrounding analyte. The Clark oxygen sensor[6] exploits such a membrane-covered electrode configuration, as do other more recent electrochemical sensors[7,8] operating on a similar principle. In the *thin-layer electrode configuration*, the third electrode configuration, two electrodes are separated by a thin layer of solution.[9-12] Provided that the reaction at the anode is the exact converse of that at the cathode, a steady state is reached. Such configurations have been used to study porous media,[9,13] to verify the laws of transport in electrolyte solutions,[14] and in the scanning electrochemical microscopy mode (SECM)[15] to probe the kinetics of solution reactions[16-18] and adsorption phenomena,[19] as well as monitoring and quantifying heterogeneous electron-transfer kinetics associated with processes on conducting surfaces.[20-23] The electrodes of the thin-layer cell may assume a planar–parallel arrangement,[9,10] or they may involve spherical or conical electrodes.[11,12] The *sufficiently convergent microelectrode*, an example of which appears in Figure 1, is the final electrode configuration at which true steady-state voltammograms may be obtained. It differs from the previous three configurations in that a steady state is not achieved by the existence of a thin transport layer adjacent to the electrode

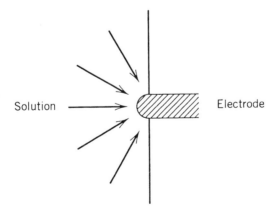

Figure 1. A microhemisphere is one example of a sufficiently convergent microelectrode. The arrows correspond to flux lines radiating from the solution to the electrode surface.

surface. Instead, the size and shape of the electrode are the important factors in the establishment of a steady state. The size of the electrode is important because this feature determines the time required for the attainment of a steady state. The importance of the shape of the electrode arises from the need for a sufficiently convergent transport field in the vicinity of the electrode. It is the recent widespread use of microelectrodes which has led to a renewed interest in steady-state voltammetry. This chapter will deal exclusively with steady states generated at microelectrodes of a sufficiently convergent geometry.

Steady-state voltammetry offers a number of advantages over the more traditionally used transient voltammetries. First, ohmic polarization, iR drop, and capacitive interferences are less of a problem.[24-44] Second, operation in the steady-state mode eliminates the requirement for timing devices or complicated input signals.[45] Third, the lack of need for rapidly sequencing current measurements eases the problems of noise associated with the small measured currents.[45] Fourth, for microelectrodes of simple shape, the steady-state current can be reliably predicted.[46-48] Fifth, the interrelationship between current and potential is unique, being totally independent of the experimental method used to achieve the steady state.[49] Harnessed with these advantages, steady-state voltammetry is the simplest form of voltammetry which has been beneficially applied to chemical analysis[45,50] and the determination of heterogeneous and homogeneous kinetic and transport constants [30-42, 50-78] under conventional and nonconventional solution conditions.

Though practiced for decades, the renewed interest in steady-state voltammetry is due to sophisticated developments in microelectrode technology.[80-95] An earlier review article emphasized steady states generated at microelectrodes.[79] A recent chapter[51] specifically addresses applications of microelectrodes in the study of heterogeneous and homogeneous kinetics under both steady-state and transient experimental conditions. This chapter focuses on developments in steady-state voltammetry since a review article by Bond et al.[79] and expands on steady-state topics presented in a chapter by Montenegro.[51]

II. STEADY-STATE MICROELECTRODES

Microelectrodes of many shapes have been described and their virtues well documented.[51,80,81,96-101] Those shapes with restricted size in *all* superficial dimensions are of special interest because of their ability to reach true steady states under diffusion control in a semi-infinite medium. Thus "band" microelectrodes, which are long narrow strips, do not engender voltammetric steady states. Furthermore, the outer diameter of a "ring" microelectrode, which may be envisaged as a band microelectrode distorted into a circular shape, must be

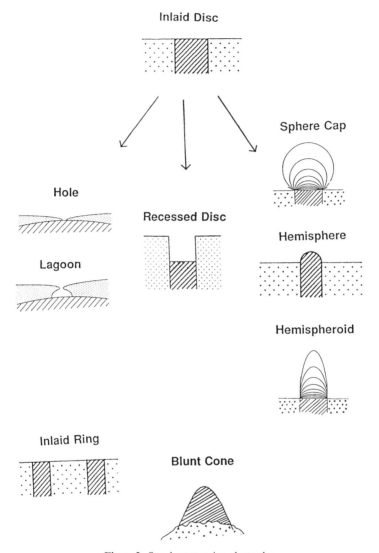

Figure 2. Steady-state microelectrodes.

small if it is to be useful in steady-state voltammetry. The narrowness of the ring is of no consequence because it is the *longest* dimension of the electrode that determines whether a satisfactory steady state will develop.

Figure 2 illustrates several steady-state engendering microelectrode shapes. Notice that the conducting microelectrode may rest on, be inlaid into, or even occupy a depression below the surface of an insulating plane of effectively infinite extent. Additionally, the diagrammed electrodes are symmetrical with respect to rotation about an axis perpendicular to the insulating plane. Thus steady-state engendering microelectrodes may be thought of as "rotationally invariant."

The inlaid-disc geometry appears at the top of Figure 2 for two reasons:

1. The inlaid disc has enjoyed the most popularity among experimentalists due to its relative ease of fabrication.
2. The inlaid structure is the foundation for the fabrication of other microelectrode geometries.

To the right of the inlaid disc are those microelectrodes fabricated by electrodepositing mercury onto inlaid discs of solid metal or carbon. Such microelectrodes conform to a sphere cap or hemispheroidal family of electrodes, both of which coalesce in the limits of the inlaid disc and hemispherical microelectrode. Directly below the inlaid-disc microelectrode lies a recessed-disc electrode which is fabricated by etching the interior of the inlaid-disc electrode. Thus the sphere cap, hemisphere, hemispheroidal, and recessed microelectrodes represent geometries which are a result of an intentional alteration to the inlaid disc electrode. In contrast, the microelectrode geometries to the extreme left in Figure 2 represent results from possible imperfections during the inlaid-disc fabrication process of sealing a fine wire into a glass tube followed by polishing. A "hole" microelectrode may result from imperfect polishing such that the insulating medium creates a small hole similar to a recessed electrode. A "lagoon"microelectrode may result from incomplete adhesion between the metal wire and the insulator such that a microcavity or lagoon forms between the orifice of the insulating material and the electrode itself. This incomplete adhesion may form around a submicroscopic pit in the metal surface during the fabrication process, or it may arise from a subsequent breakdown in the metal–insulator seal. The inlaid-ring and blunt-cone microelectrodes illustrated at the bottom of Figure 2 represent steady-state engendering microelectrodes which are not fabricated from the inlaid-disc geometry.

Though each of the microelectrodes depicted in Figure 2 is capable of attaining a steady state, they differ in the ease with which a diffusing species may reach the electrode surface. The accessibility factor,[102] a shape-specific numerical constant, reflects how easily the electrode as a whole can be reached by solute diffusing from the electrolyte solution. Generally, not all regions of the electrode surface are uniformly accessible; a heterogeneity function[102] is used to reflect this lack of uniformity. Unlike the accessibility factor, the heterogeneity function is not a single numerical constant. Instead, it is affected by both transport and kinetic factors that change with electrode potential for a given reaction. The accessibility factor and the heterogeneity function may be determined experimentally or, alternatively, calculated for simple micro-electrode shapes. These factors will be addressed again in Section VI.

Thus, the steady-state behavior of the microelectrodes diagrammed in Figure 2, regardless of geometry, depends on three factors:

1. The area of the microelectrode interface with the solution
2. The accessibility factor, a pure number
3. The heterogeneity function

The area depends on the size of the electrode, while the remaining two factors reflect its shape. None of the factors depends on the chemistry of the system, though the heterogeneity function is affected by both the kinetic and transport properties of the redox couple and hence on the electrode potential. All three factors may be determined expeimentally or calculated.

Because they do not attain true steady states, microband electrodes[103-111] and electrodes of cylindrical symmetry[112-118] will be ignored. Though micro-electrode arrays[119-127] may reach a steady state, they will not be addressed in this chapter. The dropping mercury microelectrode[83] based on heating a mer-cury reservoir to drive the droplets through a capillary will also not be addressed.

A. Inlaid-Disc Microelectrode

The term *inlaid disc*[46,79] describes a section of electrode material embedded in a surrounding insulator that forms a geometric continuation of the electrode surface. The junction between the electrode material and its surrounding insulator is called the *edge* or perimeter of the electrode. Diffusion occurs radially at the edge and is enhanced with respect to the diffusion which occurs linear to the center of the electrode. Thus the inlaid disc electrodes do not possess the feature of "uniform accessibility." This lack of diffusional uniform-ity imposes more than one distance coordinate to the diffusing electroreactant and electroproduct, thus leading to challenging mathematics in predicting the concentration profiles of diffusing species.

Fabrication procedures include glass sealing,[28,80-82,87,91] microlithography with chemical etching,[88] electrobeam lithography,[91] and scanning tunneling microscopy.[92] Recently, the development of a mercury-disc microelectrode by depositing mercury on a solid silver amalgam microdisc was described.[128]

Characterization of the microdisc surface has been achieved through tunneling spectroscopy,[92] scanning electrochemical microscopy,[86,89,94] and phase-measurement interferometric microscopy.[88]

B. Hemispherical Microelectrode

The role of the plane or "shroud" in the hemispherical microelectrode is to force diffusion to occur radially, even at the junction between the electrode and its surrounding insulator. Thus, the shrouded hemisphere, along with other similar segments of a sphere, is unique in having "uniform accessibility" to the diffusing electroreactant and electroproduct. Since all parts of the electrode surface behave similarly, voltammetry at a hemispherical electrode involves a single distance coordinate only.

Though microhemispherical electrodes may be fabricated from solid metals,[85,129,130] in most studies, liquid mercury is electroplated onto inlaid discs of solid metal[69,81,130-133] or carbon[69,81,134,135] from an electrolyte solution of mercury (I or II) ions by a cathodic reaction of the type

$$Hg_m^{n+}(soln) + ne^- \rightarrow mHg(liq) \tag{1}$$

The volume V of the deposit is proportional to the total charge q passed according to the relationship

$$V = mMq/nF\rho \tag{2}$$

where m and n are integers defined by Equation (1), F is Faraday's constant, M is the molar mass of mercury, and ρ is its density. Due to direct proportionality between V and q in the absence of competing faradaic or nonfaradaic processes, the volume may be determined coulometrically.

As many as ten different growth patterns for the mercury electrodeposit have been shown to be theoretically possible, based on the magnitudes of specific interfacial energies of the mercury–solution, mercury–disc, mercury–insulator, solution–insulator, and solution–disc interfaces.[136] For certain interrelations of the interfacial energies involved, mercury electrodeposits on the inlaid microdisc as a film almost immediately and then progresses through the shapes illustrated in Figure 3. The contact angle θ measured inside the mercury phase begins at zero and increases steadily towards π. Figure 4 shows the transition from a thin mercury film to an eventual hemisphere configuration and ultimately to an almost complete sphere. The various shapes are said to be "penned" because the perimeter of the disc serves as a barrier which restricts the basal plane of the mercury deposit. Experiments using mercury deposited from aqueous solution onto inlaid platinum microdiscs support the growth transition for microdisc radii less than 2.5 μm and deposition times less than 150 sec.[131]

It is the hemisphere geometry which has been of most interest to electrochemists due to the simplicity of the theory describing its behavior.

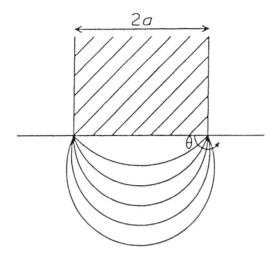

Figure 3. The shapes through which a growing mercury deposit is assumed to progress. (Modified from reference 131, with permission.)

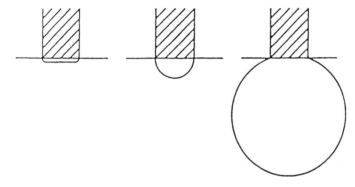

Figure 4. The transition from a thin mercury film to a hemispherical mercury deposit to a large sphere. (From reference 131, with permission.)

However, as the growth of the electrodepositing mercury on different substrates is becoming better understood, an interest in quantifying the shape of the electrodeposit both pre- and post-hemisphere has also developed. The next two subsections address this interest.

C. Hemispheroidal Microelectrode

A hemispheroid is the solid generated when half an ellipse is revolved about its axis. The geometry is characterized by an altitude b above the basal plane of the electrode and by a radius a of the circular area of a microdisc in the basal plane of the microelectrode. The ratio b/a characterizes the shape of the hemispheroid; if two hemispheroids have the same ratio of b to a, but different individual values of b and a, then they have different sizes but identical shape. If $b/a < 1$, the hemispheroid is said to be "oblate"; if $b/a > 1$, it is said to be "prolate". The hemisphere, corresponding to $b/a = 1$, represents a special case of the hemispheroid geometry. In the limit $b/a \to 0$, the hemispheroid degenerates to an inlaid disc, while in the $b/a \to \infty$ limit, a semi-infinite cylinder is approached, as Figure 2 illustrates.

The theoretical behavior of the hemispheroid in general has been addressed theoretically.[137,138] The mercury oblate hemispheroidal microelectrode has been studied in detail both experimentally and theoretically.[70,95]

D.   Sphere-Cap Microelectrode

One can also consider the growth of an electrodeposit of mercury on an inlaid-disc surface as a series of sphere-cap microelectrodes, as Figure 2 illustrates. Theoretical[138,139] and experimental[70,84,131,138] studies of the sphere-cap microelectrode geometry have been reported.

E.   Recessed-Disc Microelectrode

A recessed-disc electrode[28,140] has the geometry shown in Figure 2. At times long in comparison with L^2/D, where L is the depth of the recess and D is the diffusion coefficient of the electroactive species, a steady state is attained. A recessed microdisc electrode will achieve a steady state whether or not the solution outside the recess is stirred. Both situations have been considered theoretically,[28] but only unstirred experiments have been reported.[28]

F.   Hole and Lagoon Microelectrodes

Figure 2 portrays examples of both a microhole constructed by covering a conductor with an insulating layer leaving a small aperture and a pronounced microlagoon formed between an aperture and a conductor. Even more extreme, but realistic, examples of such electrodes appear in Figure 5, which depicts microelectrodes possessing cavities with a narrow opening and a wide opening. A narrow gap on the order of 0.1 µm between the insulating glass and conducting metal, a tiny crack, or a gas bubble becomes important to the behavior of the electrode as its diameter approaches a nanometer or less. Recognizably, a wide-open cavity may be converted into a narrow-neck cavity by adsorption of a colloidal particle or lodging of polishing material. Thus it is the narrow-neck aperture which has been treated theoretically[141,142] More-

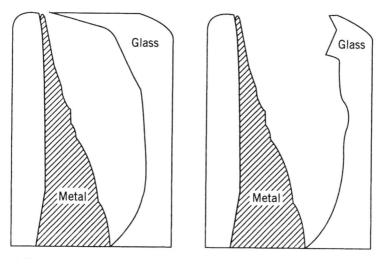

Figure 5. Two types of cavities which may form during the fabrication or use of a microelectrode.

over, it is the dimensions of the aperture that primarily determine the response of the lagooned electrode. As long as the linear dimensions of the lagoon are at least twice as wide as the aperture connecting the lagoon to the bulk solution, and the thickness of the aperture is infinitesimal compared to its width, a lagooned microelectrode behaves, under diffusion control, substantially as if the orifice itself were the electrode,[142] with mass transport properties similar to an inlaid disc. When the thickness of the aperture is greater than its width, then it is the diffusion inside this "channel" which determines the ultimate mass transport.[141]

G. Inlaid-Ring Microelectrode

Inlaid-ring microelectrodes arose from the desire to achieve a larger but more uniform current density than possible at an inlaid-disc microelectrode. An inlaid ring, as shown in Figure 2, is surrounded by an insulator, both interior and exterior to the ring, which forms a geometric continuation of the electrode plane. Thus there are two "edges" to consider: one formed by an inner radius and the other by an outer radius. The radial diffusion which occurs to *both* circular edges of a thin ring leads to an enhanced, and perhaps more uniformly distributed, current density than possible at an inlaid disc of similar area.

Inlaid-ring electrodes are generally fabricated by applying a conductor to the walls of an insulating cylindrical support,[52,72,80,98,143–145] which is most often a glass rod.[72,143,145] For smaller-diameter rings on the order of 1 μm, a flame-drawn glass rod[52,72,80,98,144,145] is used. For metal rings, the insulating support is either painted with organometallic compounds[52,80,98,143,145] or coated by vapor deposition or sputtering of metal onto a rotating glass rod.[72,80,143,145] The vapor deposition method ensures a more uniform coating of the metal; rings of thickness ranging from 10 nm to 5 μm have been fabricated. The coated support is then insulated from solution by sealing into a larger glass tube with epoxy resin or by collapsing the glass around the rod. The structure is then sectioned and polished to expose the inlaid ring.

Noninsulated carbon rings have been fabricated by the pyrolysis of a methane jet through a small, heated capillary.[144,146] Carbon forms on the inner walls and the tip is filled with epoxy. Ultrasmall carbon-ring electrodes with tip diameters as small as 1 μm have been fabricated in this way. Because the resulting ring electrode is not surrounded by an insulating plane, diffusion of electroactive species from behind the plane of the electrode enhances the steady-state faradaic current significantly compared to that of an inlaid ring whose structural diameter is at least ten times greater than the electrode surface diameter.

Fully insulated carbon microrings have been constructed by suspending a fully fabricated noninsulated ring in a glass tube and filling with epoxy.[147]

Microring electrodes have also been the key element in the development of an optoelectrochemical microprobe.[148] A gold-jacketed optical fiber is polished to a flat surface such that the gold forms a microring electrode immediately adjacent to the optical fiber.

Both experimental[72,143,144,146-148] and theoretical[48,72,149,150] steady-state studies have been reported at microring electrodes.

H. Conical Microelectrodes

The interest in conical microelectrodes[11,15,85,94,151-153] can largely be attributed to the development of techniques which allow the examination of interfaces and reactions that occur at them on an atomic and/or cellular scale under electrochemical conditions. Scanning tunnelling microscopy (STM),[154-158] scanning electrochemical microscopy (SECM),[15] and single-cell investigations[152] require microsized electrodes shaped such that they can be brought very close to another surface or inserted directly into a single cell. A conical microelectrode meets these requirements.

Platinum and carbon conical microelectrodes have been used experimentally[85,94,152,153] Etching of the carbon or metal is the common feature in the fabrication process. However, the electrochemical etching bath varies from laboratory to laboratory, as does the final tip coating. Detailed fabrication procedures are described in references 15, 85, 94, 152, and 153. The final tip shape has been evaluated using SECM.[94]

Theoretical investigations are few, but nonetheless include references 11, 94, and 151.

III. PATHWAY OF THE ELECTRODE REACTION

An overall electrode reaction is generally composed of a series of steps that cause the conversion of a dissolved reactant R to a dissolved product P with an exchange of electrons at the electrode surface.

The current (or reaction rate) is governed by the rate of processes such as

- Mass transfer of the electoactive species
- Electron transfer at the electrode surface
- Chemical reactions preceding and/or following the electron transfer

as diagrammed in Figure 6. The rates of all the reactions steps are the same when a steady state is reached.

Other possible surface reactions, such as adsorption/desorption, may also occur but will not be addressed in this chapter.

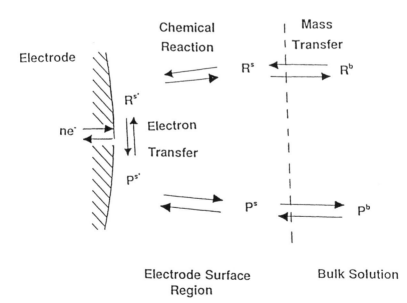

Figure 6. The pathway of the electrode reaction $R(soln) + ne^- \rightleftarrows P(soln)$.

IV. EXPERIMENTAL APPROACHES TO THE STEADY STATE

A steady state is a condition that is approached as a limit and is never theoretically achieved. Nonetheless, in practice, states that are experimentally indistinguishable from the steady state can be attained quite rapidly, especially with small working electrodes.[48,49,79,159] Because steady-state voltammetry is time-independent, the shape of the voltammogram does not depend on the experimental method used for its measurement. In practice, there are four types of experiment that lead to a steady state with suitable electrodes:

- Potentiostatic
- Galvanostatic
- Capped ramp
- Potentiodynamic

The potentiostatic steady-state experiment[48,49,79] involves a sudden jump to a constant potential. Simultaneously, the current jumps to a large, and possible infinite, value after which it decays, approaching a steady-state value.

In galvanostatic steady-state voltammetry,[48,49,79] the current is kept constant. For a reduction, the electrode potential drifts steadily towards more negative values and ultimately approaches a steady-state potential, provided that the applied constant current is not too large.

In the capped-ramp experiment,[79,159] one ramps the potential to a final potential, which is then held constant. The current will generally peak before approaching its steady-state value. In this method, the rate at which the potential is ramped is not crucial because no measurement is made until the potential is constant and the current is almost constant.

In the potentiostatic, galvanostatic and capped-ramp experiments, a steady-state voltammogram is recorded in a point-by-point fashion by performing a series of experiments. This is not the case in potentiodynamic steady-state voltammetry,[48,79,159] the most popular method for recording steady-state voltammograms. As shown in Figure 7, the waveform generally used resembles that used in traditional cyclic voltammetry—a slow sweep to negative potentials, for a reduction, followed by a reverse scan. The resulting voltammogram is, however, very different from a classical cyclic voltammogram, and it retraces itself as long as the scan rate is slow enough to encounter a continuous succession of steady states. This retracing feature is characteristic of the steady state and is independent of the kinetics of the electrode reaction. The coincidence of forward and backward curves is used as an experimental criterion that a steady state has been attained. If the scan rate is too fast, there will be a gap between the two branches as Figure 7 illustrates. However,

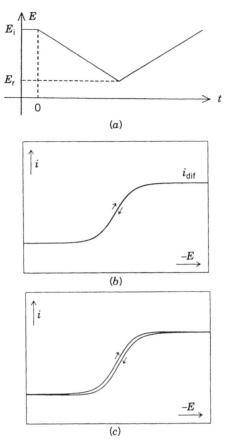

Figure 7. The details of a potentiodynamic steady-state experiment: (**a**) Potential waveform where E_i and E_r represent initial and reversal potentials, respectively. (**b**) Steady-state voltammogram. (**c**) Near-steady-state voltammogram.

methods have been published[159] to correct such "near-steady-state" voltammograms to produce the true steady-state voltammogram.

It is a valuable characteristic of steady-state voltammetry that the resulting current–voltage curves are totally independent of the details of the method used to obtain them.

V.   THE STEADY-STATE VOLTAMMOGRAM

The majority of steady-state voltammograms resemble that of Figure 7b— that is, wave-shaped with a horizontal plateau. Features of the steady-state

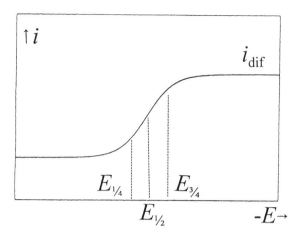

Figure 8. The features of a steady-state voltammogram used in obtaining quantitative information.

voltammogram used in obtaining quantitative information about an electrode process include

- Wave height: i_{dif}
- Wave position: $E_{1/2}$
- Wave steepness: $E_{1/4} - E_{3/4}$

These features are illustrated in Figure 8. Changes in these features will be due to one or any combination of four factors:

- Electrode geometry
- Heterogeneous kinetics
- Supporting electrolyte level
- Homogeneous kinetics

Under strictly diffusion conditions and ignoring homogeneous kinetics for the present, the detailed shape of a steady-state voltammogram reflects the fact that three distinct processes are required to bring about the overall reaction

$$R^{bulk} + ne^- \rightarrow P^{bulk} \qquad (3)$$

when the reactant R, but not product P, is present in the bulk solution. First, the electroreactant R must diffuse to the electrode with diffusion coefficient D_R:

$$R^{bulk} \xrightarrow{D_R} R^{surface} \tag{4}$$

Second, the electron(s) must be transferred:

$$R^{surface} + ne^- \underset{k_b(E)}{\overset{k_f(E)}{\rightleftharpoons}} P^{surface} \tag{5}$$

where $k_f(E)$ and $k_b(E)$ are the potential-dependent heterogeneous rate constants. Third, the product P must diffuse away from the electrode surface

$$P^{surface} \xrightarrow{D_P} P^{bulk} \tag{6}$$

with its own diffusion coefficient D_P. To examine this situation, one must first determine what the current would be if each of these processes alone were solely responsible for determining the current.

If process (4) were the slow step, then the overall reaction is controlled by the rate of diffusion of R to the electrode surface, and this diffusion-limiting current would be given by

$$i = vnFc_R^b A^{1/2} = i_{dif} \tag{7}$$

where v is the shape-specific accessibility factor[102] referred to in Section II and discussed in more detail in Section VI, A is the area of the electrode, c_R^b is the bulk concentration of the electroreactant, and F is Faraday's constant.

If the kinetic process (5) is rate-determining, then the reaction obeys the simple kinetic law

$$i/nAF = \text{rate} = k_f(E)c_R^s - k_b(E)c_P^s \tag{8}$$

where c^s denotes a surface concentration. However, assuming that processes (4) and (6) impose no restriction on the current, the surface concentration of R would have its bulk value, whereas that of P would be zero. It then follows that the kinetic-limiting current is given by

$$i = nAFc_R^b k_f(E) = i_{kin} \tag{9}$$

If the third process (6), the diffusion of P away from the electrode surface, determines the current, then

$$i = vnFc_P^s D_P A^{1/2} \tag{10}$$

by analogy with equation (7). The surface concentration c_P^s can be established by the following argument. Assuming that the kinetic processes (5) are fast in both directions, then a thermodynamic equilibrium exists between the two species at the electrode surface, so that

$$c_P^s = k_f(E)c_R^s/k_b(E) = k_f(E)c_R^b/k_b(E) \qquad (11)$$

where the concentration of R at the surface of the electrode and in the bulk solution have been set equal because process (4) is treated as infinitely fast. Combining (10) and (11) we obtain

$$i = vnFc_R^b D_P A^{1/2}[k_f(E)/k_b(E)] = i_{t/d} \qquad (12)$$

The subscript "t/d" stands for "thermodynamic/diffusion" and reflects the fact that the current, in this case, is controlled jointly by thermodynamics and by the transport of product P by diffusion.

In summary, if the overall process

$$R^{bulk} \rightarrow R^{surface} \underset{}{\overset{ne^-}{\rightleftarrows}} P^{surface} \rightarrow P^{bulk} \qquad (13)$$

were controlled solely by the first step, the current would be given by equation (7). If the second step alone controlled the current, it would be given by equation (9). If the third step were solely controlling, Eq. (12) would describe the current. The joint control of the current by all three steps is considered in Section VI. For completeness, a glossary of symbols appears as Section IX.

VI.   STEADY-STATE VOLTAMMETRY AT MICROELECTRODES OF ARBITRARY SHAPE

The current passed when all three steps in the overall process

$$R^{bulk} \rightarrow R^{surface} \underset{}{\overset{ne^-}{\rightleftarrows}} P^{surface} \rightarrow P^{bulk} \qquad (14)$$

are in joint control is given by the reciprocal-sum formula[102]

$$\frac{1}{i} = \frac{1}{i_{dif}} + \frac{1}{i_{t/d}} + \frac{1 + h(\xi)}{i_{kin}} \qquad (15)$$

where

$$i_{dif} = vnF c_R^b D_R A^{1/2} \tag{16}$$

$$i_{t/d} = vnF c_R^b D_P A^{1/2}[k_f(E)/k_b(E)]$$
$$= vnF c_R^b D_P A^{1/2} \exp\{-nF(E - E^0)/RT\} \tag{17}$$

$$i_{kin} = nAF c_R^b k_f(E) = \exp\{-\alpha nF(E - E^0)/RT\} \tag{18}$$

and the Butler–Volmer relationships

$$k_f(E) = k^0 \exp\{-\alpha nF(E - E^0)/RT\} \tag{19}$$

and

$$k_b(E) = k^0 \exp\{(1 - \alpha)nF(E - E^0)/RT\} \tag{20}$$

between the heterogeneous rate constants and kinetics have been assumed. The parameter v is a shape-specific numerical constant called the *accessibility factor*.[102] The function h(ξ) represents a heterogeneity function of the dimensionless potential-dependent variable

$$\xi = A^{1/2}\left[\frac{k_f(E)}{D_R} + \frac{k_b(E)}{D_P}\right] \tag{21}$$

that depends on the geometry, not the size, of the electrode. Only for simple microelecrode shapes can v and h(ξ) be calculated, but both may be determined experimentally as will be explained in Section VI.F.

A. The Accessibility Factor

A diffusion-limiting current flows when the electrode is polarized at such negative or positive extremes that further polarization has no effect. For extreme negative potentials, this diffusion-limiting current is proportional to n and F and also depends on the bulk concentration c_R^b of reactant R and its diffusion coefficient D_R, as well as on the geometry of the microelectrode and its surroundings. Thus

$$i_{dif} = nF f(c_R^b, D_R, \text{geometry}) \tag{22}$$

where f( ) denotes a shape-dependent function. Comparison of Eqs. (22) and (16) leads to the conclusion that the "geometry" factor must reflect some "characteristic length"

$$i_{\text{dif}} = nF c_R^b D_R(\text{length}) \tag{23}$$

However, the exact size and shape of a microelectrode becomes less certain as its size decreases due to increases in fabrication difficulties. For both irregular-

Table 1. Accessibility Factors for Electrode Geometries[a]

Microelectrode Geometry	Accessibility Factor v	
	Analytical	Numerical
Sphere[160]	$2\pi^{1/2}$	3.545
Thin ring,[150] $a + b \gg a - b$	$\sqrt{\dfrac{a+b}{a-b}} \dfrac{\pi^{3/2}}{\ln\{16(a+b)/(a-b)\}}$	3.47
Prolate hemispheroid,[137,138] $b > a$	$\dfrac{2\sqrt{\pi}(b^2 - a^2)^{3/4}}{\text{arcosh}\left\{\dfrac{b}{a}\right\}\sqrt{a^2\sqrt{b^2 - a^2} + ab^2 \arccos\left\{\dfrac{a}{b}\right\}}}$	2.522
Hemisphere[47]	$(2\pi)^{1/2}$	2.507
Oblate hemispheroid,[95,138] $a > b$	$\dfrac{2\sqrt{\pi}(a^2 - b^2)^{3/4}}{\arccos\left\{\dfrac{b}{a}\right\}\sqrt{a^2\sqrt{a^2 - b^2} + ab^2\, \text{arcosh}\left\{\dfrac{a}{b}\right\}}}$	2.495
Three-quarter sphere[139]	$\dfrac{6}{\pi^{1/2}} \displaystyle\int_0^\infty \dfrac{d\omega}{1 + \text{sech}\,\omega \cosh 2\omega}$	2.493
Quarter sphere[102]	$12\left(\dfrac{3}{\pi}\right)^{1/2} \displaystyle\int_0^\infty \dfrac{d\omega}{1 + \text{sech}\,\omega \cosh 5\omega}$	2.485
Disc[161]	$4/\pi^{1/2}$	2.257
Lagoon[142]	$\dfrac{2\lambda\sqrt{\pi}}{[\pi - \arcsin \lambda]\sqrt{1 + [(1/\lambda) - \lambda]\text{arctanh}\,\lambda}}$	0.501

[a]Excluding the sphere, all listed electrodes rest on or are inlaid into an infinite insulating plane. Values given for the hemispheroids relate to a major:minor axis ratio of 2:1. The analytical expressions are exact, except for the thin ring which is a limiting expression valid when the sum of the outer and inner diameters is much greater than their difference; the ratio of the sum to the difference was set to 10.0 in calculating the numerical value of v. "Quarter sphere" and "three-quarter sphere" refer to spherical caps of heights equal, respectively, to one-half and three-halves of the radius of curvature. The lagoon microelectrode assumes a shape of an oblate hemispheroid with major and minor semiaxes a and b, respectively, symmetrically positioned behind a hole of radius r_h, where $r_h/a = \lambda$. It was assumed that the radius a of the lagoon was twice that of the aperture r_h in calculating v.

ly and regularly shaped microelectrodes, selecting $vA^{1/2}$ as the "characteristic length" reflects the area A of the electrode. Thus, Eq. (16) provides a definition of the accessibility factor v of a microelectrode of any geometry. Specifically, v reflects how easily the electrode as a whole can be reached by an electroactive solute diffusing from the electrolyte solution.

Analytical expressions for v have been determined for several microelectrode shapes. A selection of these is listed in Table 1, along with corresponding numerical values of v. The table is arranged in order of decreasing accessibility.

B.   The Heterogeneity Function

Generally, not all regions of the microelectrode are uniformly accessible. The heterogeneity function $h(\xi)$ reflects the extent to which different portions of the electrode surface have different accessibilities to solutes diffusing from the bulk solution. Unique microelectrodes possessing uniform accessibility at all points on the electrode surface include the isolated sphere, the shrouded hemisphere, and the hemispheroid; $h(\xi) = 0$ exactly. All other microelectrodes have heterogeneities exceeding zero.

Contrary to v, a single numerical value cannot be assigned to $h(\xi)$ due to the effects of transport and kinetic factors that change with electrode potential for a given reaction. One can think of the function h as a property of the microelectrode's shape, whereas ξ depends on the potential and size of the microelectrode in addition to the chemistry of the electrode reaction.

C.   The Reversible Steady-State Voltammogram

Under reversible steady-state conditions, $k_f(E)$ approaches infinity so that $i_{kin} \rightarrow \infty$, allowing the third term of Eq. (15) to be ignored:

$$\frac{1}{i} = \frac{1}{i_{dif}} + \frac{1}{i_{t/d}} \qquad \text{(reversible)} \qquad (24)$$

Expression (24) demonstrates that the behavior at any potential can be written in terms of the limiting behaviors at extreme potentials for a reversible steady-state voltammogram. Substituting for $i_{t/d}$ from Eq. (17) leads to

$$i = \frac{i_{dif}}{1 + \dfrac{D_R k_b(E)}{D_P k_f(E)}} = \frac{i_{dif}}{1 + \dfrac{D_R}{D_P} \exp\left\{ \dfrac{nF}{Rt}(E - E^0) \right\}} \qquad \text{(reversible)} \qquad (25)$$

and further reformulation leads to the classical equation of a polarographic wave:

$$E = E^0 - \frac{RT}{nF}\ln\left\{\frac{D_R}{D_P}\right\} = E^0 - \frac{RT}{nF}\ln\left\{\frac{i_{dif} - i}{i}\right\} \quad \text{(reversible)} \quad (26)$$

The reversible half-wave potential, corresponding to the steady-state current i equaling one-half of i_{dif}

$$E_{1/2} = \frac{RT}{nF}\ln\left\{\frac{D_R}{D_P}\right\} \quad \text{(reversible)} \quad (27)$$

demonstrates that the half-wave potential of a reversible steady-state voltammogram lies twice as far from E^0 than in classical voltammetry because the diffusion coefficient ratio is raised to the half power in the latter case. Thus a reversible steady-state voltammogram possesses a shape and position that is totally independent of the microelectrode.

The "log plot" of the reversible steady-state voltammogram is linear

$$\ln\left\{\frac{i_{dif} - i}{i}\right\} = \frac{nF}{RT}(E - E_{1/2}) \quad \text{(reversible)} \quad (28)$$

with a slope of nF/RT. Another measure of the slope is provided by the Tomeš criterion,[162] the difference between the one-quarter wave and three-quarter wave potentials:

$$E_{1/4} = E_{3/4} = \frac{RT}{nF}\ln 9 \quad \text{(reversible)} \quad (29)$$

which is identical with that for a classical reversible voltammetric wave.

D.   The Irreversible Steady-State Voltammogram

Irreversibility is characterized by the absence of any participation in the electrode process by the reverse electron-transfer reaction $P \rightarrow R + ne^-$. Under these circumstances, $k_b(E) \rightarrow 0$ and subsequently $i_{t/d} \rightarrow 0$, allowing Eq. (15) to be recast as

$$\frac{1}{i} = \frac{1}{i_{dif}} + \frac{1 + h(\xi)}{i_{kin}} \quad \text{(irreversible)} \quad (30)$$

for an irreversible steady-state wave.[102] ξ takes the form

$$\xi_{irr} = A^{1/2}k_f(E)/D_R \quad \text{(irreversible)} \quad (31)$$

For irreversible processes, the steady-state current approaches the i_{kin} or i_{dif} limiting expressions at extreme potentials. The relationship which applies at intermediate potentials bridging these two limits is more complicated than in expression (24) for the reversible case. The additional complexity is due to the interplay between transport and kinetic factors that change with electrode potential for a given reaction. The dimensionless parameter ξ, which depends on the potential and size of the microelectrode as well as on the kinetics of the electrode reaction as demonstrated Eq. (31), encapsulates this interplay. Thus, the extent of heterogeneity is expressed by a function $h(\xi)$ of this parameter, where the function h is a property of the shape of the microelectrode, as will be demonstrated in the following subsections.

1. Hemispherical Microelectrode

For the shrouded microhemisphere of radius r_0 and area

$$A = 2\pi r_0^2 \quad \text{(hemisphere)} \tag{32}$$

resting on an infinite plane,

$$h(\xi) = 0 \quad \text{(hemisphere)} \tag{33}$$

so that on rearrangement, Eq. (30) may be rewritten[47]

$$i^{\text{hemi}} = \frac{i_{\text{dif}}^{\text{hemi}}}{1 + \dfrac{(2\pi)^{1/2}}{\zeta_{\text{irr}}^{\text{hemi}}}} \quad \text{(irreversible)} \tag{34}$$

in terms of $\zeta_{\text{irr}}^{\text{hemi}}$, where from equations (31) and (32) we obtain

$$\zeta_{\text{irr}}^{\text{hemi}} = \frac{A^{1/2}k_{\text{f}}(E)}{D_{\text{R}}} = \frac{(2\pi)^{1/2}r_0 k_{\text{f}}(E)}{D_{\text{R}}} \quad \text{(irreversible)} \tag{35}$$

under irreversible kinetics, and it can be rewritten further as

$$i^{\text{hemi}} = \frac{i_{\text{dif}}^{\text{hemi}}}{1 + \dfrac{D_{\text{R}}}{r_0 k_{\text{f}}(E)}} = \frac{2\pi n F D_{\text{R}} c_{\text{R}}^b r_0}{1 + \dfrac{D_{\text{R}}}{r_0 k^0} \exp\left\{\dfrac{\alpha n F}{RT}(E - E^0)\right\}} \quad \text{(irreversible)} \tag{36}$$

assuming Butler–Volmer kinetics in defining $k_{\text{f}}(E)$ from Eq. (19). Equation (36) represents the equation of an irreversible steady-state voltammogram at a hemispherical microelectrode. It is in agreement with earlier work concerning the theory of voltammetry at electrodes of spherical symmetry published first by Delmastro and Smith[160] and later by Bond and Oldham.[163] The four equations

$$E_{1/2}^{\text{hemi}} = E^0 + \frac{RT}{\alpha nF} \ln\left\{\frac{r_0 k^0}{D_R}\right\} \qquad \text{(irreversible)} \qquad (37)$$

$$i^{\text{hemi}} = \frac{i_{\text{dif}}^{\text{hemi}}}{1 + \exp\left\{\frac{\alpha nF}{RT}(E - E_{1/2}^{\text{hemi}})\right\}} \qquad \text{(irreversible)} \qquad (38)$$

$$E^{\text{hemi}} = E_{1/2}^{\text{hemi}} + \frac{RT}{\alpha nF} \ln\left\{\frac{i_{\text{dif}}^{\text{hemi}} - i^{\text{hemi}}}{i^{\text{hemi}}}\right\} \qquad \text{(irreversible)} \qquad (39)$$

and

$$E_{1/4}^{\text{hemi}} - E_{3/4}^{\text{hemi}} = \frac{RT}{\alpha nF} \ln 9 = \frac{56.45\,\text{mV}}{\alpha n} \quad \text{at } 25°C \qquad \text{(irreversible)} \qquad (40)$$

follow directly from Eq. (36). The superscript "hemi" is used because microelectrodes do not behave exactly the same when the voltammetry is irreversible.[47] The Tomeš potential separation given by Eq. (40) provides a method of evaluating the transfer coefficient α.

Unlike most other irreversible voltammetric waves,[46,164] the irreversible steady-state voltammogram at a hemispherical electrode is symmetrical in the sense that its halfwave is a point of inversion symmetry. This is borne out by the fact that the "log plot" is linear, as can be deduced from Eq. (39).

2.   Inlaid-Disc Microelectrode

Steady-state voltammetry at an inlaid disc has been the subject of three independent theoretical studies.[46,62,160] Though formally distinct, they have been shown to be equivalent,[161] and all three approaches ultimately lead to expressions which must be evaluated numerically. However, these numerical results have been demonstrated to be matched to an empirical formula to better than 0.3%.[47,161] To describe irreversible behavior at an inlaid-disc microelectrode of radius a and area

$$A = \pi a^2 \qquad \text{(disc)} \qquad (41)$$

the empirical formula takes the form[102]

$$i^{\text{disc}} = \frac{i_{\text{dif}}^{\text{disc}}}{1 + \dfrac{2\pi^{1/2}}{\zeta_{\text{irr}}^{\text{disc}}}\left(\dfrac{\zeta_{\text{irr}}^{\text{disc}} + 6\pi^{1/2}}{\zeta_{\text{irr}}^{\text{disc}} + 3\pi^{3/2}}\right)} = \frac{4nFD_{\text{R}}c_{\text{R}}^{\text{b}}a}{1 + \dfrac{2\pi^{1/2}}{\zeta_{\text{irr}}^{\text{disc}}}\left(\dfrac{\zeta_{\text{irr}}^{\text{disc}} + 6\pi^{1/2}}{\zeta_{\text{irr}}^{\text{disc}} + 3\pi^{3/2}}\right)} \qquad \text{(irreversible)}$$

$$(42)$$

where, from Eq. (31) and (41), we obtain

$$\zeta_{\text{irr}}^{\text{disc}} = \frac{A^{1/2}k_{\text{f}}(E)}{D_{\text{R}}} = \frac{\pi^{1/2}ak_{\text{f}}(E)}{D_{\text{R}}} \qquad \text{(irreversible)} \qquad (43)$$

and $\zeta_{\text{irr}}^{\text{disc}}$ is related to potential via Eq. (19) for $k_{\text{f}}(E)$. Equation (42) may be cast as the general irreversible equation [Eq. (30)] in which $h(\xi)$ is[102]

$$h(\xi) = \left(\frac{\pi}{2} - 1\right)\left(\frac{\zeta_{\text{irr}}^{\text{disc}}}{\zeta_{\text{irr}}^{\text{disc}} + 3\pi^{3/2}}\right) \qquad \text{(irreversible)} \qquad (44)$$

Inverting Eq. (42) and (43) leads to an explicit equation for the potential[47]

$$E^{\text{disc}} = E^0 - \frac{RT}{\alpha nF}\ln\left\{\frac{D_{\text{R}}\zeta_{\text{irr}}^{\text{disc}}}{\pi^{1/2}ak^0}\right\}$$

$$= E^0 - \frac{RT}{\alpha nF}\left[\ln\left\{\frac{D_{\text{R}}}{2ak^0}\right\}\right.$$

$$\left. + \ln\left(\sqrt{\frac{4i^2}{(i_{\text{dif}} - i)^2} + \frac{12(4 - \pi)i}{i_{\text{dif}} - i} + 9\pi^2} + \frac{2i}{i_{\text{dif}} - i} - 3\pi\right)\right] \qquad \text{(irreversible)}$$

$$(45)$$

in terms of current. Equation (42) or (45) describes the shape of an irreversible steady-state voltammogram at an inlaid-disc electrode. An example is graphed in Figure 9, where the wave shape is compared to that of an "equivalent" hemisphere, as given by Eq. (36). A microdisc and microhemisphere are considered equivalent in size when their "superficial diameters" are equal. The superficial diameter is the distance d from one edge of the electrode to the other, measured along the electrode surface,[47] as Figure 10 illustrates. Thus, a microdisc and microhemisphere are of equivalent size when

$$\pi r_0 = d = 2a \qquad (46)$$

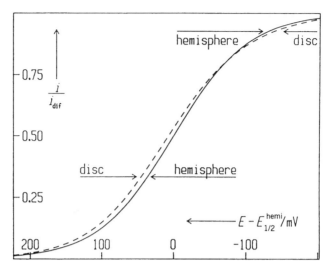

Figure 9. A comparison of irreversible steady-state voltammograms for microdisc and micro-hemispherical electrodes of equivalent size. (From reference 47, with permission.)

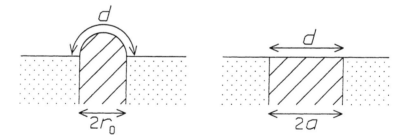

Figure 10. A microhemispherical and an inlaid-microdisc electrode. The two electrodes are said to be of equivalent size when their superficial diameters d are equal. (From reference 47, with permission.)

Figure 9 shows that the disc steady-state voltammogram is less steep than its hemispherical equivalent and is displaced towards positive potentials for the most part. Unlike that at a microhemisphere, the irreversible wave at a microdisc is mildly asymmetric in that it is steeper above the half-wave point than at a corresponding point below.[46,47]

Setting $i^{\text{disc}}/(i^{\text{disc}}_{\text{dif}} - i^{\text{disc}})$ equal to unity in Eq. (45) leads to the half-wave potential expression

$$E^{\text{disc}}_{1/2} = E^0 + \frac{RT}{\alpha nF} \ln\left\{\frac{2k^0 a}{2.730 D_R}\right\} \qquad \text{(irreversible)} \qquad (47)$$

Similarly, setting $i^{\text{disc}}/(i^{\text{disc}}_{\text{dif}} - i^{\text{disc}})$ equal to 1/3 or 3 in Eq. (45) followed by subtraction leads to the Tomeš criterion of wave steepness at the microdisc:

$$E^{\text{disc}}_{1/4} - E^{\text{disc}}_{3/4} = \frac{2.34RT}{\alpha nF} = \frac{60.2\,\text{mV}}{\alpha n} \quad \text{at } 25°\text{C} \qquad \text{(irreversible)} \qquad (48)$$

3. Oblate Hemispheroidal Microelectrode

A hemispheroidal microelectrode is characterized by an altitude b above the basal radius a of the electrode formed by a circular microdisc encased in an insulator. When $b < a$, the microelectrode is described as an "oblate hemispheroid," where a represents the major semiaxis and b represents the minor semiaxis of the resulting oblate hemispheroidal microelectrode. For such an electrode, the area is given as[138]

$$A = \pi a^2 + \frac{\pi ab^2 \operatorname{arcosh}\{a/b\}}{\sqrt{a^2 - b^2}} \qquad \text{(oblate hemispheroid)} \qquad (49)$$

and the heterogeneity function may be shown to be

$$h(\xi) = 0 \qquad \text{(oblate hemispheroid)} \qquad (50)$$

allowing Eq. (30) to be recast[70,95,138] as

$$i^{\text{oblate}} = \frac{i^{\text{oblate}}_{\text{dif}}}{1 + \dfrac{v}{\zeta^{\text{oblate}}_{\text{irr}}}} \qquad \text{(irreversible)} \qquad (51)$$

The accessibility factor v is listed in Table 1 and, using equations (31) and (32), $\zeta^{\text{oblate}}_{\text{irr}}$ may be written

$$\zeta^{\text{oblate}}_{\text{irr}} = A^{1/2}\frac{k_{\text{f}}(E)}{D_R} = \frac{\pi^{1/2}k_{\text{f}}(E)\sqrt{a^2\sqrt{(a^2 - b^2)} + ab^2\operatorname{arcosh}(a/b)}}{D_R(a^2 - b^2)^{1/4}}$$
$$\text{(irreversible)} \qquad (52)$$

so that the equation [Eq. (51)] of an irreversible steady-state voltammogram at an oblate hemispheroidal microelectrode may be written

$$i^{\text{oblate}} = \cfrac{i_{\text{dif}}^{\text{oblate}}}{1 + \cfrac{2(a^2 - b^2)D_R}{k_f(E) \arccos\left\{\dfrac{b}{a}\right\}\left[a^2\sqrt{a^2 - b^2} + ab^2 \operatorname{arcosh}\left\{\dfrac{a}{b}\right\}\right]}}$$

$$= \cfrac{2\pi n F D_R c_R^b \sqrt{a^2 - b^2}/\arccos\{b/a\}}{1 + \cfrac{2(a^2 - b^2)D_R \exp\left\{\dfrac{\alpha n F}{RT}(E - E^0)\right\}}{k^0 \arccos\left\{\dfrac{b}{a}\right\}\left[a^2\sqrt{a^2 - b^2} + ab^2 \operatorname{arcosh}\left\{\dfrac{a}{b}\right\}\right]}}$$

(irreversible)   (53)

in agreement with Birke and Huang.[70] The assumption of Butler–Volmer kinetics led to the final equality in expression (53). The four equations

$$E_{1/2}^{\text{oblate}} = E^0 + \frac{RT}{\alpha n F}\ln\left\{\frac{k^0 \arccos\{b/a\}[a^2\sqrt{a^2 - b^2} + ab^2 \operatorname{arcosh}\{a/b\}]}{(2a^2 - b^2)D_R}\right\}$$

(irreversible)   (54)

$$i^{\text{oblate}} = \frac{i_{\text{dif}}^{\text{oblate}}}{1 + \exp\left\{\dfrac{\alpha n F}{RT}(E - E_{1/2}^{\text{oblate}})\right\}}$$

(irreversible)   (55)

$$E^{\text{oblate}} = E_{1/2}^{\text{oblate}} + \frac{RT}{\alpha n F}\ln\left\{\frac{i_{\text{dif}}^{\text{oblate}} - i^{\text{oblate}}}{i^{\text{oblate}}}\right\}$$

(irreversible)   (56)

$$E_{1/4}^{\text{oblate}} - E_{3/4}^{\text{oblate}} = \frac{RT}{\alpha n F}\ln 9 = \frac{56.45\,\text{mV}}{\alpha n}$$

(irreversible)   (57)

follow directly from Eq. (53). Similar to the microhemisphere, the irreversible steady-state voltammogram at an oblate hemispheroidal microelectrode is symmetrical around the half-wave potential, as can be deduced from a linear "log plot" analysis involving Eq. (56).

4. Inlaid-ring Microelectrode

The most comprehensive theory of voltammetry at ring electrodes is that by Szabo.[150] He developed an approximate, yet analytically tractible, approach

to the calculation of the long-time behavior of the current under quasireversible conditions at thin microring electrodes. From this work, steady-state and
near-steady-state equations have been derived.[48]

Underlying Szabo's "approximate" approach is the assumption that the
microring surface is uniformly accessible. This assumption, while simplifying
the mathematics, is technically incorrect as shown by Fleischmann and
Pons.[80,165] However, as the thickness of the ring is decreased, their work
demonstrates that the microring surface becomes markedly more uniformly
accessible. Later work by Cope and Tallman,[66] using an integral equation
method, found agreement between their numerical results and those calculated using Szabo's approximate theory.

To describe the irreversible steady-state behavior at an inlaid-microring
electrode of outer radius a and inner radius b of area and perimeter

$$A = \pi(a^2 - b^2) \qquad \text{(microring)} \tag{58}$$

$$P = 2\pi(a + b) \qquad \text{(microring)} \tag{59}$$

Eq. (30) takes the form

$$i^{\text{ring}} = \frac{i^{\text{ring}}_{\text{dif}}}{1 + \dfrac{\pi^2(a + b)}{A^{1/2}\zeta^{\text{ring}}_{\text{irr}} \ln\left\{16\left(\dfrac{a + b}{a - b}\right)\right\}}}, \qquad a - b \ll a \qquad \text{(irreversible)} \tag{60}$$

Due to the uniform accessibility assumption, we have

$$h(\xi) = 0 \qquad \text{(uniformly accessible microring)} \tag{61}$$

From Eqs. (31) and (58) we obtain

$$\zeta^{\text{ring}}_{\text{irr}} = \frac{A^{1/2}k_{\text{f}}(E)}{D_{\text{R}}} = \frac{\pi^{1/2}(a^2 - b^2)^{1/2}k_{\text{f}}(E)}{D_{\text{R}}} \qquad \text{(irreversible)} \tag{62}$$

allowing Eq. (60) to be recast as

$$i^{\text{ring}} = \frac{i^{\text{ring}}_{\text{dif}}}{1 + \dfrac{\pi P D_{\text{R}}}{2A k_{\text{f}}(E) \ln\left\{4P^2/\pi A\right\}}} = \frac{\pi n F c^{\text{b}}_{\text{R}} D_{\text{R}} P/2 \ln\left\{4P^2/\pi A\right\}}{1 + \dfrac{\pi P D_{\text{R}} \exp\left\{\alpha n F(E - E^0)/RT\right\}}{2A k^0 \ln\left\{4P^2/\pi A\right\}}}$$

$$\text{(irreversible)} \tag{63}$$

where $P^2 \gg 4\pi A$, in terms of the electrode area A and perimeter P. This reformulation demonstrates that the diffusion-limiting current for a thin microring is a much stronger function of the perimeter than of the area. This is the expected result for a current mainly carried by the circular edges of the electrode, despite the assumption of uniform accessibility.

As with the hemisphere, inlaid disc, and oblate hemispheroid, the irreversible steady-state voltammogram at a thin microring is less steep than, and negatively displaced from, the reversible wave. The uniform accessibility assumption results in an irreversible steady-state voltammogram which is symmetrical about the half-wave potential:

$$E_{1/2}^{\text{ring}} = E^0 + \frac{RT}{\alpha nF}\ln\left\{\frac{2Ak^0\ln\{4P^2/\pi A\}}{\pi PD_{\text{R}}}\right\} \qquad \text{(irreversible)} \qquad (64)$$

has a linear "log plot" that can be deduced from

$$E^{\text{ring}} = E_{1/2}^{\text{ring}} + \frac{RT}{\alpha nF}\ln\left\{\frac{i_{\text{dif}}^{\text{ring}} - i^{\text{ring}}}{i^{\text{ring}}}\right\} \qquad \text{(irreversible)} \qquad (65)$$

and a Tomeš potential difference of

$$E_{1/4}^{\text{ring}} - E_{3/4}^{\text{ring}} = \frac{RT}{\alpha nF}\ln 9 = \frac{56.45\,\text{mV}}{\alpha n} \quad \text{at } 25°C \qquad \text{(irreversible)} \qquad (66)$$

Equations (64)–(66) follow directly from Eq. (63). Equations (60)–(66) are in agreement with those previously published.[79]

E.   The Quasireversible Steady-State Voltammogram

Quasireversibility is the most general voltammetric regime, incorporating reversibility and irreversibility as special cases. Thus, when steady-state voltammetry is applied under quasireversible conditions, all of the right-hand terms of the reciprocal-sum equation [Eq. (15)] are required, so that no simplification is possible. The definition of ξ from Eq. (21) is retained in its entirety. Because it is not possible to invert the reciprocal-sum equation [Eq. (15)] to express E in terms of i/i_{dif}, explicit equations for $E_{1/2}$ and $E_{1/4}-E_{3/4}$ cannot be written. Nevertheless, this equation may be used to construct voltammograms numerically, as was done in generating the middle voltammogram of Figure 11 for a microdisc.

The general result (15) reduces to the reversible case, equation (25), if $i_{\text{kin}} \gg i_{\text{t/d}}$, and to the irreversible case, equation (30), if $i_{\text{kin}} \ll i_{\text{t/d}}$. From equations (17) and (18), these two inequalities require that

$$k_{\text{b}}(E)A^{1/2} \gg vD_{\text{P}} \qquad \text{(reversible)} \qquad (67)$$

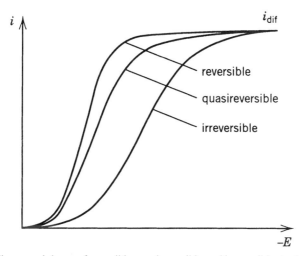

Figure 11. The general shapes of reversible, quasireversible, and irreversible steady-state voltammograms at a microelectrode.

or

$$k_b(E)A^{1/2} \ll vD_P \quad \text{(irreversible)} \quad (68)$$

at all points along the voltammogram. For aqueous electrochemistry, the product vD_P will have a numerical value on the order of $10^{-9}\,\text{m}^2/\text{sec}$. This result arises from assuming a typical diffusion coefficient of $1 \times 10^{-9}\,\text{m}^2/\text{sec}$ in aqueous solution and use of the numerical values of v listed in Table 1. With these limits in place, the reversible and irreversible behavior at a steady-state microelectrode of any shape have been listed,[102] as approximations, in Table 2. Perusal of Table 2 demonstrates that the smaller the microelectrode of

Table 2. Values of the Backward Rate Constant $k_b(E)$ Needed to Ensure Reversibility or Irreversibility of Steady-State Voltammograms Recorded at a Microelectrode in Aqueous Solution.[a]

Area $A(\text{m}^2)$	$k_b(E)(\text{m sec}^{-1})$	
	Reversibility	Irreversibility
10^{-18}	>10	<0.1
10^{-16}	>1	<0.01
10^{-14}	>0.1	$<10^{-3}$
10^{-12}	>0.01	$<10^{-4}$
10^{-10}	$>10^{-3}$	$<10^{-5}$

[a]Because $k_b(E)$ is potential–dependent, the tabulated inequalities must be satisfied at all potentials along the escarpment of the steady-state voltammogram.

any shape, the more irreversibly it behaves. There is interest in using quasireversible steady-state voltammograms obtained at microelectrodes to determine the kinetics of fast electron-transfer reactions; this topic is the subject of Section VIII.A.

Figure 11 compares the typical shapes of reversible, quasireversible, and irreversible steady-state voltammograms that result from experiments at microelectrodes. The same diffusion-limiting current expressed by Eq. (7) is attained irrespective of the degree of reversibility. However, the slope of the voltammetric wave diminishes and the wave shifts along the potential axis as one proceeds from reversibility to irreversibility.

1. Hemispherical Microelectrode

For a shrouded microchemisphere of radius r_0, and with area and heterogeneity functions defined in Eqs. (32) and (33), respectively, the reciprocal-sum formula (15) may be written as

$$i = \frac{i_{\text{dif}}^{\text{hemi}}}{\left[1 + \dfrac{D_R k_b(E)}{D_P k_f(E)}\right]\left[1 + \dfrac{(2\pi)^{1/2}}{\xi}\right]} = \frac{2\pi n F D_R c_R^b r_0}{\left[1 + \dfrac{D_R}{D_P}\exp\left\{\dfrac{nF}{RT}(E - E^0)\right\}\right]\left[1 + \dfrac{(2\pi)^{1/2}}{\xi}\right]}$$

$$\text{(quasireversible)} \quad (69)$$

where ξ is defined by Eq. (21). Under reversible conditions where $\xi \to \infty$, Eq. (69) reduces to Eq. (25). Similarly, under irreversible conditions, where $k_b(E) \to 0$, Eq. (69) reduces to Eq. (34).

Substituting for ξ in Eq. (69) leads to

$$i^{\text{hemi}} = \frac{i_{\text{dif}}^{\text{hemi}} k_f(E)}{D_R\left[\dfrac{k_f(E)}{D_R} + \dfrac{k_b(E)}{D_P} + \dfrac{1}{r_0}\right]} = \frac{2\pi n F c_R^b k_f(E)}{\dfrac{k_f(E)}{D_R} + \dfrac{k_b(E)}{D_P} + \dfrac{1}{r_0}} \quad \text{(quasireversible)} \quad (70)$$

which is in agreement with previous work.[47,160,163]

2. Inlaid-Disc Microelectrode

The empirical formula, based on numerical results, describing the quasireversible behavior at an inlaid microelectrode takes the form[102]

$$i^{\text{disc}} = \frac{i^{\text{disc}}_{\text{dif}}}{\left[1 + \dfrac{D_R k_b(E)}{D_P k_f(E)}\right]\left[1 + \dfrac{2\pi^{1/2}}{\xi}\left(\dfrac{\xi + 6\pi^{1/2}}{\xi + 3\pi^{3/2}}\right)\right]}$$

$$= \frac{4nFD_R c_R^b a}{\left[1 - \dfrac{D_R}{D_P}\exp\left\{\dfrac{nF}{RT}(E - E^0)\right\}\right]\left[1 + \dfrac{2\pi^{1/2}}{\xi}\left(\dfrac{\xi + 6\pi^{1/2}}{\xi + 3\pi^{3/2}}\right)\right]}$$

(quasireversible)   (71)

where ξ is defined by Eq. (21). Equation (71) may be cast as the reciprocal-sum Eq. (15) in which h(ξ) is[102]

$$h(\xi) = \left(\frac{\pi}{2} - 1\right)\left(\frac{\xi}{\xi + 3\pi^{3/2}}\right) \qquad \text{(quasireversible)} \qquad (72)$$

Equation (71) reduces to Eq. (25) under reversible conditions and reduces to Eq. (42) when the electrode reaction is irreversible.

Figure 12 compares the shapes of quasireversible steady-state voltammograms for a microdisc and a microhemispherical electrode of equivalent size. As would be expected, the microdisc steady-state voltammogram is somewhat less steep than for the microhemispherical counterpart and lies at slightly more positive potentials. Descrepancies between quasireversible voltammograms at equivalent microhemispheres and microdiscs are less than for

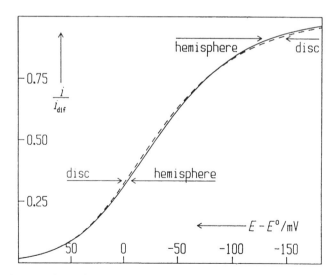

Figure 12. A comparison of quasireversible steady-state voltammograms for microdisc and microhemispherical electrodes of equivalent size. (From reference 47, with permission.)

the irreversible case, as expected from the intermediacy of quasireversibility between full reversibility and total irreversibility.[47,79]

3. Oblate Hemispheroidal Microelectrode

With the area and heterogeneity function defined by Eqs. (49) and (50), respectively, the reciprocal-sum formula (15) becomes

$$
i^{\text{oblate}} = \frac{i_{\text{dif}}^{\text{oblate}}}{\left[1 + \dfrac{D_R k_b(E)}{D_P k_f(E)}\right]\left[1 + \dfrac{2\pi\sqrt{a^2 - b^2}}{\xi A^{1/2} \arccos\{b/a\}}\right]}
$$

$$
= \frac{2\pi n F D_R c_R^b \sqrt{a^2 - b^2}/\arccos\{b/a\}}{\left[1 + \dfrac{D_R}{D_P}\exp\left\{\dfrac{nF}{RT}(E - E^0)\right\}\right]\left[1 + \dfrac{2\pi\sqrt{a^2 - b^2}}{\xi A^{1/2} \arccos\{b/a\}}\right]}
$$

$$\text{(quasireversible)} \quad (73)$$

where ξ is defined by Eq. (21), $k_f(E)$ and $k_b(E)$ are defined by equations (19) and (20), respectively, and $i_{\text{dif}}^{\text{oblate}}$ results from use of Eq. (16) and Table 1. Equation (73) is in agreement with the quasireversible equation reported by Birke and Huang.[70]

When the electrode reaction is reversible, Eq. (73) reduces to Eq. (25). Similarly, under irreversible conditions, Eq. (73) may be shown to be identical to Eq. (53).

4. Inlaid-Ring Microelectrode

Under quasireversible conditions, the reciprocal-sum formula (15) takes the form

$$
i^{\text{ring}} = \frac{i_{\text{dif}}^{\text{ring}}}{\left[1 + \dfrac{D_R k_b(E)}{D_P k_f(E)}\right]\left[1 + \dfrac{\pi^{3/2}\sqrt{(a+b)/(a-b)}}{\xi \ln\left\{16\left(\dfrac{a+b}{a-b}\right)\right\}}\right]}
$$

$$
= \frac{\pi^2 n F D_R c_R^b(a+b)/\ln\{16(a+b)/(a-b)\}}{\left[1 + \dfrac{D_R}{D_P}\exp\left\{\dfrac{nF}{RT}(E - E^0)\right\}\right]\left[1 + \dfrac{\pi^{3/2}\sqrt{(a-b)/(a-b)}}{\xi \ln\left\{16\left(\dfrac{a+b}{a-b}\right)\right\}}\right]},
$$

$$a - b \ll a \quad \text{(quasirversible)} \quad (74)$$

making use of Eqs. (19)–(21), (58), and (61), in addition to Table 1.

Equation (74) reduces to equation (25) when reversible electrode kinetics prevail; Eq. (60) results when irreversible kinetics are applied to Eq. (74).

F. Microelectrode Shape Sensitivity

The term *conforming microelectrode*[138] has been used to describe any small electrode that conforms to the four constraints:

1. They rest on an infinite insulating plane.
2. They are symmetrical with respect to rotation about an axis perpendicular to the insulating plane.
3. The line of intersection of the electrode with the insulating plane is a single concentric circle.
4. The electrode stands proud of the insulating plane by a maximum distance which will often, but not necessarily, lie on the symmetry axis.

Criterion 3 excludes the ring electrode which intersects the plane in two concentric circles. However, the remaining steady-state microelectrodes depicted in Figure 2 may be referred to as conforming microelectrodes.

The diffusion-limiting steady-state current at the more popular conforming microelectrodes diagrammed in Figure 2 may be written in the following forms:

$$i_{dif} = 4nFDc^b a = (\sqrt{8}/\pi)nFDc^b\sqrt{2\pi A} \qquad \text{(inlaid disc)} \qquad (75)$$

$$i_{dif} = 2\pi nFDc^b r_0 = nFDc^b\sqrt{2\pi A} \qquad \text{(hemisphere)} \qquad (76)$$

$$i_{dif} = \alpha(b/a)nFDc^b\sqrt{2\pi A} \qquad \text{(sphere-cap)} \qquad (77)$$

$$i_{dif} = \beta(b/a)nFDc^b\sqrt{2\pi A} \qquad \text{(hemispheroid)} \qquad (78)$$

Equation (77) for the sphere-cap family of electrodes contains the shape-dependent factor $\alpha(b/a)$ defined as[138]

$$\alpha(b/a) = \frac{\sqrt{2}a}{\sqrt{a^2 + b^2}} \int_0^\infty \frac{\cosh\{\omega \arctan\{b/a\}\}}{\cosh\{\omega \operatorname{arccot}\{b/a\}\}\cosh\{\pi\omega/2\}} d\omega \qquad \text{(sphere-cap)}$$

$$(79)$$

for all values of the b/a ratio. It has been reported[138] to lie in the ranges

$$\begin{cases} 0.9003 = \sqrt{8/\pi} \leqslant \alpha(b/a) \leqslant 1, & 0 \leqslant b/a \leqslant 1 \\ 1 \geqslant \alpha(b/a) > \sqrt{2}\ln\{2\} = 0.9803, & 1 \leqslant b/a < \infty \end{cases} \qquad \text{(sphere-cap)} \quad (80)$$

Similarly, equation (6.65) for the hemispheroid contains the shape-dependent factor $\beta(b/a)$[138]

$$\beta(b/a) = \begin{cases} \dfrac{\sqrt{2}(a^2 - b^2)^{3/4}}{\arccos\{b/a\}\,[a^2\sqrt{a^2 - b^2} + ab^2\,\mathrm{arcosh}\{a/b\}]^{1/2}}, & b < a \\[4mm] \dfrac{\sqrt{2}(b^2 - a^2)^{3/4}}{\mathrm{arcosh}\{b/a\}\,[a^2\sqrt{b^2 - a^2} + ab^2\,\arccos\{a/b\}]^{1/2}}, & b > a \end{cases}$$

(hemispheroid)   (81)

each of which approaches unity, the shape factor for the hemisphere, as the b/a ratio approaches unity. The $b/a \to \infty$ limit gives an unbounded result. However, as long as $b < 6.3a$, we obtain $\beta(b/a) = (1.0 \pm 0.1)$.

Thus Eqs. (75)–(81) suggest that the steady-state diffusion-limiting current is given by

$$i_{\mathrm{dif}} = (1.0 \pm 0.1)nFDc^{\mathrm{b}}\sqrt{2\pi A} \tag{82}$$

for those microelectrodes which may be classified as conforming microelectrodes. This significant result implies that the steady-state behavior of a conforming microelectrode is insensitive to its shape and therefore to unintended geometric imperfections during fabrication.

VII.   GENERAL THEORY OF STEADY-STATE VOLTAMMETRY

Traditionally, voltammetry is carried out with excess supporting electrolyte. In practice, an excess of supporting electrolyte is added to provide ions that are electrochemically and chemically inert, thereby increasing the solution conductivity, lowering the ohmic potential drop across the electrochemical cell, and diminishing the migratory flux of the electroactive species relative to its diffusion flux. Additional beneficial effects include the following[167]:

- A double layer populated almost exclusively by supporting ions becomes less of an uncontrolled variable in the experiment. This is true as long as the electroactive species does not specifically adsorb on the electrode surface.
- A swamping excess of supporting electrolyte renders the activity coffi-cients of the electroactive species constant.
- The presence of excess supporting electrolyte makes the solution density much less dependent on the concentration of the electroactive species with beneficial consequences of inhibiting natural convection.

Nevertheless, there are circumstances in which it is undesirable or impossible to provide such support. For example, with solvents of low permittivity[31,38,41] the solubility of supporting salts may be inadequate to provide excess support. A valuable property of microelectrodes in steady-state voltammetry is their capacity to deliver useful and predictable results in media of low electrical conductivity. This property arises because the electrode process causes a redistribution of ions in the vicinity of the microelectrode in response to ionic charge that is either removed or created during the electrode reaction. The ionic redistribution will either increase or decrease the local conductivity.[27] Thus oxidations of cations or neutral species, as well as reductions of anions or neutral species, usually generate an increased ionic strength in the solution at the electrode interface. Because of the convergent geometry around microelectrodes, extra ions close to the electrode are effective in reducing cell resistance. In this way, the ohmic polarization becomes much less than would be calculated from the conductivity of the bulk solution. Conversely, oxidations of anions or reductions of cations lead to an increase in ohmic polarization, though less pronounced than the decrease in polarization that occurs in an electrode process that consumes neutral species. This is because counterions must initially be present at a concentration at least comparable to that of the electroactive ion.[24,27,79]

Experimental[30,31,33-36,38-41,42,44,168,169] and theoretical[25,27,29,32,56,65,170,172] studies of steady-state voltammetry at microelectrodes have shown that there is a great diversity in the shape and position of the steady-state voltammogram to support level due to the critical dependence of the steady-state current on the charge numbers of the reactant, its counterion, and the product. However, the majority of steady-state voltammograms are wave-shaped, with the usual horizontal plateau.[171] Theoretically, diffusion-limiting steady-state currents are sometimes absent when no supporting electrolyte is present in the cell solution.[56,170,171]

The most comprehensive theory has been developed for the microhemisphere due to its convenient feature of uniform accessibility, leading to simplification of mathematics as described in Section II.B. Moreover, as discussed in Sections VI.C, VI.D.2, and VI.3.2, the steady-state behavior is close to that of the inlaid-microdisc electrode which is experimentally more popular due to ease of fabrication.

A. Cell Conditions

A hemispherical microelectrode of radius r_0, resting on an infinite plane, is assumed. The cell has two electrodes only, the second electrode being large and remote so that its precise size and shape are irrelevant. The reaction

$$R(\text{soln}) \rightarrow n e^- + P(\text{soln}) \tag{82}$$

at the microelectrode surface occurs as a single step. Though written as an oxidation with n being positive, reductions may also be considered with n then assuming negative values. The reactant R, with charge number z, may be a cation, an anion, or a neutral species. Thus z may be any positive or negative integer, or it may be zero. The charge number of the product P is $n + z$; n may be any positive or negative integer, but it cannot be zero.

The solution initially contains a bulk concentration of R, but P is absent initially. Unless R is a neutral species, it is usually introduced into the solution in the form of a binary electrolyte with its counterion being a univalent anion A or cation C. Thus, in addition to c_R^b, the solution also contains a concentration $|z|c_R^b$ of A if z is positive or C if z is negative.

The solution also contains a level of supporting electrolyte, containing a univalent cation C and a univalent anion A. The ions of the supporting electrolyte need not be identical with the corresponding counterion of the electroactive species. The ratio of the bulk concentration of the supporting electrolyte, c_S^b, to the reactant bulk concentration defines the support ratio ρ

$$\rho = c_S^b/c_R^b \tag{83}$$

which approaches infinity when supporting electrolyte is in excess.[171] The support ratio is useful in defining the concentrations of the electroinactive ions and the total bulk concentration c^b of all solutes. Electroinactive ions arise as counterions accompanying the electroreactant, unless $z = 0$, and from the separately added supporting electrolyte. Table 3 summarizes the properties of the four solutes in solution.

Each of the three solutes contributes to the total solute concentration in the bulk solution. The symbol x_i^b is used to denote the solute fraction of solute i in the bulk for each component[171]:

$$x_R^b = \frac{c_R^b}{c^b} = \frac{c_R^b}{c_R^b + c_C^b + c_A^b} = \frac{1}{2\rho + |z| + 1} \tag{84}$$

$$x_C^b = \frac{c_C^b}{c^b} = \frac{c_S^b + \frac{|z| - z}{2}c_R^b}{c_R^b + c_C^b + c_A^b} = \frac{2\rho + |z| - z}{2(2\rho + |z| + 1)} \tag{85}$$

$$x_A^b = \frac{c_A^b}{c^b} = \frac{c_S^b + \frac{|z| + z}{2}c_R^b}{c_R^b + c_C^b + c_A^b} = \frac{2\rho + |z| + z}{2(2\rho + |z| + 1)} \tag{86}$$

Table 3. Solute Properties

Species	Charge Number	Bulk Concentraton	Role
R	z	c_R^b	Electroreactant
P	$n+z$	0	Electroproduct
C	1	$\dfrac{(2\rho + |z| - z)c_R^b}{2}$	Inert cation
A	-1	$\dfrac{(2\rho + |z| + z)c_R^b}{2}$	Inert anion
All		$c^b = (2\rho + |z| + 1)c_R^b$	Solutes

The possibility of homogeneous reactions is ignored in the presentation of the theory to follow. Additionally, ion pairing is not considered, though it is acknowledged that there will be experimental instances in which this assumption deviates from reality.

B. Transport Relationships

When both migration and diffusion contribute to the transport of species i, the general equation

$$j_i = j_i^{\text{dif}} + j_i^{\text{mig}} = -D_i \frac{\partial c_i}{\partial r} - \frac{z_i u_i c_i}{|z_i|} \frac{\partial \phi}{\partial r} \tag{87}$$

expresses the diffusive flux density through Fick's first law, while the migratory flux density is expressed as proportional to the local field strength $-\partial \phi / \partial r$, the local concentration c_i of species i, and its mobility u_i. The Nernst–Einstein equation[173]

$$RTu_i = |z_i|FD_i \tag{88}$$

provides a relationship between the mobility of an ion and its diffusivity, which may be used to replace Eq. (87) by

$$\frac{\partial c_i}{\partial r} + \frac{z_i c_i F}{RT} \frac{\partial \phi}{\partial r} = -\frac{j_i}{D_i} = -\frac{J_i}{2\pi r^2 D_i} \tag{89}$$

Equation (89), a form of the Nernst–Planck law, applies throughout the solution of range $r_0 \leqslant r < \infty$, where r is the radial coordinate measured from the center of the hemisphere. Because the flux density $j_i (\text{mol m}^{-2} \text{sec}^{-1})$ is uniform at all points on a hemisphere located at the center of a microelectrode, it may be replaced by $J_i/2\pi r^2$. J_i is the total flux (mol sec^{-1}) across and $2\pi r^2$ is the area of that hemisphere. In its present form, Eq. (89) applies to either a transient or a steady state. In a steady state, the total flux J_i is constant with time and uniform in space. Due to space uniformity, J_i must be independent of r and equal to its special value J_i^s at the electrode surface, denoted by a superscript s. Restricted to the steady state, Eq. (89) takes the form

$$\frac{dc_i}{d(1/r)} + \frac{z_i c_i F}{RT}\frac{d\phi}{d(1/r)} = \frac{J_i^s}{2\pi D_i} \tag{90}$$

in which the independent distance variable r is replaced by its reciprocal. The right-hand side of the Nernst–Planck equation [Eq. (7.9)] is a constant. Equation (7.9) is a general relationship, applicable to each of the four solute species R, P, C, and A, with z_i replaced by the corresponding charge number. For electroactive species R and P, J_i^s is proportional to the current i from Faraday's law so that Eq. (90) becomes

$$\frac{dc_R}{d(1/r)} + \frac{z c_R F}{RT}\frac{d\phi}{d(1/r)} = -\frac{i}{2\pi n F D_R} \approx -\frac{i}{2\pi n F D} \tag{91}$$

and

$$\frac{dc_P}{d(1/r)} + \frac{(n+z)c_P F}{RT}\frac{d\phi}{d(1/r)} = \frac{i}{2\pi n F D_P} \approx \frac{i}{2\pi n F D} \tag{92}$$

The current i is positive when anodic. Because species A and C are electroinactive, J_i^s is zero, leading to

$$\frac{dc_C}{d(1/r)} + \frac{c_C F}{RT}\frac{d\phi}{d(1/r)} = 0 \tag{93}$$

and

$$\frac{dc_A}{d(1/r)} - \frac{c_A F}{RT}\frac{d\phi}{d(1/r)} = 0 \tag{94}$$

It is assumed that the electroactive species R and P share a common diffusion coefficient D in the final approximations of Eqs. (91) and (92).

The electroneutrality condition

$$\sum_i z_i c_i = z c_R + (n+z)c_P + c_C - c_A = 0 \tag{95}$$

establishes that when Eqs. (91)–(94) are summed, the potential gradient term disappears. Because the right-hand terms also sum to zero, the only remaining terms are

$$\frac{d(c_R + c_P + c_C + c_A)}{d(1/r)} = 0 \tag{96}$$

Equation (96) establishes that the sum of the four concentrations must be independent of r and therefore a constant

$$c_R + c_P + c_C + c_A = \text{constant} = c^b \tag{97}$$

as defined by the final entry in Table 3.

Equation (95) stipulates that the electroneutrality condition holds at all points in the solution. However, it has been demonstrated[29,56,172,174–176] that electroneutrality may fail in the depletion layer around the microelectrode even in the presence of excess supporting electrolyte. The effect of the supporting electrolyte level on the double layer and depletion layer surrounding the microelectrode has been difficult to incorporate into voltammetric theory and thus was ignored in the theory[170,171] outlined throughout Section VII.

Equations (91)–(94) relate the concentration of each of the four solutes to the electrostatic potential. Equations (95) and (97) permit the interrelationship of these five parameters at some specific value of the radial coordinate r in solution extending from the hemispherical surface into the bulk solution, thus covering the range $r_0 \leqslant r \leqslant \infty$. At each fixed position r in solution, one speaks of "localized" concentrations and potentials. Working with dimensionless counterparts, x_i has been used[171] to denote the fraction that species i contributes locally to the total solute concentration

$$x_i = \frac{c_i}{c^b} \tag{98}$$

while the potential has been undimensionalized through the definition

$$\psi = \frac{F}{RT}(\phi - \phi^b) \tag{99}$$

where ϕ^b is the potential in the bulk solution. Thus the electroneutrality relationship (95) may be recast as

$$zx_R + (n + z)x_P + x_C - x_A = 0 \tag{100}$$

and Eq. (97) may be rewritten as

$$x_R + x_P + x_C + x_A = 1 \tag{101}$$

Similarly, the transport equations (92)–(94) become

$$dx_P + (n + z)x_P d\psi = \frac{i}{2\pi n F D c^b} d\left(\frac{1}{r}\right) \tag{102}$$

$$dx_C + x_C d\psi = 0 \tag{103}$$

and

$$dx_A - x_A d\psi = 0 \tag{104}$$

From definition (99), the normalized potential is expressed as $\psi \to \psi^b = 0$ as $\phi \to \phi^b$. This allows Eqs. (103) and (104) to be integrated to

$$x_C = x_C^b \exp\{-\psi\} \tag{105}$$

and

$$x_A = x_A^b \exp\{\psi\} \tag{106}$$

thus providing a link between the solute fractions of inert ions and the normalized potential. Use of Eqs. (100) and (101) extend these links to the electroactive species

$$x_R = \frac{(n + z) + (1 - n - z)x_C - (1 + n + z)x_A}{n}$$
$$= \frac{n + z}{n} + \frac{(1 - n - z)x_C^b}{n} \exp\{-\psi\} - \frac{(1 + n + z)x_A^b}{n} \exp\{\psi\} \tag{107}$$

and

$$x_P = \frac{-z - (1 - z)x_C + (1 + z)x_A}{n}$$
$$= \frac{-z}{n} - \frac{(1 - z)x_C^b}{n} \exp\{-\psi\} + \frac{(1 + z)x_A^b}{n} \exp\{\psi\} \tag{108}$$

Equations (102)–(108) may then be used to generate the expression[171]

$$\frac{i}{2\pi F D c^b r} = (1 - n - z)(1 - z)x_C^b[1 - \exp\{-\psi\}]$$
$$+ (1 + n + z)(1 + z)x_A^b[\exp\{\psi\} - 1] - z(n + z)\psi \tag{109}$$

which allows the calculation of the dimensionless group $i/2\pi FDc^b r$ for any value of the normalized potential ψ. Equation (109) applies to a steady state at all points in the solution. At the surface of the microelectrode, this equation may be recast as

$$\frac{i}{2\pi FDc^b r_0} = (1 - n - z)(1 - z)x_C^b[1 - \exp\{-\psi^s\}]$$

$$+ (1 + n + z)(1 + z)x_A^b[\exp\{\psi^s\} - 1] - z(n + z)\psi^s \quad (110)$$

where the superscript s is used to specify conditions at the electrode surface, and the radius r_0 of the microhemisphere has been used in place of r^s. Equation (110) expresses the current in terms of the normalized surface potential ψ^s for any steady-state voltammogram in a solution of any level of support.

For steady-state voltammetry in the presence of supporting electrolyte, a diffusion-limiting current always arises because the concentration c_R^s of the electroreactant eventually reaches zero. The corresponding value of the normalized potential ψ_L can be found from Eq. (107) by setting x_R to zero, thus generating a quadratic equation in $\exp\{\psi_L\}$ which has been solved to produce the result[171]

$$\exp\{\psi_L\} = \frac{n + z + \sqrt{(n + z)^2 + 4[1 - (n + z)^2]x_C^b x_A^b}}{2(1 + n + z)x_A^b} \quad (111)$$

or, alternatively,

$$\exp\{\psi_L\} = \frac{2(1 - n - z)x_C^b}{-n - z + \sqrt{(n + z)^2 + 4[1 - (n + z)^2]x_C^b x_A^b}} \quad (112)$$

The x^b terms in both equations are defined in Eqs. (85) and (86), permitting ψ_L to be determined for any support ratio. The subscript L accompanying ψ is used to denote this limiting condition. Equation (111) is indeterminant when the product species is a univalent anion, whereas Eq. (112) is indeterminant when the product is a univalent cation.

The current coordinate of the steady-state voltammogram described by Eq. (110) is determined by the single parameter ψ^s which varies monotonically from a value of zero at the foot of the wave to its limiting value, from Eq. (111),

$$\psi_L = \ln\left\{\frac{n + z + \sqrt{(n + z)^2 + 4[1 - (n + z)^2]x_C^b x_A^b}}{2(1 + n + z)x_A^b}\right\} \quad (113)$$

on the voltammetric plateau. The current may be found from the ψ^s parameter by the equation

$$
\frac{i}{i_L^{\rho=\infty}} = \frac{1}{nx_R^b} \{(1-n-z)(1-z)x_C^b[1-\exp\{-\psi^s\}]
$$
$$
+ (1+n+z)(1+z)x_A^b[\exp\{\psi^s\}-1] - z(n+z)\psi^s\} \quad (114)
$$

which follows from Eq. (110) after multiplication by the factor $(1/nx_R^b)$ and normalization by $i_L^{\rho=\infty}$ where

$$
i_L^{\rho=\infty} = 2\pi nFDc_R^b r_0 = i_{dif}^{hemi} \quad (115)
$$

Equation (115) is the result for the diffusion-limiting steady-state current at a microhemisphere in the presence of excess supporting electrolyte; the superscript $\rho = \infty$ is used to indicate this excess. The current is zero at the foot of the steady-state wave, increasing in magnitude to approach the asymptotic limit given by

$$
\frac{i_L}{i_L^{\rho=\infty}} = \frac{1}{nx_R^b} \left[(n+z)(1-z^2)x_R^b - z + z\sqrt{(n+z)^2 + 4[1-(n+z)^2]x_C^b x_A^b} \right.
$$
$$
\left. - z(n+z)\ln\left\{\frac{n+z+\sqrt{(n+z)^2+4[1-(n+z)^2]x_C^b x_A^b}}{2(1+n+z)x_A^b}\right\} \right] \quad (116)
$$

on the voltammetric plateau. This equation results from Eqs (110)–(112). To be useful in the case of a univalent anionic product, the argument of the logarithm in Eq. (116) must be replaced by the right-hand side of Eq. (112).

As long as the support ratio satisfies the inequality[171]

$$
\rho \geqslant 12.5|nz| \quad (117)
$$

the limiting current i_L will be within 2% of $i_L^{\rho=\infty}$.

C.   Electrode Potential and Cell Resistance

Contributions to the microelectrode potential will include the standard potential E^0 and three overvoltages arising from ohmic, concentration, and activation polarizations.

The ohmic voltage η_{ohm} represents the potential difference $\phi^s - \phi^b$ required to drive the current through the solution by the migration of ions. ϕ^s and ϕ^b correspond to the local potentials established in solution at points immediately adjacent to the microelectrode and the second electrode, respectively. The ohmic overvoltage depends on the current i, and the ratio $(\phi^s - \phi^b)/i$ represents the effective resistance of the cell. This resistance has been referred to as the "steady-state resistance," R_{ss},[27,79] and may differ from the better known "static resistance," R_{st},[27,79] which is measurable by alternating current (ac) techniques in an unpolarized cell. The product iR_{st} is the quantity most often reported as the "iR drop."

The static resistance is found by integrating the resistances of an infinite number of hemispherical shells of area $2\pi r^2$ and thickness dr

$$R_{st} = \int_{r_0}^{\infty} \frac{dr}{2\pi r^2 \kappa} = \frac{1}{2\pi r_0 \kappa} \tag{118}$$

where κ is the conductivity of the bulk ionic solution. The conductivity may be expressed in terms of the concentrations and mobilities of all ionic species present

$$\kappa = F \sum_i |z_i| u_i c_i^b = \left(\frac{F^2}{RT} \right) \sum_i z_i^2 D_i c_i^b \tag{119}$$

or in terms of diffusion coefficients with the help of the Nernst–Einstein relationship (88) as demonstrated in the second identity of Eq. (119). Combination of Eqs. (118) and (119) produces the expression

$$R_{st} = \frac{RT}{2\pi F^2 r_0 \sum_i z_i^2 D_i c_i^b} \tag{120}$$

for the static resistance of the microhemispherical cell.

Mathematically, the "steady-state resistance" R_{ss} is the ratio

$$R_{ss} = \frac{\phi^s - \phi^b}{i} \tag{121}$$

of the ohmic polarization in the solution between the microelectrode and the second electrode to the current. Being a potential difference inside the solution and not across the electrodes, $\phi^s - \phi^b$ cannot be measured directly; hence, neither can R_{ss}. Both must be determined by solving the transport equations for the specific electrode geometry, electrode reaction, and solution composition.

Because most of the cell resistance occurs close to the microelectrode surface,[79] the conductivity of the layers of solution in the immediate vicinity of the microelectrode will affect the magnitude of R_{ss}. Thus the steady-state resistance is influenced by the chemistry taking place at the microelectrode surface and any consequent changes which may occur in the ion population surrounding the surface. For example, the oxidation of cations or neutral species and the reduction of anions or neutral species increase the ionic strength in solution at the electrode interface because ions are produced from neutral precursors or ions of low charge are replaced by ions of greater absolute charge. In an attempt to maintain electroneutrality, electrogenerated ions are expelled from the vicinity of the microelectrode, and counterions are drawn or "scavenged" towards the electrode. This scavenging effect occurs during the transient phase preceding the steady state; once the steady state is reached, the electroinactive ions remain stationary.

The redistribution of ions close to the interface is dramatically enhanced because of the convergent geometry surrounding the microelectrode; consequently, the redistribution is effective in reducing R_{ss}. This ionic redistribution can be so effective that a bulk concentration of electrolyte no greater than the electroactive bulk concentration can behave as "excess" supporting electrolyte. Under such circumstances, iR_{ss} may become much less than iR_{st} which is calculated from the conductivity of the bulk solution. When the electrode reaction occurs in a solution containing a bulk concentration of supporting electrolyte greatly exceeding that of the electroactive species, any redistribution of ions which may occur in the vicinity of the microelectrode surface will have little effect on the already large ionic strength; under these conditions, $R_{ss} = R_{st}$.

Thus, for any level of support, one must compute the ohmic overvoltage by integrating the electrostatic potential ϕ across the cell:

$$\eta_{ohm} = \int_{r=\infty}^{r=r_0} d\phi = \phi^s - \phi^b = \frac{RT\psi^s}{F} \tag{122}$$

where the last equality is a result of definition (99).

In the absence of activation polarization, the effect of concentration polarization is embodied in the Nernst equation:

$$E^s - E^0 = \frac{RT}{nF} \ln\{x_P^s/x_R^s\} \tag{123}$$

where the surface concentrations of the electroactive species are related to the ψ^s parameter by

$$x_R^s = \frac{n+z}{n} + \frac{(1-n-z)x_C^b}{n} \exp\{-\psi^s\} - \frac{(1+n+z)x_A^b}{n} \exp\{\psi^s\} \tag{124}$$

and

$$x_P^s = \frac{-z}{n} - \frac{(1-z)x_C^b}{n}\exp\{-\psi^s\} + \frac{(1+z)x_A^b}{n}\exp\{\psi^s\} \qquad (125)$$

which follow from Eqs. (107) and (108).

When the microelectrode reaction is not reversible, the interfacial potential depends on the concentrations of the electroactive species at the electrode surface in addition to the current density there. Assuming Butler–Volmer kinetics via Eqs. (19) and (20), we obtain

$$\frac{i}{nAFk^0} = x_R^s \exp\left\{\frac{(1-\alpha)nF}{RT}(E^s - E^0)\right\} - x_P^s \exp\left\{\frac{\alpha nF}{RT}(E^s - E^0)\right\} \qquad (126)$$

which may be normalized by Eq. (115), the diffusion-limiting current under excess supporting electrolyte conditions for a microhemisphere:

$$\frac{iDx_R^b}{r_0 k^0 i_L^{\rho=\infty}} = x_R^s \exp\left\{\frac{(1-\alpha)nF}{RT}(E^s - E^0)\right\} - x_P^s \exp\left\{\frac{\alpha nF}{RT}(E^s - E^0)\right\} \qquad (127)$$

When $\alpha = 1/2$, Eq. (127) may be inverted to give

$$E^s - E^0 = \frac{2nF}{RT}\ln\left\{\frac{iDx_R^b}{r_0 k^0 i_L^{\rho=\infty}(x_R^s - x_P^s)}\right\} \qquad (128)$$

in this case.

If the reaction is totally irreversible, the second term on the right-hand side of Eq. (127) will be negligible in comparison with the first, leading to

$$E^s - E^0 = \frac{RT}{(1-\alpha)nF}\ln\left\{\frac{iDx_R^b}{r_0 k^0 i_L^{\rho=\infty}x_R^s}\right\} \qquad (129)$$

From Eqs. (122), (123), (128), and (129), the total electrode potential may be written

$$\eta_{ohm} + E^s - E^0 = \frac{RT\psi^s}{F} + \text{a term involving } x_R^s, \text{ sometimes } x_P^s \qquad (130)$$

where Eqs. (124) and (125) define x^s.

D. Classifications of Steady-State Voltammograms When $\rho < \infty$

Changes in electrolyte support shift the steady-state voltammetric wave along both the i and E axes. The position of the voltammetric wave is characterized by its half-wave potential $E_{1/2}$, which is found from Eq. (114) by setting $i/i_L^{\rho = \infty}$ to one-half. The steepness of the wave is quantified by the Tomeš separation $E_{3/4} - E_{1/4}$, also accessible from Eq. (114). These shape characteristics are so dependent on the precise values of the charge numbers of the reactants and products that electrode reactions in the presence of varying supporting-electrolyte concentrations have been described in terms of seven broad classifications.[170,171]

1. Class I: Neutral Reactant

Figure 13 illustrates the steady-state voltammograms for a reversible oxidation of a neutral species:

$$R(\text{soln}) \to 3e^- + P^{3+}(\text{soln}) \tag{131}$$

Because migration does not affect the rate of supply of the uncharged reactant to the electrode surface, the limiting current is not affected by the support ratio.

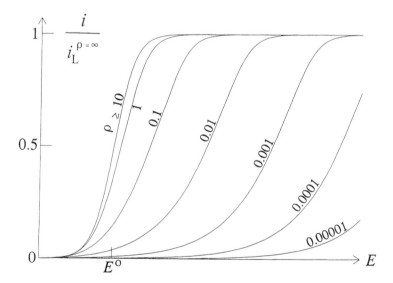

Figure 13. Class I: Neutral reactant. Steady-state voltammograms for the reversible reaction $R(\text{soln}) \to 3e^- + P^{3+}(\text{soln})$ with the support ratios indicated. (Modified from reference 171, with permission.)

Thus

$$i_L^{\rho = \infty} = \pm 2\pi n F D c_R^b r_0 = i_{dif}^{hemi} \tag{132}$$

However, because there are no counterions, ρ has an unusually large effect on η_{ohm}. This effect manifests as a dependence of the wave position and shape on ρ, as demonstrated in Figure 13. For the $n = +3$ reaction (131), $\rho \geqslant 10$ corresponds to a steady-state wave shape that is typical of a reversible steady-state voltammogram recorded under conditions of excess supporting electrolyte. As ρ begins to diminish, two overvoltage effects interact to shift $E_{1/2}$ and affect the wave steepness. First, ohmic overvoltage pushes the wave increasingly positive of E^0. Second, there is a counteraction to this "ohmic push" caused by the migration of P^{3+} away from the electrode surface, thus decreasing its concentration there and shifting the potential in a negative direction. The net effect results in a small positive shift in $E_{1/2}$ and a small increase in the Tomeš potential separation. These effects magnify for $\rho < 1$; $E_{1/2}$ shifts approximately 59 mV for each tenfold reduction in supporting electrolyte concentration, and the wave steepness decreases. These predictions may be quantified through the following expressions[171]:

$$E_{1/2} = E^0 + \frac{3RT}{8F\rho}, \qquad\qquad \rho \gg 1 \tag{133}$$

$$E_{3/4} - E_{1/4} = \frac{RT}{3F} \ln\{9\} + \frac{3RT}{8F\rho}, \qquad\qquad \rho \gg 1 \tag{134}$$

$$E_{1/2} = E^0 - \frac{RT}{3F} \ln\left\{\frac{2048}{27}\right\} + \frac{RT}{F} \ln\{\rho\}, \qquad \rho \ll 1 \tag{135}$$

$$E_{3/4} - E_{1/4} = \frac{5RT}{6F} \ln 9, \qquad\qquad \rho \ll 1 \tag{136}$$

A well-studied system in Class I both theoretically[27,171] and experimentally[30] is the oxidation of ferrocene:

$$(C_5H_5)_2Fe(soln) \rightarrow e^- + (C_5H_5)_2Fe^+(soln) \tag{137}$$

2. Class II: Neutral Product

Class II represents the case of a multiply charged reactant producing a neutral product; for example,

$$R^{3-}(soln) \rightarrow 3e^- + P(soln) \tag{138}$$

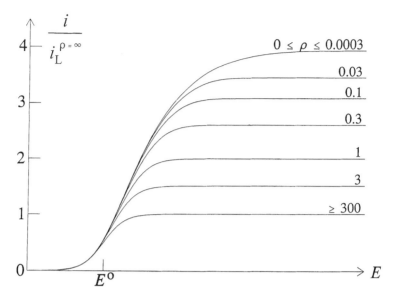

Figure 14. Class II: Neutral product. Steady-state voltammograms for the reversible reaction $R^{3-}(soln) \to 3e^- + P(soln)$ with the support ratios indicated. (Modified from reference 171, with permission.)

The electric field which drives current through the solution causes migration of the R^{3-} anion towards the electrode surface. Withdrawal of supporting electrolyte enhances this mode of transport, thereby increasing the limiting current i_L, as demonstrated in Figure 14, which eventually reaches a definite limit as $\rho \to 0$.

There is a decrease in the symmetry of the steady-state voltammograms corresponding to low electrolyte levels. The upper reaches of the oxidation wave are displaced increasingly towards positive potentials due to the lack of a charged product in the vicinity of the electrode surface as the voltammetric plateau is approached.

3. Class III: Sign Reversal/Univalent Product

Class III incorporates those electrode reactions in which "sign reversal" occurs. This means that z_P is opposite in sign to z_R. Specifically comprising Class III are those oxidations producing univalent cations from an anionic reactant and reductions giving univalent anions from a cationic reactant.

Similar to Class II, as ρ is diminished the waveheight increases from the $\rho = \infty$ case due to the facilitated migration of the reactant anion R^{2-} in the

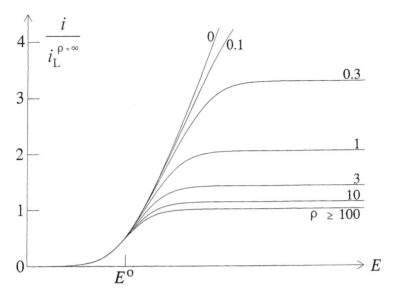

Figure 15. Class III: Sign reversal/univalent product. Steady-state voltammograms for the reversible reaction $R^{2-}(soln) \rightarrow 3e^- + P^+(soln)$ with the support ratios indicated. (Modified from reference 171, with permission.)

example reaction

$$R^{2-}(soln) \rightarrow 3e^- + P^+(soln) \tag{139}$$

towards the electrode surface. As Figure 15 illustrates, when the support ratio actually reaches 0, the voltammetric wave assumes a ramp shape due to the facilitated migration of the cationic P^+ away from the electrode surface.

4. *Class IV: Sign Reversal/Multicharged Product*

Those cases of sign-reversal voltammetry not included in Class III fall into Class IV. Specifically, electrode reactions in which a univalent reactant is converted to a multiply charged product of opposite sign comprise Class IV. Figure 16 shows the steady-state voltammograms predicted to be generated by the reaction

$$R^-(soln) \rightarrow 3e^- + P^{2+}(soln) \tag{140}$$

For reasons similar to those discussed in context of Class III reactions, the wave height increases limitlessly as $\rho \rightarrow 0$.

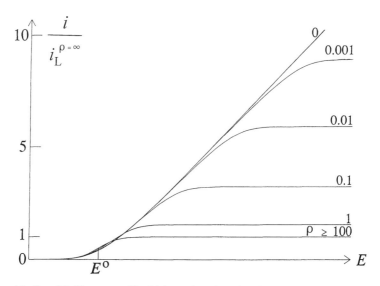

Figure 16. Class IV: Sign reversal/multicharged product. Steady-state voltammograms for the reversible reaction $R^-(soln) \rightarrow 3e^- + P^{2+}(soln)$ with the support ratios indicated. (Modified from reference 171, with permission.)

5.   Classes V and VI: Sign Retention/Univalent and Lesser Multiply Charged Product

Classes V and VI incorporate those reactions in which "sign retention" occurs. This means that z_R and z_P share the same sign.

In Class V electrode reactions, a univalent product is formed from a more highly charged reactant of the same sign, as in

$$R^{4-}(soln) \rightarrow 3e^- + P^-(soln) \tag{141}$$

As Figure 17 demonstrates, the effect of changing the support ratio ρ is largely limited to the wave height. The limiting current increases when supporting electrolyte is decreased because the migration of the anionic reactant to the electrode surface is facilitated; however, the hindrance of the removal of the anionic product away from the electrode surface prevents the wave height from increasing limitlessly as $\rho \rightarrow 0$. The position and steepness of the steady-state voltammograms are virtually unchanged.

Class VI represents those electrode reactions in which a multiply charged product is produced from a multiply charged reactant of the same sign. Qualitatively, the voltammograms for Classes V and VI are similar.

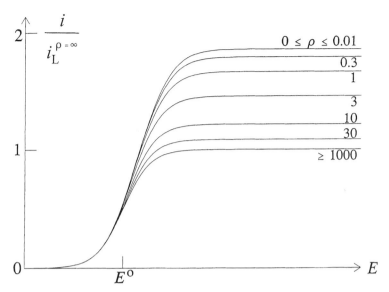

Figure 17. Class V: Sign retention/univalent-charged product. Steady-state voltammograms for the reversible reaction $R^{4-}(\text{soln}) \rightarrow 3e^- + P^-(\text{soln})$ with the support ratios indicated. (Modified from reference 171, with permission.)

6. Class VII: Sign Retention/Greater Multiply Charged P from Univalent R

Class VII comprises electrode reactions in which a univalent ion is converted to another ion of the same sign, but higher charge, as in

$$R^+(\text{soln}) \rightarrow 3e^- + P^{4+}(\text{soln}) \tag{142}$$

Class VII steady-state voltammograms illustrated in Figure 18 are similar to those generated in Classes V and VI: the effect of diminishing ρ is mostly limited to the wave height with the wave position and steepness virtually unchanged. However, in Class VII the limiting current decreases as supporting electrolyte is removed because migration of both the cationic reactant and product away from the electrode surface is facilitated.

E. Steady-State Voltammetry in the Absence of Deliberately Added Supporting Electrolyte

Theoretical studies[56,170,171] have shown that, in the total absence of supporting electrolyte, steady-state voltammograms for sign-reversal electrode reac-

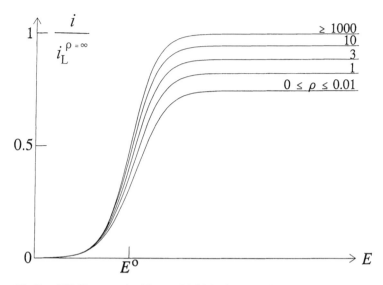

Figure 18. Class VII: Sign retention/Greater Multiply charged P from univalent R. Steady-state voltammograms for the reversible reaction $R^-(soln) \rightarrow 3e^- + P^{4+}(soln)$ with the support ratios indicated. (Modified from reference 171, with permission.)

tions are ramp-shaped. Also demonstrated[33,56,170,171] is that all steady-state voltammograms will have a wave shape if any supporting electrolyte, no matter how little, is present. The presence of ionic impurities is difficult to prevent experimentally due to impurities introduced with the redox species, ions resulting from the autoionization of the solvent or water, or problems with reference electrode function.

Minute quantities of adventitious electrolyte have a dramatic effect on the shape of the steady-state voltammogram because the ion having the same sign as the reactant migrates from the surrounding solution during the pre-steady-state regime and accumulates close to the electrode in a surprisingly large concentration. Its presence causes the voltammogram to revert to the typical wave shape[33] in the absence of deliberately added supporting electrolyte.

F. Benefits of Steady-State Voltammetry with Meager Support

Numerous endeavors in the use of steady-state voltammetry in the presence of low concentrations of supporting electrolyte have resulted from a general interest in extending the scope of electrochemical measurements to solvents in which electrolytes are insoluble or to systems in which electrolyte ions interfere with quantitative analyses. Additionally, the often unexpected results of such experiments have raised fundamental questions regarding the role of the supporting electrolyte in charge-transport mechanisms as well as in establishing electric fields at the electrode–electrolyte interface.

A valuable aspect of low support steady-state voltammetry is that more information concerning the product is present in the resulting steady-state voltammogram than in a traditional wave.[31,32,34,39,42,168] The voltammetric shape under conditions of low supporting electrolyte reflects the charge number of the product and can therefore aid in the identity of a doubtful product; no such information is available from a classical voltammogram.

Other benefits include greater freedom from activity-coefficient and ion-pairing effects, in addition to less uncertainty concerning speciation. The possibility of using larger-than-traditional electroactive concentrations is an asset.

G. Activation Overvoltage Effects in Low Support Steady-State Voltammetry

Figures 13–18 pertain to reversible steady-state voltammograms. Figure 19 illustrates the inclusion of the effect of electron-exchange kinetics on steady-state voltammograms recorded at a support ratio of unity. This inclusion is possible through the combination of Eqs. (124), (125), and (127). As is found in

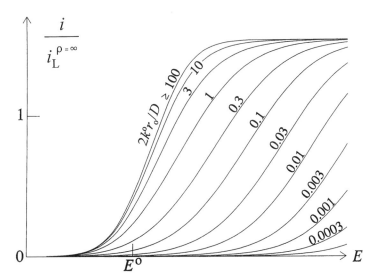

Figure 19. Steady-state voltammogram for the Class IV reaction $R^- \rightarrow 3e^- + P^{2+}$ at unity support ratio for the $2k^0 r_0/D$ values shown. (Modified from reference 171, with permission.)

steady-state voltammograms where $\rho = \infty$, the wave steepness decreases as the magnitude of the ratio $2k^0 r_0/D$ decreases, and a kinetic transition from a reversible through a quasireversible and eventually to an irreversible regime occurs.

VIII.   APPLICATIONS OF STEADY-STATE VOLTAMMETRY

Steady-state voltammetry has been used successfully in the determination of

- Heterogeneous electrode kinetics
- Homogeneous electrode kinetics
- Diffusion coefficients
- Concentration

The reasons for this are attributed to the much greater freedom from interferences arising from capacitive currents, uncompensated resistance, and instrumental imperfections[79] compared with transient voltammetries. Microelectrodes provide one of the simplest methods for achieving steady-state voltammetry.

A.   Determination of Heterogeneous Electrode Kinetics

As in other voltammetries, the shape of a steady-state voltammogram depends on the degree of reversibility of the electrode process. In steady-state voltammetry with microelectrodes, the degree of reversibility is determined by comparing a "kinetic distance scale," D/k^0, of the electrode reaction with a "characteristic length"

$$\lambda = vA^{1/2} \tag{143}$$

of the microelectrode. As discussed in Section VI.A, v refers to the accessibility factor. If the dimension of the microelectrode sufficiently exceeds this kinetic distance, the electrode process behaves reversibly and its kinetic parameters are not measurable in the steady-state. However, if the size of the microelectrode is sufficiently small compared with λ, quasireversibility is ensured and the kinetic parameters can be measured in the steady state without any interference from charging currents. Thus, one has

$$
\begin{aligned}
D/k^0 \ll \lambda \quad &\text{(reversible)} \\
D/k^0 \approx \lambda \quad &\text{(quasireversible)} \\
D/k^0 \gg \lambda \quad &\text{(irreversible)}
\end{aligned}
\tag{144}
$$

Table 4. Classification of Voltammetric Regimes and the Kinetic Parameters That Are Experimentally Accessible

Regime	α	k^0	E^0
Reversible	Inaccessible	Inaccessible	Accessible
Quasireversible	Accessible	Accessible	Accessible
Irreversible	Accessible	Inaccessible, though a parameter involving both k^0 and E^0 is calculable.	

Microelectrodes of micrometer sizes are necessary; such sizes are readily accessible using modern fabrication technology, as will be found in references cited throughout Section II. By employing a range of microelectrode sizes, one is able to scan a range of kinetic distances which embrace reversible, quasireversible, and irreversible voltammetric behavior. As illustrated in Figure 11, the steady-state waves become less steep and shift along the potential axis as one moves from reversible to irreversible kinetics.

Determining the kinetics of an electrode reaction involves measuring the standard heterogeneous rate constant k^0 and the transfer coefficient α. In addition, knowledge of the standard potential E^0 may be necessary. The kinetic regime of the steady-state experiment determines which kinetic parameters, if any, are accessible, as detailed in Table 4.

Several methods have been reported for determining k^0, E^0, and α from steady-state voltammograms.[50,52,54,57–59,70,71,177] One method[57,58] relies on employing a range of microdisc radii in order to scan the reversible, quasireversible, or irreversible kinetic regimes. The experimentally determined quartile potentials $E_{1/4}$, $E_{1/2}$, and $E_{3/4}$ are either graphed or used to locate points on so-called "kinetic indicator diagrams," two of which are illustrated in Figure 20. The procedures in this method are not based on the usual assumptions of "uniform assessibility" to the microdisc or equal diffusion coefficients.

In a second method,[50] a steady-state voltammogram recorded at a microhemisphere is analyzed by plotting the logarithm of $[(i_{dif}/i) - 1]$ versus potential, as shown in Figure 21. The graph should be linear at either extremity. The transfer coefficient α is determined from the slope of the negative branch, while extrapolating the positive branch leads to E^0. The rate constant k^0 is then calculated from the gap between the two intercepts at $E = E^0$. An advantage of this method is that all three parameters may be found from one graph.

A third method[71] involves the use of tables for given values of $(E_{1/4} - E_{1/2})$ and $(E_{1/2} - E_{3/4})$. Experimentally, only one steady-state voltammogram is required from which $E_{1/4}$, $E_{1/2}$, and $E_{3/4}$ potentials have been determined.

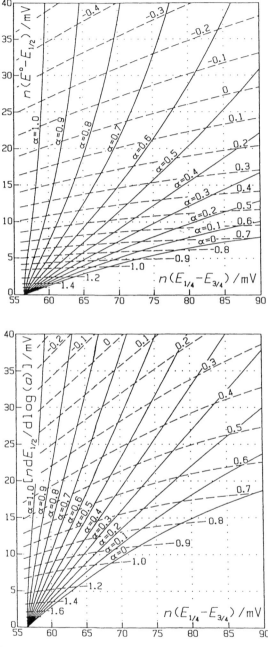

Figure 20. Kinetic-indicator diagrams which show the relationship between $n(E^0 - E_{1/2})$ and $nd E_{1/2}/d \log\{a\}$ with $n(E_{1/4} - E_{3/4})$ or $\log\{\pi k^0 a/4 D_R\}$ at constant α for steady-state voltammograms at microdisc electrodes. (From reference 57, with permission.)

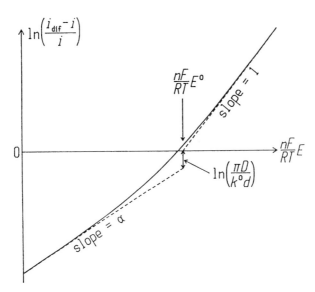

Figure 21. Method of analyzing a steady-state voltammogram of wave height i_{dif} for a quasireversible reaction at a microhemisphere of superficial diameter d. The kinetic parameters α, k^0, and E^0 may be obtained. (From reference 50, with permission.)

A fourth method involves curve-fitting procedures to calculate optimized values of α, k^0, and E^0.[59,70]

All of these procedures have been used successfully in extracting kintic information from steady-state voltammograms; a comprehensive list of specific examples may be found in Table 1 of reference 51.

B. Determination of Homogeneous Electrode Kinetics

Just as the kinetics of an electron-transfer reaction are accessible from steady-state studies at microelectrodes of a variety of sizes, kinetics of preceding or succeeding homogeneous reactions may be elucidated in a similar manner.[63,179,188] CE and EC mechanisms, catalytic follow-up processes, and ECE or DISP reactions have been categorized in Table 2 of reference 51. The analyses are most successful for first-order rate constants whose magnitude is comparable with D/r_0^2 for a microhemisphere and should be ideal for reactions with first-order rate constants in the 10^2–10^4 sec^{-1} range.

The simplest examples of the determination of homogeneous kinetics via steady-state voltammetry include those which take place at a microhemisphere of radius r_0 under conditions of excess supporting electrolyte and with reversible heterogeneous kinetics.[63] Analyses of steady-state voltammograms are shown in Figure 22 for the following mechanisms:

$$CE_r: \quad B \underset{k_{-1}}{\overset{k_1}{\rightleftarrows}} R \rightleftarrows P \tag{145}$$

$$E_rC: \quad R \rightleftarrows P \underset{k_{-1}}{\overset{k_1}{\rightleftarrows}} Z \tag{146}$$

$$E_rCE_r: \quad R \rightleftarrows P \underset{k_{-1}}{\overset{k_1}{\rightleftarrows}} Q \rightleftarrows Z \tag{147}$$

$$E_rC': \quad R \underset{\underset{k_{-1}}{\overset{\longleftarrow}{\rule{1.2cm}{0pt}}}}{\rightleftarrows} P \tag{148}$$

In the CE_r reaction (145), an electroinactive species B forms electroactive species R through first-order/pseudo-first-order kinetics. Electroactive R then undergoes a reversible electron transfer to the product P. R and P are initially absent from solution. The upper left graph of Figure 22 illustrates how the linear dependence of the limiting current i_L^{CE} on the radius of the microhemisphere can be analyzed to obtain homogeneous rate constants k_1, k_{-1}, and $k = k_1 - k_{-1}$ from the slope and intercept.

In the E_rC homogeneous reaction (146), the product P, formed by a preceding reversible electron transfer, undergoes first-order/pseudo-first-order kinetics to species Z. Only the electroreactant R is initially present in solution. The homogeneous reaction does not affect the height or shape of the reversible steady-state voltammogram; it does, however, influence the position of the voltammogram along the potential axis. The upper right graph of Figure 22 illustrates how the half-wave potential $E_{1/2}$ may be plotted as a logarithmic function of r_0 to determine the homogeneous rate constants k_1 and k_{-1} and their quotient $K = k_1/k_{-1}$.

For the E_rCE_r mechanism (147), the product P of a reversible electron-transfer reaction undergoes a homogeneous reaction to electroactive species Q which subsequently undergoes a further reversible electron-transfer reaction of species Z. Only the electroreactant R is initially present in solution. For this mechanism, both the limiting current i_L^{ECE} and half-wave potential are affected as the radius of the microhemisphere is varied. The central graphs of Figure 22 illustrate how this behavior can be linearized so that either approach may be used to determine the homogeneous rate constants k_1 and k_{-1}.

The E_rC' reaction (148) is known alternatively as "catalytic regeneration"[189] in which the product P of a reversible electron transfer is reconverted to reactant R by a homogeneous pseudo-first-order reaction of rate constant k_{-1}. As shown in the lower graph of Figure 22, the limiting current varies

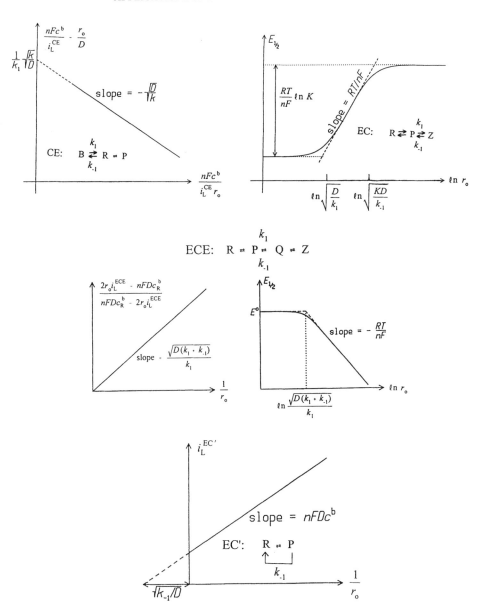

Figure 22. Methods of analyzing steady-state voltammograms at microhemispheres of various sizes in order to determine the kinetics of homogeneous reactions coupled to the electrode reaction. (Modified from reference 63, with permission.)

linearly with the reciprocal of the microhemispherical radius, thus allowing the catalytic rate constant k_{-1} to be determined.

C. Determination of Diffusion Coefficients

The diffusion coefficient of an electroreactant can be calculated directly from the steady-state diffusion-limiting current[66,78] defined by Eq. (16) with a value of the accessibility factor v appropriate for the microelectrode geometry. Diffusion coefficients measured in this way accrue greater accuracy due to the proportionality of i_{dif} to $DA^{1/2}$, in contrast to those measured using transient voltammetries in which the proportionality is with $AD^{1/2}$. An additional asset arises from the precision with which a constant current can be measured when compared to one that is changing with time.

D. Determination of Concentration

The diffusion-limiting current, Eq. (16), has also been used in the measurement of concentration.[45,190,191] Figure 23 demonstrates that the predicted linearity between the steady-state diffusion-limiting current and concentration holds

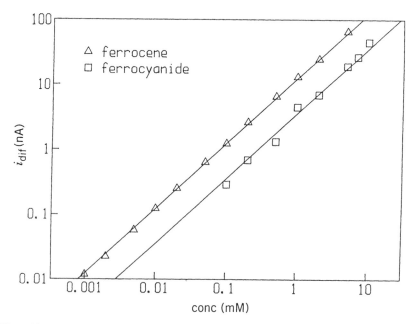

Figure 23. Wave heights of ferrocene and ferrocyanide oxidations versus concentration. (From reference 45, with permission.)

down to millimolar and micromolar levels.[45] Other studies[190,191] have achieved still lower detection limits.

IX. NOMENCLATURE

A	univalent anion
A	electrode surface area (m^2)
a	radius of inlaid disc electrode (m)
a	radius of lagoon (m)
a	outer radius of inlaid microring (m)
b	altitude above the basal plane of an electrode (m)
b	inner radius of inlaid microring (m)
c	concentration ($mol\,m^{-3}$)
D	diffusion coefficient ($m^2\,sec^{-1}$)
d	superficial diameter (m)
E	electrode potential (V)
e^-	an electron
F	Faraday's constant ($C\,mol^{-1}$)
$f(\ )$	shape-dependent function
$h(\ )$	heterogeneity function
i	current (A)
J	total flux ($mol\,sec^{-1}$)
j	flux density ($mol\,m^{-2}\,sec^{-1}$)
k	heterogeneous rate constant ($m\,sec^{-1}$)
k	homogeneous rate constant (sec^{-1})
L	depth of recess (m)
M	molar mass of mercury
m	integer
n	number of electrons
P	product species
P	perimeter of inlaid microring
Q	intermediate species in a homogeneous reaction
q	charge (C)
R	reactant species
R	gas constant ($8.3144\,JK^{-1}\,mol^{-1}$)
R	resistance (Ω)
r	radial distance measured from the center of an electrode (m)
T	thermodynamic temperature (K)
u	mobility ($m^2\,V^{-1}\,sec^{-1}$)
V	volume (m^3)
x	solute fraction

Z	arbitrary species in homogeneous reaction
z	charge number of R
α	transfer coefficient
$\alpha(\ )$	shape-dependent factor of sphere-cap electrode
$\beta(\ )$	shape-dependent factor of hemispheroid electrode
η	overvoltage (V)
κ	conductivity ($S\,m^{-1}$)
λ	characteristic length of a microelectrode (m)
ν	accessibility factor
ξ	potential dependent function
π	Archimedes constant (3.1416)
ρ	support ratio
ρ	density of mercury
ϕ	potential in solution (V)
ψ	undimensionalized potential

Superscripts

b	relates to the bulk solution
bulk	relates to the bulk solution
CE	pertaining to a CE homogeneous mechanism
disc	denotes inlaid disc
EC	pertaining to a EC homogeneous mechanism
ECE	pertaining to a ECE homogeneous mechanism
hemi	denotes hemisphere
0	relates to standard constants
oblate	denotes oblate hemispheroid
s	relates to the electrode surface
surface	relates to the electrode surface
$\rho = \infty$	denotes excess supporting electrolyte

Subscripts

A, C, P, R	denotes a property of that species
b	denotes backward electron transfer
dif	denotes a diffusion-limiting condition
f	denotes forward electron transfer
h	relates to aperature of lagoon electrode
i	denotes any solute species
irr	denotes irreversible
kin	denotes a kinetic-limiting condition
o	relates to radius of hemisphere

L	relates to a limiting condition
ohm	denotes ohmic
r	denotes reversible electron transfer
ring	denotes inlaid microring
S	denotes supporting electrolyte
ss	refers to steady-state condition
st	refers to static condition
t/d	denotes a joint thermodynamic and diffusion limiting condition
1	denotes forward direction in a homogeneous reaction
−1	denotes backward direction in a homogeneous reaction
1/4, 1/2, 3/4	denotes quartile values

ACKNOWLEDGMENT

The finanical assistance of the donors of the Petroleum Research Fund administered by the American Chemical Society is greatly appreciated.

REFERENCES

1. A. J. Bard and L. R. Faulkner, *Electrochemical Methods: Fundamentals and Applications*, John Wiley & Sons, New York (1980), Chapter 8.

2. C. M. A. Brett and A. M. C. F. O. Brett, in *Comprehensive Chemical Kinetics*, Vol. 26, C. H. Bamford and R. G. Compton, eds., Elsevier, New York (1986), Chapter 5.

3. R. S. Stojanovic, A. M. Bond, and E. C. V. Butler, *Anal. Chem.*, **62**, 2692–2697 (1990).

4. R. G. Compton, A. C. Fisher, M. H. Latham, C. M. A. Brett, and A. M. C. F. O. Brett, *J. Phys. Chem.*, **96**, 8363–8367 (1992).

5. J. V. Macpherson, S. Marcar, and P. R. Unwin, *Anal. Chem.*, **66**, 2175–2179 (1994).

6. I. Fatt, *Polarographic Oxygen Sensors*, CRC Press, Boca Raton, FL (1976).

7. P. Arquint, M. Koudelka-Hep, N. F. de Rooij, H. Buhler, and W. E. Morf, *J. Electroanal. Chem.*, **378**, 177–183 (1994).

8. O. C. Keller, S. Poitry, and J. Buffle, *J. Electroanal. Chem.*, **378**, 165–175 (1994).

9. M. A. Armitage, G. D. Brydon, P. J. J. Connelly, C. J. E. Farina, H. L. Gordon, K. B. Oldham, and T. Yanagi, *Can. J. Chem.*, **58**, 1966–1972 (1980).

10. A. T. Hubbard and F. C. Anson, in *Electroanalytical Chemistry*, Vol. 4, A. J. Bard, ed., Marcel Dekker, New York (1970), p. 129.

11. J. M. Davis, F.-R. F. Fan, and A. J. Bard, *J. Electroanal. Chem.*, **238**, 9–31 (1987).

12. M. V. Mirkin and A. J. Bard, *J. Electrochem. Soc.*, **139**, 3535–3539 (1992).

13. L. E. Topol and K. B. Oldham, *J. Phys. Chem.*, **73**, 1455 (1969).

14. C. L. Colyer, P. J. J. Connelly, H. L. Gordon, and K. B. Oldham, *Can. J. Chem.*, **66**, 1333–1341 (1988).

15. A. J. Bard, F.-R. F. Fan, and M. V. Mirkin, in *Electroanalytical Chemistry*, Vol. 18, A. J. Bard ed., Marcel Dekker, New York (1994), pp. 243–373.

16. P. R. Unwin and A. J. Bard, *J. Phys. Chem.*, **95**, 7814–7824 (1991).

17. F. Zhou, P. R. Unwin, and A. J. Bard, *J. Phys. Chem.*, **96**, 4917–4924 (1992).

18. D. A. Treichel, M. V. Mirkin, and A. J. Bard, *J. Phys. Chem.*, **98**, 5751–5757 (1994).

19. P. R. Unwin and A. J. Bard, *J. Phys. Chem.*, **96**, 5035–5045 (1992).

20. D. O. Wipf and A. J. Bard, *J. Electrochem. Soc.*, **138**, 469–474 (1991).

21. A. J. Bard, M. V. Mirkin, P. R. Unwin, and D. O. Wipf, *J. Phys. Chem.*, **96**, 1861–1868 (1992).

22. B. R. Horrocks, M. V. Mirkin, and A. J. Bard, *J. Phys. Chem.*, **98**, 9106–9114 (1994).

23. M. V. Mirkin, T. C. Richards, and A. J. Bard, *J. Phys. Chem.*, **97**, 7672–7677 (1993).

24. K. B. Oldham, in *Microelectrodes: Theory and Applications*, M. I. Montenegro, M. A. Queiros, and J. L. Daschbach, eds., NATO ASI Series, Vol. 197, Kluwer Academic Publishers, Boston (1991), pp. 83–99.

25. K. B. Oldham, *J. Electroanal. Chem.*, **237**, 303–307 (1987).

26. S. Bruckenstein, *Anal. Chem.*, **59**, 2098 (1987).

27. K. B. Oldham, *J. Electroanal. Chem.*, **250**, 1–21 (1988).

28. A. M. Bond, D. Luscombe, K. B. Oldham, and C. G. Zoski, *J. Electroanal. Chem.*, **249**, 1–14 (1988).

29. J. D. Norton, H. S. White, and S. W. Feldberg, *J. Phys. Chem.*, **94**, 6772–6780 (1990).

30. S. M. Drew, R. M. Wightman, and C. A. Amatore, *J. Electroanal. Chem.*, **317**, 117–124 (1991).

31. J. B. Cooper and A. M. Bond, *J. Electroanal. Chem.*, **315**, 143–160 (1991).

32. J. D. Norton, W. E. Benson, H. S. White, B. D. Pendley, and H. D. Abruna, *Anal. Chem.*, **63**, 1909–1914 (1991).

33. B. D. Pendley, H. D. Abrunã; J. D. Norton, W. E. Benson, and H. S. White, *Anal. Chem.*, **63**, 2766–2771 (1991).

34. M. Ciszkowska, Z. Stojek, S. E. Morris, and J. G. Osteryoung, *Anal. Chem.*, **64**, 2372–2377 (1992).

35. A. M. Bond and V. B. Pfund, *J. Electroanal. Chem.*, **335**, 281–295 (1992).

36. J. B. Cooper, A. M. Bond, and K. B. Oldham, *J. Electroanal. Chem.*, **331**, 877–895 (1992).

37. C. Lee and F. C. Anson, *J. Electroanal. Chem.*, **323**, 381–389 (1992).

38. A. M. Bond, S. W. Feldberg, H. B. Greenhill, P. J. Mahon, R. Colton, and T. Whyte, *Anal. Chem.*, **64**, 1014–1021 (1992).

39. J. B. Cooper and A. M. Bond, *Anal. Chem.*, **65**, 2724–2730 (1993).

40. M. Ciszkowska and Z. Stojek, *J. Electroanal. Chem.*, **344**, 135–143 (1993).

41. M. F. Bento, M. J. Medeiros, M. I. Montenegro, C. Beriot, and D. Pletcher, *J. Electroanal. Chem.*, **345**, 273–286 (1993).

42. C. Beriet and D. Pletcher, *J. Electroanal. Chem.*, **375**, 213–218 (1994).

43. M. Ciszkowska and J. G. Osteryoung, *J. Phys. Chem.*, **98**, 3194–3201 (1994).

44. R. T. Robertson and B. D. Pendley, *J. Electroanal. Chem.*, **374**, 173–177 (1994).

45. R. D. Lafleur, J. C. Myland, and K. B. Oldham, *Electroanalysis*, **2**, 223–228 (1990).

46. A. M. Bond, K. B. Oldham, and C. G. Zoski, *J. Electroanal. Chem.*, **245**, 71–104 (1988).

47. K. B. Oldham and C. G. Zoski, *J. Electroanal. Chem.*, **256**, 11–19 (1988).

48. C. G. Zoski, *J. Electroanal. Chem.*, **296**, 317–333 (1990).

49. C. G. Zoski, A. M. Bond, E. T. Allinson, and K. B. Oldham, *Anal. Chem.*, **62**, 37–45 (1990).

50. K. B. Oldham, in M. I. Montenegro, M. A. Queiros, and J. L. Daschbach, eds., *Microelectrodes: Theory and Applications*, NATO ASI Series, Vol. 197, Kluwer Academic Publishers, Boston (1991), pp. 35–50.

51. M. I. Montenegro, in R. G. Compton and G. Hancock, eds., *Research in Chemical Kinetics*, Elsevier, Amsterdam (1994), pp. 1–80.

52. A. Russell, K. Repka, T. Dibble, J. Ghoroghchian, J. J. Smith, M. Fleischmann, C. H. Pitt, and S. Pons, *Anal. Chem.*, **58**, 2961–2964 (1986).

53. Z. Galus, J. Golas, and J. Osteryoung, *J. Phys. Chem.*, **92**, 1103–1107 (1988).

54. T. Abe, K. Itaya, I. Uchida, K. Aoki, and K. Tokuda, *Bull. Chem. Soc. Jpn.*, **61**, 3417–3420 (1988).

55. R. B. Morris, K. F. Fisher, and H. S. White, *J. Phys. Chem.*, **92**, 5306–5313 (1988).

56. C. Amatore, B. Fosset, J. Bartelt, M. R. Deakin, and R. M. Wightman, *J. Electroanal. Chem.*, **256**, 255–268 (1988).

57. K. B. Oldham, J. C. Myland, C. G. Zoski, and A. M. Bond, *J. Electroanal. Chem.*, **270**, 79–101 (1989).

58. K. B. Oldham, C. G. Zoski, A. M. Bond, and D. A. Sweigart, *J. Electroanal. Chem.*, **248**, 467–473 (1988).

59. A. Owlia and J. F. Rusling, *Electroanalysis*, **1**, 141 (1989).

60. C. G. Phillips, *J. Electroanal. Chem.*, **296**, 255–258 (1990).

61. M. D. Pritzker, *J. Electroanal. Chem.*, **296**, 1–18 (1990).

62. D. R. Baker and M. W. Verbrugge, *J. Electrochem. Soc.*, **137**, 3836–3845 (1990).

63. K. B. Oldham, *J. Electroanal. Chem.*, **313**, 3–16 (1991).

64. A. S. Baranski, *J. Electroanal. Chem.*, **307**, 287–292 (1991).

65. D. R. Baker, M. W. Verbrugge, and J. Newman, *J. Electroanal. Chem.*, **314**, 23–44 (1991).

66. J. E. Baur and R. M. Wightman, *J. Electroanal. Chem.*, **305**, 73–81 (1991).

67. G. Denuault and D. Pletcher, *J. Electroanal. Chem.*, **305**, 131–134 (1991).

68. L. K. Safford and M. J. Weaver, *J. Electroanal. Chem.*, **312**, 69–96 (1991).

69. M. M. C. Dos Santos and M. L. S. Goncalves, *Electrochim. Acta*, **37**, 1413–1416 (1992).

70. R. L. Birke and Z. Huang, *Anal. Chem.*, **64**, 1513–1520 (1992).

71. M. V. Mirkin and A. J. Bard, *Anal. Chem.*, **64**, 2293–2302 (1992).

72. U. Kalapathy, D. E. Tallman, and S. Hagen, *J. Electroanal. Chem.*, **325**, 65–81 (1992).

73. O. R. Tutty, *J. Electroanal. Chem.*, **377**, 39–51 (1994).

74. O. R. Tutty, *J. Electroanal. Chem.*, **379**, 519–521 (1994).

75. T. Carofiglio, M. Magno, and I. Lavagnini, *J. Electroanal. Chem.*, **373**, 11–17 (1994).

76. M. D. Pritzker, *J. Electroanal. Chem.*, **373**, 39–52 (1994).

77. S. A. Olsen and D. E. Tallman, *Anal. Chem.*, **66**, 503–509 (1994).

78. M. B. Moressi and H. Fernandez, *J. Electroanal. Chem.*, **369**, 153–159 (1994).

79. A. M. Bond, K. B. Oldham, and C. G. Zoski, *Anal. Chim. Acta*, **216**, 177–230 (1989).

80. M. Fleischmann, S. Pons, D. R. Rolison, and P. P. Schmidt, *Ultramicroelectrodes*, Datatech Systems, Science Publishers (1987).

81. R. M. Wightman and D. O. Wipf, in *Electroanalytical Chemistry*, Vol. 15, A. J. Bard, ed., Marcel Dekker, New York (1989), pp. 267–353.

82. C. D. Baer, N. J. Stone, and D. A. Sweigart, *Anal. Chem.*, **60**, 188–191 (1988).

83. J. W. Pons, J. Daschbach, S. Pons, and M. Fleischmann, *J. Electroanal. Chem.*, **239**, 427–431 (1988).

84. Z. Stojek and J. Osteryoung, *Anal. Chem.*, **61**, 1305–1308 (1989).

85. R. M. Penner, M. J. Heben, and N. S. Lewis, *Anal. Chem.*, **61**, 1630–1636 (1989).

86. A. J. Bard, G. Denuault, C. Lee, D. Mandler, and D. O. Wipf, *Acc. Chem. Res.*, **23**, 357–363 (1990).

87. B. D. Pendley and H. D. Abruna, *Anal. Chem.*, **62**, 782–784 (1990).

88. C. P. Smith, H. L. Kennedy, H. J. Kragt, H. S. White, and J. F. Biegen, *Anal. Chem.*, **62**, 1135–1138 (1990).

89. D. O. Wipf and A. J. Bard, *J. Electrochem. Soc.*, **138**, L4–L6 (1991).

90. S. T. Yau, D. Saltz, A. Wriekat, and M. H. Nayfeh, *J. Appl. Phys.*, **69**, 2970–2974 (1991).

91. C. Lee, C. J. Miller, and A. J. Bard, *Anal. Chem.*, **63**, 78–83 (1991).

92. N. Casillas, S. R. Snyder, and H. S. White, *J. Electrochem. Soc.*, **138**, 641–642 (1991).

93. S. P. Kounaves and W. Deng, *J. Electroanal. Chem.*, **301**, 77–85 (1991).

94. M. V. Mirkin, F. R.-F. Fan, and A. J. Bard, *J. Electroanal. Chem.*, **328**, 47–62 (1992).

95. R. L. Birke, *J. Electroanal. Chem.*, **274**, 297–304 (1989).

96. *Microelectrodes: Theory and Applications*, M. I. Montenegro, M. A. Queiros, and

J. L. Daschbach, eds., NATO ASI Series, Vol. 197, Kluwer Academic Publishers, Boston (1991).

97. R. M. Wightman, *Anal. Chem.*, **53**, 1125A–1134A (1981).

98. A. M. Bond, M. Fleischmann, S. B. Khoo, S. Pons, and J. Robinson, *Indian J. Technol.*, **24**, 492–500 (1986).

99. S. Pons and M. Fleischmann, *Anal Chem.*, **59**, 1391A–1399A (1987).

100. J. O. Howell, *Curr. Sep.*, **8**, 2–16 (1987).

101. R. M. Wightman, *Science*, **240**, 415–420 (1988).

102. K. B. Oldham, *J. Electroanal. Chem.*, **323**, 53–76 (1992).

103. K. R. Wehmeyer, M. R. Deakin, and R. M. Wightman, *Anal. Chem.*, **57**, 1913 (1985).

104. P. M. Kovach, W. L. Caudill, D. G. Peters, and R. M. Wightman, *J. Electroanal. Chem.*, **185**, 285 (1985).

105. A. M. Bond, T. L. E. Henderson, and W. Thorman, *J. Phys. Chem.*, **90**, 2911 (1986).

106. S. Coen, D. K. Cope, and D. E. Tallman, *J. Electroanal. Chem.*, **215**, 29 (1986).

107. M. R. Deakin, R. M. Wightman, and C. A. Amatore, *J. Electroanal. Chem.*, **215**, 49 (1986).

108. A. Szabo, D. K. Cope, D. E. Tallman, P. M. Kovach, and R. M. Wightman, *J. Electroanal. Chem.*, **217**, 417 (1987).

109. D. K. Cope, C. H. Scott, U. Kalapathy, and D. E. Tallman, *J. Electroanal. Chem.*, **280**, 27–35 (1990).

110. J. D. Seibold, E. R. Scott, and H. S. White, *J. Electroanal. Chem.*, **264**, 281–289 (1989).

111. M. Samuelsson, M. Armgarth, and C. Nylander, *Anal. Chem.*, **63**, 931–936 (1991).

112. J.-L. Ponchon, R. Cespuglio, F. Gonon, M. Jouvet, and J. F. Pujol, *Anal. Chem.*, **51**, 1483 (1979).

113. K. Aoki, K. Honda, K. Tokuda, and H. Matsuda, *J. Electroanal. Chem.*, **182**, 267 (1985).

114. C. A. Amatore, M. R. Deakin, and R. M. Wightman, *J. Electroanal. Chem.*, **206**, 23 (1986).

115. K. Aoki, K. Honda, K. Tokuda, and H. Matsuda, *J. Electroanal. Chem.*, **195**, 51 (1985).

116. K. Aoki, K. Tokuda, and H. Matsuda, *J. Electroanal. Chem.*, **206**, 47 (1986).

117. S. T. Singleton, J. J. O'Dea, and J. Osteryoung, *Anal. Chem.*, **61**, 1211–1215 (1989).

118. M. M. Murphy, J. J. O'Dea, and J. Osteryoung, *Anal. Chem.*, **63**, 2743–2750 (1991).

119. C. Amatore, J. M. Saveant, and D. Tessier, *J. Electroanal. Chem.*, **147**, 39 (1983).

120. R. M. Penner and C. R. Martin, *Anal. Chem.*, **59**, 2625 (1987).

121. W. Thormann, P. van den Bosch, and A. M. Bond, *Anal. Chem.*, **57**, 2764–2770 (1985).

122. C. G. Phillips, *J. Electrochem.*, **139**, 2222–2230 (1992).

123. A. Szabo and R. Zwanzig, *J. Electroanal. Chem.*, **314**, 307–311 (1991).

124. K. Aoki, M. Morita, O. Niwa, and H. Tabei, *J. Electroanal. Chem.*, **256**, 269–282 (1988).

125. K. Tokuda, K. Morita, and Y. Shimizu, *Anal. Chem.*, **61**, 1763–1768 (1989),

126. H. P. Wu, *Anal. Chem.*, **65**, 1643–1646 (1993).

127. W. Peng and E. Wang, *Anal. Chem.*, **65**, 2719–2723 (1993).

128. M. Ciszkowska, M. Donten, and Z. Stojek, *Anal. Chem.*, **66**, 4112–4115 (1994).

129. G. A. Brydon and K. B. Oldham, *J. Electroanal. Chem.*, **122**, 353 (1981).

130. G. Hills, A. K. Pour, and B. Scharifker, *Electrochim. Acta*, **28**, 891–898 (1983).

131. C. L. Colyer, D. Luscombe, and K. B. Oldham, *J. Electroanal. Chem.*, **283**, 379–387 (1990).

132. S. P. Kounaves and W. Deng, *J. Electroanal. Chem.*, **301**, 77–85 (1991).

133. K. R. Wehmeyer and R. M. Wightman, *Anal. Chem.*, **57**, 1989–1993 (1985).

134. J. Golas and J. Osteryoung, *Anal. Chim. Acta*, **181**, 211–218 (1986).

135. J. Golas and J. Osteryoung, *Anal. Chim. Acta*, **186**, 1–9 (1986).

136. C. L. Colyer, K. B. Oldham, and S. Fletcher, *J. Electroanal. Chem.*, **290**, 33–48 (1990).

137. K. B. Oldham, *J. Electroanal. Chem.*, **284**, 491–497 (1990).

138. J. C. Myland and K. B. Oldham, *J. Electroanal. Chem.*, **288**, 1–14 (1990).

139. P. A. Bobbert, M. M. Wind, and J. Vlieger, *Physica*, **141A**, 58 (1987).

140. A. C. West and J. Newman, *J. Electrochem. Soc.*, **138**, 1620–1625 (1991).

141. A. S. Baranski, *J. Electroanal. Chem.*, **307**, 287–292 (1991).

142. K. B. Oldham, *Anal. Chem.*, **64**, 646–651 (1992).

143. D. R. MacFarlane and D. K. Y. Wong, *J. Electroanal. Chem.*, **185**, 197–202 (1985).

144. Y.-T. Kim, D. M. Scarnulis, and A. G. Ewing, *Anal. Chem.*, **58**, 1782–1786 (1986).

145. U. Kalapathy, D. E. Tallman, and D. K. Cope, *J. Electroanal. Chem.*, **285**, 71–77 (1990).

146. R. A. Saraceno and A. G. Ewing, *Anal. Chem.*, **60**, 2016–2020 (1988).

147. R. A. Saraceno and A. G. Ewing, *J. Electroanal. Chem.*, **257**, 83–93 (1988).

148. L. S. Kuhn, A. Weber, and S. G. Weber, *Anal. Chem.*, **62**, 1631–1636 (1990).

149. M. Fleischmann, S. Bandyopadhyay, and S. Pons, *J. Phys. Chem.*, **89**, 5537–5541 (1985).

150. A. Szabo, *J. Phys. Chem.*, **91**, 3108–3111 (1987).

151. K. Aoki, *J. Electroanal. Chem.*, **281**, 29–40 (1990).

152. K. T. Kawagoe, J. A. Jankowski, and R. M. Wightman, *Anal. Chem.*, **63**, 1589–1594 (1991).

153. M. V. Mirkin, F. R. F. Fan, and A. J. Bard, *Science*, **257**, 364–366 (1992).

154. H. Y. Liu, F. R. F. Fan, C. W. Lin, and A. J. Bard, *J. Am. Chem. Soc.*, **108**, 3838–3839 (1986).

155. A. J. Arvia, *Surf. Sci.*, **181**, 78–91 (1987).

156. F. R. F. Fan and A. J. Bard, *J. Phys. Chem.*, **94**, 3761–3766 (1990).

157. H. Chang and A. J. Bard, *Langmuir*, **7**, 1143–1153 (1991).

158. L. Vazquez, A. H. Creus, P. Carro, P. Ocon, P. Herrasti, C. Palacio, J. M. Vara, R. C. Salvarezza, and A. J. Arvia, *J. Phys. Chem.*, **96**, 10454–10460 (1992).

159. C. G. Zoski, A. M. Bond, C. L. Colyer, J. C. Myland, and K. B. Oldham, *J. Electroanal. Chem.*, **263**, 1–21 (1989).

160. J. R. Delmastro and D. E. Smith, *J. Phys. Chem.*, **71**, 2138 (1967).

161. Y. Saito, *Rev. Polarogr. Jpn.*, **15**, 177 (1968).

162. J. Tomeš, *Collect. Czech. Commun.*, **9**, 50 (1937).

163. A. M. Bond and K. B. Oldham, *J. Electroanal. Chem.*, **158**, 193 (1983).

164. M. Goto and K. B. Oldham, *Anal. Chem.*, **48**, 1617 (1976).

165. M. Fleischmann and S. Pons, *J. Electroanal. Chem.*, **222**, 107 (1987).

166. D. K. Cope and D. E. Tallman, *J. Electroanal. Chem.*, **285**, 79 (1990).

167. K. B. Oldham and C. G. Zoski, in *Comprehensive Chemical Kinetics*, Vol. 26, C. H. Bamford and R. G. Compton, eds., Elsevier, New York (1986), Chapter 2.

168. Z. Stojek, M. Ciszkowska, and J. G. Osteryoung, *Anal. Chem.*, **66**, 1507–1512 (1994).

169. M. Ciszowska and J. G. Osteryoung, *J. Phys. Chem.*, **98**, 3194–3201 (1994).

170. K. B. Oldham, *J. Electroanal. Chem.*, **337**, 91–126 (1992).

171. J. C. Myland and K. B. Oldham, *J. Electroanal. Chem.*, **347**, 49–91 (1993).

172. C. P. Smith and H. S. White, *Anal. Chem.*, **65**, 3343–3353 (1993).

173. J. Koryta and J. Dvořák, in *Principles of Electrochemistry*, John Wiley & Sons, Chichester (1987), p. 91.

174. S. Bruckenstein, *Anal. Chem.*, **59**, 2098 (1987).

175. N. Ibl, in *Comprehensive Treatise of Electrochemistry*, Vol. 6, E. Yeager, J. O'M. Bockris, B. E. Conway, and S. Saragapani, eds., Plenum, New York (1986), p. 42.

176. T. R. Brumleve and R. P. Buck, *J. Electroanal. Chem.*, **90**, 1 (1978).

177. Z. Galus, J. Golas, and J. Osteryoung, *J. Phys. Chem.*, **92**, 1103–1107 (1988).

178. M. A. Dayton, A. G. Ewing, and R. M. Wightman, *Anal. Chem.*, **52**, 2392 (1980).

179. M. Fleischmann, F. Lassere, J. Robinson, and D. Swan, *J. Electroanal. Chem.*, **177**, 97 (1984).

180. M. Fleischmann, F. Lassere, and J. Robinson, *J. Electroanal. Chem.*, **177**, 115 (1984).

181. G. Denuault, M. Fleischmann, D. Pletcher, and O. R. Tutty, *J. Electroanal. Chem.*, **280**, 243–254 (1990).

182. C. G. Phillips, *J. Electroanal. Chem.*, **296**, 255–258 (1990).

183. Q. Zhuang and H. Chen, *J. Electroanal. Chem.*, **346**, 29–51 (1993).

184. I. Lavaganini, P. Pastore and F. Magno, *J. Electroanal. Chem.*, **358**, 193–201 (1993).

185. T. Carofiglio, F. Magno, and I. Lavagnini, *J. Electroanal. Chem.*, **373**, 11–17 (1994).

186. S. Dong, and G. Che, *Electrochim. Acta*, **37**, 2587–2589 (1992).

187. G. Che and S. Dong, *Electrochim. Acta*, **37**, 2695–2699 (1992).

188. G. Che and S. Dong, *Electrochim. Acta*, **37**, 2701–2705 (1992).

189. A. J. Bard and L. R. Faulkner, *Electrochemical Methods: Fundamentals and Applications*, John Wiley & Sons, New York (1980), Chapter 11.

190. J. W. Bixler, A. M. Bond, P. A. Lay, W. Thormann, P. van den Bosch, M. Fleischmann, and S. Pons, *Anal. Chim. Acta*, **187**, 67 (1986).

191. J. W. Bixler and A. M. Bond, *Anal. Chem.*, **52**, 2859 (1986).

CHAPTER

7

SIMULATION ANALYSIS OF ELECTROCHEMICAL MECHANISMS

DAVID K. GOSSER, Jr.

Department of Chemistry
City College of the City University of New York
New York, New York 10031

Cyclic voltammetry and other electrochemical experiments can be utilized for qualitative and quantitative analysis of mechanisms involving electron transfer. An exploration of the EC mechanism (electrode electron transfer followed by chemical reaction) illustrates the basic principles underlying most electrochemical mechanisms. The notion of reversibility in electrochemistry relates the rates of electrode electron transfer, diffusion, and chemical reacion. Mechanistic analysis employing computer simulation of electrochemical experiments can overcome the limitations of analytic methods. The principles of

Modern Techniques in Electroanalysis, Edited by Petr Vanýsek, Chemical Analysis Series, Vol. 139.
ISBN 0-471-55514-2 © 1996 John Wiley & Sons, Inc.

simulation on the basis of the method of fractional steps are outlined, and a sample tutorial program is provided. After a brief description of CVSIM,[1-3] a general program for the simulation and analysis of electrochemical experiments, the stability and efficiency of the program is discussed. The program CVFIT utilizes nonlinear curve fitting in conjuction with CVSIM. An application of a CVFIT analysis to an actual EC mechanism reveals that although the voltammogram appears irreversible, it is in fact possible to determine the reduction potential. Many of the points raised in this discussion will be applicable to the use of simulation programs that use other mathematical approaches.[4,5]

I. COMPONENTS OF ELECTROCHEMICAL MECHANISMS

Figure 1 shows a schematic of the fundamental physical processes occurring during an electrochemical experiment: electron transfer at the electrode, diffusion, and coupled chemical reactions. Although in the scheme discussed here adsorption is disregarded, it is not presumed to be unimportant. To date, simulation methods lack a physically realistic and general model (one which includes both competitive adsorption and a realistic isotherm), a situation which should be remedied.

Each of the processes taking place near at or near the electrode (electron transfer, diffusion, and chemical reactions) can be characterized by a rate term. In the case of electron transfer and diffusion, the rate has the nature of a flux

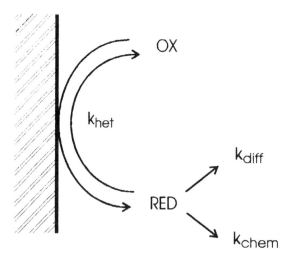

Figure 1. Scheme for electrode reactions.

and has units commonly expressed as $cm\,sec^{-1}$. The chemical rate term is a volume-based rate, the units of which depend on whether the reaction is first- (sec^{-1}) or second-order ($M\,sec^{-1}$). The shape of a cyclic voltammetric response reflects the processes and their relative rates. The terms reversible, irreversible, and quasireversible are used to characterize the essential character of cyclic voltammograms. In general, these terms refer to the relative rate of the electrode electron transfer as compared to either the diffusion rate or the rate of a coupled chemical reaction.

Let us take the simple EC mechanism to illustrative the essential character of electrode mechanisms. All electrode mechamisms (excluding adsorption), however complex, could be reduced to components which are one electron transfer and/or a chemical reaction:

$$Ox + e^- \rightleftarrows Red \qquad E^{0'}, k^0, \text{ and } \alpha$$

$$Red \Rightarrow Products \qquad k_{chem}$$

$$D_{ox} \approx D_{red} = D$$

The reduction potential, which might well be called the *solution phase electron affinity*, is characteristic of each substance and is determined by the lowest occupied molecular orbital energy* and the solvation energy change upon addition of an electron. For a reduction, the reduction potential with respect to the hydrogen reference is estimated as

$$FE^0 = -\Delta G_{EA} - \delta\Delta G_{Solv} - 4.4F \qquad \text{(see Figure 2)}$$

the contribution from the electron affinities [reflecting the lowest unoccupied molecular orbitals (LUMOs)], the change in solvation energies, and a term to relate the hydrogen reference to the vacuum scale (with which the electron affinity and solvation energy terms are taken in reference to).

In a potential controlled electrochemical experiment, a potential controlled rate of electron transfer is effected through the Butler–Volmer equation:

$$\frac{i}{nFA} = k^0 C_{ox} \exp\left[\frac{-\alpha nF}{RT}(E - E^{0'})\right] - C_{red}\exp\left[\frac{(1-\alpha)nF}{RT}(E - E^{0'})\right] \quad (1)$$

The Butler–Volmer equation is a linear free energy relationship very similar to such relationships established for rates of organic reactions.[6] The Marcus theory,[7] which predicts a quadratic free energy relationship for electron

---

* Actually, one is making a Koopman-like assumption here—that is, that the energy levels of the molecule do not change upon reduction.

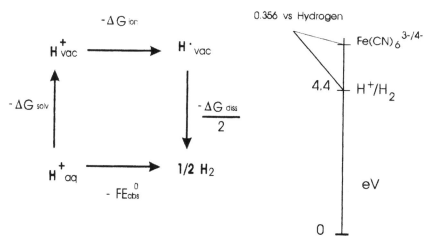

Figure 2. Cycle to establish the absolute reduction potential for hydrogen.

transfer, reduces to the Butler–Volmer equation by taking only linear terms. In fact, the *inverted region*, where the electron transfer rate is predicted to decrease with increasing negative free energy change, is not expected to occur for electrode reactions because of the availability of electrons from the continuim of energy levels in the electrode (Figure 3). The transfer coefficient, α, is particularly informative for reactions in which the electron transfer is concerted with bond breaking.[8] In this case the heterogeneous rate constant is slow, because the bond breaking part of the activation process, and the transfer coefficient is smaller than 0.5. Such cases have been observed, for instance, in the reduction of alkyl halides[8] and the reduction of peroxide species.[9]

It can be shown[3] that the following semiquantitative formulas allow us to compare the rates of electrode electron transfers and the rates of diffusion and chemical reaction:

$$k_D = 2.18(vD)^{1/2} \tag{2}$$

$$k_c^* = \frac{k_c(2D)^{1/2}}{10^3} \tag{3}$$

where k_c^* is the heterogeneous equivalent of the rate of a homogeneous chemical reaction. Consider an EC mechanism where the heterogeneous rate constant is 10^{-5} cm sec^{-1}, the diffusion coefficient is $D = 10^{-5}$ cm^2 sec^{-1}, the scan rate is 1.0 V sec^{-1}, and the following chemical reaction rate is 1000 sec^{-1}; then the electrode mechanism is deemed irrisible by virtue of the fact that the

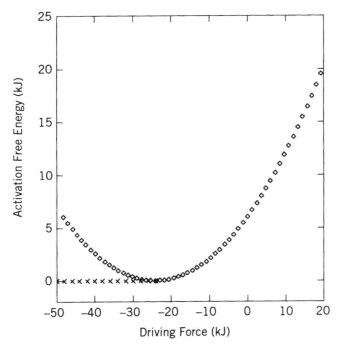

Figure 3. Marcus theory predicts an inverted region (diamonds) but not for electrode reactions (x's).

chemical rate and diffusion rate are both larger than the electron transfer rate by several orders of magnitude.

II. SIMULATION OF ELECTROCHEMICAL PROCESSES BY THE METHOD OF FRACTIONAL STEPS

The method of fractional steps, or operator splitting, is an approach to the numerical solution of partial differential equations.[10] It is suited to the solution of problems that have several terms, such as the diffusive-reactive problem of electrochemistry.

Consider an equation of the type

$$\frac{\partial u}{\partial t} = Pu \tag{4}$$

where P is a generalized set of m additive operators

$$P = P_1 u + P_2 u + P_3 u \cdots P_m u$$

If, for each of the operators, there is a differencing scheme

$$u^{n+1} = F_1(u^n, \Delta t)$$
$$u^{n+1} = F_2(u^n, \Delta t)$$
$$u^{n+1} = F_3(u^n, \Delta t)$$
$$\cdots$$
$$u^{n+1} = F_m(u^n, \Delta t)$$

a fractional step method realizes the $n+1$ value of u with the following sequence:

$$u^{n+(1/m)} = F_1(u^n, \Delta t)$$
$$u^{n+(2/m)} = F_2(u^{n+(1/m)}, \Delta t)$$
$$\cdots$$
$$u^{n+1} = F_m(u^{n+(m-1)/m}, \Delta t)$$

Let us examine how this applies to the simulation of the cyclic voltammetric experiment by using the explicit finite differences for the diffusion term and a second-order Runge–Kutta method for the chemical kinetic term.

The partial differential equation that we want to solve has Fick's second law describing diffusion and a generalized term for chemical reaction kinetics:

$$\frac{\partial C_q}{\partial t} = D_q \frac{\partial^2 C_q}{\partial x^2} \pm \sum k_g C_{g_1} C_{g_2} \tag{5}$$

where C_q is the concentration of species q, k_n is the chemical rate constant for the nth solution chemical reaction, and C_{g_1} and C_{g_2} are the concentrations for the first and second species for the nth solution chemical reaction rate expression.

Fick's second law can be converted to discrete form through and application of Taylor series approximations of a function around a point in time. The "forward" and "backward" series are

$$C(t + \Delta t) = C(t) + \Delta t \frac{\partial C}{\partial t} + \frac{1}{2} \Delta t^2 \frac{\partial^2 C}{\partial C^2} \tag{6}$$

$$C(t - \Delta t) = C(t) - \Delta t \frac{\partial C}{\partial t} + \frac{1}{2} \Delta t^2 \frac{\partial^2 C}{\partial C^2} \tag{7}$$

Adding the two series, we arrive at

$$\frac{\partial^2 C}{\partial x^2} = \frac{D[C_{i-1} - 2C_i + C_{i+1}]}{\Delta x^2} \tag{8}$$

This is the discrete version of the right side of Fick's second law, where i is an index that keeps track of the spatial points.

Considering only the first two terms of the forward series and introducing the time index j, we obtain

$$\frac{\partial C}{\partial t} = \frac{[C_{j+1} - C_j]}{\Delta t} \tag{9}$$

the left side of Fick's law. Taken together, we have

$$C_{i,j+1} = C_{i,j} + \frac{D\delta t}{\delta x^2}[C_{i+1,j} - 2C_{i,j} + C_{i-1,j}] \tag{10}$$

the concentration in the next time element given the previous concentration and its neighbors—the discrete version of Fick's law. A grid of points can be constructed on the basis equation (see Figure 4). Each spatial point represents the center of a spatial element of width δx. In any one time increment, δt, diffusion only propagates to neighboring volume elements. This brings in

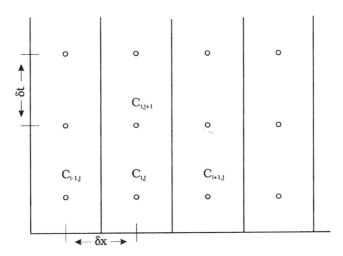

Figure 4. Grid established by the explicit finite difference method.

a constraint to Eq. (10)—that is,

$$\frac{D_k \delta t}{\delta x^2} \leqslant 0.5 \tag{11}$$

The experiment takes place over a total time (t_{exp}), so that the value of δt

$$\delta t = \frac{t_{exp}}{nt} \tag{12}$$

will depend on the number of time increments, nt. For a given value of D, if nt is chosen, time δx is fixed.

t_{exp} also fixes the total size (X) of the space grid. Nearly the entire diffusion layer can be enclosed in about four times the mean diffusion length.

$$X = 6\sqrt{Dt_{exp}} \tag{13}$$

Then the number of spatial elements, ns, is

$$ns = \frac{X}{\delta x} \tag{14}$$

The distance between the center of the first gridpoint and the surface of the electrode is set here to $\frac{1}{2}\delta x$ (Figure 5). The current flux, $J = -i/nFA$, is determined by the concentration gradients of the oxidized and reduced species by the Butler–Volmer equation:

$$J_{ox} = \frac{-2D_{ox}[C_{ox,1} - C_{ox,0}]}{\delta x} \tag{15}$$

$$J_{red} = -\frac{2D_{red}[C_{red,1} - C_{red,0}]}{\delta x} \tag{16}$$

$$-J_{ox} = k_1 C_{ox,0} - k_{-1} C_{red,0} \tag{17}$$

where $J_{ox} = -J_{red}$.

Solving Eqs. (15) and (16) for C_{ox} and C_{red} and substituting these values into the Butler–Volmer equation results in the following expression for the flux:

$$-J_{ox} = \frac{k_1 C_{ox,1} - k_{-1} C_{red,1}}{1 + k_1 \dfrac{\delta x}{2D_{ox}} + k_{-1} \dfrac{\delta x}{2D_{red}}} \tag{18}$$

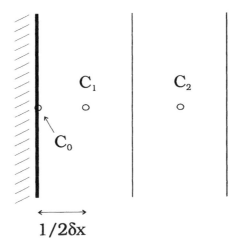

Figure 5. Grid near the electrode surface.

The flux expression can be incorporated into an expression similar to Eq. (6). The flux ($mol\,cm^{-2}\,sec^{-1}$) is converted to a concentration change by multiplying through by the appropriate dimensional factors:

$$C_{1,j+1} = C_{1,j} + \frac{D\delta t}{\delta x^2}[C_{1,j} - C_{0,j}] + J\frac{\delta t}{\delta x} \tag{19}$$

All the variables can be made dimensionless by normalizing them to a standard value. For instance (denoting the dimensionless unit with a double dagger),

$$t^{\ddagger} = \frac{t}{\delta t} \tag{20}$$

$$\delta x^{\ddagger} = \frac{x}{\delta x} \tag{21}$$

$$C^{\ddagger} = \frac{C}{C_n} \tag{22}$$

$$D^{\ddagger} = \frac{D}{D_n} < 0.5 \tag{23}$$

where δx and δt are the time and space increments of the simulation, C_n is a chosen normalizing concentration (typically the concentration of the principle electroactive species), and D_n is a chosen normalizing diffusion coefficient.

In the calculations, all units must be consistent. The heterogeneous rate constant is converted to dimensionless form by multiplying through by the approprate simulation units:

$$k^{\ddagger} = k \frac{\delta t}{\delta x} \tag{24}$$

The equation for diffusion from the first spatial cell written in dimensionless units is

$$C_{i,j+1}^{\ddagger} = C_{i,j}^{\ddagger} + D^{\ddagger} \frac{\delta t^{\ddagger}}{\delta x^{\ddagger 2}} = [C_{i-1,j}^{\ddagger} - C_{i,j}^{\ddagger}] + J^{\ddagger} \frac{\delta t^{\ddagger}}{\delta x^{\ddagger}} \tag{25}$$

We get back to real units by dimensional analysis, by noting that

$$J = J^{\ddagger} C_n \frac{\delta x}{\delta t} \tag{26}$$

According to the method of fractional steps, we can solve the differential equation corresponding to the chemical reaction kinetics, separately, utilizing the results (concentrations) obtained from the solution of the diffusion equation. Runge–Kutta methods provide general and robust solutions for sets of ordinary differential equations. We employ an iteration on the second-order Runge–Kutta method to solve a first-order chemical reaction. The maximum useful iteration is about 4. For the rate law

$$\frac{\delta C}{\delta t} = kC$$

the solution is

$$C_n = C_0 + \frac{k\delta t}{2} [C_0 + C_{n-1}]$$

where $n = 0, 1, 2, 3\ldots$ is the iteration number, C_0 is the initial concentration, and C_n is the nth concentration. This approach can be generalized for sets of first- or second-order chemical reactions.

A.   A Simple Simulation Program

CV, an interactive program in the Turbo Pascal program, will simulate cyclic voltammograms based upon the method of fractional steps, which includes a second-order Runge–Kutta method for the solution of the chemical kinetics.

```pascal
Program CV;
{A sample program, using the method of fractional steps, with Runge–Kutta for
chemical kinetics}
{$N + } {enable numeric coprocessor}
Uses Crt;
Var
nt, ns, k, a, b, r, s, i: longint;
pot, ipot, spot, fpot, T, X: extended;
delx, delp, delt, E, scanr, area: extended;
Current, khet, kf, kr, kchem, rsc: extended;
C, Ct, Ctemp: array[1..3,1..200] of extended;
J: array[1..3] of extended;
outfile: text;

Procedure Setup;
begin
    write ('What is the reduction potential in Volts?:');
    readln (E);
    write ('What is the heterogeneous rate constant?');
    readln(khet);
    writeln('transfer coefficient set to 0.5')
    write ('What is the k (sec − 1) of the following reaction?:');
    readln(kchem);
    write ('What is the initial potential?:');
    readln(ipot);
    write ('What is the switching potential?:');
    readln (spot);
    writeln ('final potential = initial potential');
    fpot:= ipot;
    write ('What is the scan rate?:');
    readln (scanr);
    T:= 2*abs(spot-ipot)/scanr; {Time of experiment}
    X:= 6*sqrt(1E-5*T); {The diffusion layer assuming D = 1E-5}
    nt:= 800; {A default number of time increments}
    area:= 0.01;
    delt:= T/nt;
```

```
    if kchem > 100 then begin
    delt:= 0.3/kchem; {if kchem high, reset time increment}
    nt:= trunc(T/delt);
    end;
    kchem:= kchem*delt; {make it dimensionless}
    delx:= sqrt(1E-5*delt/0.45);
    ns:= trunc(X/delx);
    khet:= khet*delt/delx; {make it dimensionless}
For a:= 1 to ns + 1 do
    begin
    C[1,a]:= 1.0;
    C[2,a]:= 0.00;
    C[3,a]:= 0.00;
    Ctemp[1,a]:= 1.00;
    Ctemp[2,a]:= 0.00;
    Ctemp[3,a]:= 0.00;
    end;
    for a:= 1 to 3 do
    begin
    j[a]:= 0.0
    end;
Delp:= 2*(Spot − ipot)/nt;
pot:= ipot + delp;
assign (outfile, 'data.pas');
rewrite (outfile);
end;


Procedure Electrode;
Begin
kf:= khet*exp(− 19.46*(pot − E));
kr:= khet*exp(19.46*(pot − E));
J[1]:= (kf*C[1,1] − kr*C[2,1])/(1 + kf/0.9 + kr/0.9);
J[2]:= − J[1]; J[3]:= 0.00;
Current:= J[1]*(delx/delt)*96484*1E-6*area; {This is i/Area}
writeln(outfile,pot:9:5,' ',current:12);
end;


Procedure Diffusion:
Begin
For k:= 1 to 3 do
begin
    Ctemp[k,1]:= C[k,1] + 0.45* (C[k,2] − C[k,1]) − J[k];
    For b:= 2 to ns do
begin
```

```
Ctemp[k,b]:= C[k,b] + 0.45*(C[k,b − 1] − 2.0*C[k, b] + C[k,b + 1]);
end;
end;
end;

Procedure Chemreact;
Begin
If kchem > 0.00 then begin
   for i:= 1 to ns do
   ct[2,i]:= ctemp[2,i];
   For k:= 1 to ns do
      begin
         Ctemp[2,k]:= Ct[2,k] − 0.5*kchem*(Ct[2,k] + ctemp[2,k]);
      end; end;
      end;

{****************** The Main Program **************}
begin
clrcr;
Writeln('CV Simulation of EC Mechanism');
writeln;

setup;
for a:= 1 to nt do
   begin
      for s:= 1 to 3 do
         begin
            for r:= 1 to ns do
               begin
         rsc:= 550/ns;
      C[s,r]:= Ctemp[s,r];
      end;
      end;
   electrode;
   diffusion;
   chemreact;
   if a < nt/2 then pot:= pot + delp;
   if a > = nt/2 then pot:= pot − delp;
   end; close(outfile); writeln('simulation finished');
   end.
```

B. CVSIM

CVSIM is a general program for the simulation of cyclic voltammetric experiments that is based upon the method of fractional steps. It is a flexible and accessible program that has been utilized to aid in the analysis of a wide variety of electrochemical mechanisms, including those with very fast chemical reactions. [12-21] The features and use of the program have been described in detail elsewhere* and are only summarized here.

1. Heterogeneous Electron Transfers

An unlimited number of heterogeneous single-electron transfers can be included in the simulation. The maximum number of consecutive electron transfers is two. Electron transfer is described by the Butler–Volmer equation, with $E^{0\prime}$, k^0, and α.

2. Coupled Chemical Reactions

An unlimited number of coupled chemical reactions of first or second order can be included. This introduces a tremendous flexibility in the program. The simulation is not restricted by a menu of choices of mechanisms, but can include any mechanism that is a combination of electron transfers and homogeneous chemical reactions.

3. Diffusion Coefficients

Individual diffusion coefficients may be a specified, as well as a chosen, normalizing diffusion coefficient. The correct choice of the normalizing diffusion coefficient can optimize the time of the simulation.[3]

4. IR Effects

Resistance effects are included as an option. An experimental data file is needed. This option is mostly useful in the regression analysis program CVFIT.

5. Capacitive Current

The capacitive current calculation is included and utilizes the relationship between the capacitive current and the resistance. Only constant capacitance (not as a function of potential) is available.

---

*CVSIM and companion programs are available with the book *Cyclic Voltammetry: Simulation and Analysis of Reaction Mechanisms*, VCH, New York (1993).

6. *Stability*

In any numerical method the issue of stability arises. A numerical method is said to be unstable when individual roundoff errors propagate through a calculation with increasing effect[11]. Even the widely used workhorse, the Runge–Kutta method, can have problems with stiff sets of differential equations. The Runge–Kutta solution of a stiff set of equations can be unstable, due to the different time scales of different kinetic terms. For instance,

$$y = Ae^{-20x} + Be^{20x}$$

Near $x = 0$, an error in the second term will dominate the solution. Roundoff error is directly related to *machine accuracy*—that is, the smallest (in magnitude) floating point number which, when added to 1.0, gives a number different from 1.0. Turbo Pascal 7.0 (the language used here) has the data type *extended*, with 19–20 significant digits, which results in a machine accuracy on the order to magnitude 10^{-19}. Truncation error is the error associated with the approximations inherent in the numerical method—that is, an error that would appear even if one used a hypothetical perfect computer with no roundoff error.

The physical nature of the problems associated with electrochemical simulations are such that differences in time scales large enough to cause roundoff error will not occur. The effects of truncation error can be controlled by the appropriate choice of increment size of the numerical method.

Instability is not a significant problem in CVSIM, as can be easily demonstrated by performing simulations with decreasing time step size and checking for acceptable convergence. We give as an example the simulation of an EC (catalytic) mechanism

$$Ox + e^- = Red$$

$$Red \Rightarrow Ox$$

For large rate constants and first-order conditions, a steady-state current is expected. The equation for current is

$$i_{t=\infty} = nFAC(k_f D)^{1/2}$$

The results of the simulations shown in Figure 6, for a pseudo-first-order rate constant $k = 1000\,\mathrm{sec}^{-1}$, are visually identical. The plateau current values are listed as ratioed to the analytical result in Table 1.

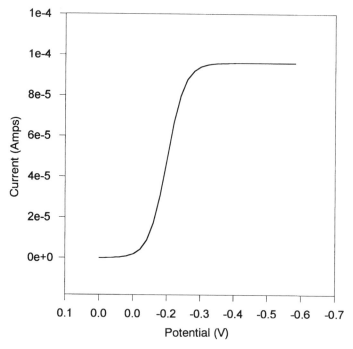

Figure 6. Simulations for different numbers of time increments indicate stability of the numerical method.

**Table 1. Simulated Enrents for an EC Catalytic
Mechanism with Different Numbers of Time Increments.**

Time Increments	$\dfrac{i_{\text{simulated}}}{i_{\text{theory}}}$
1,200	0.9995
12,000	1.0001
24,000	1.0002

7. Efficiency

An issue of great concern to simulators is computation time, particularly for cases where many simulations must be performed, such as in a regression analysis. The most significant factors influencing computation time are the maximum rate constant in the simulation and the scan rate. These two effectively control the size of the time increment of the simulation and the time of the experiment, and thus they control the number of time increments to be

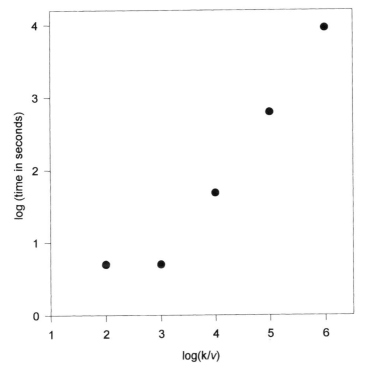

Figure 7. Computation time vs. k/v for a first-order reaction.

included in the simulation as well as the number of spatial grip points. CVSIM and companion programs have been optimized through the use of the fractional step method and the use of an expanding spatial grid.

A series of simulations of the EC mechanism with increasing value of the parameter k/v (chemical rate constant divided by scan rate) were performed and the elapsed simulation time was recorded, for first- and second-order chemical reactions (Figures 7 and 8). The simulation time in each case is seen as constant and small, until a certain value of k/v is reached, at which point the simulation time becomes proportional to k/v. It is clear from these results that the great majority of mechanisms of interest can be simulated in a matter of seconds, and that even diffusion-controlled reactions can be simulated at moderate scan rates.

C.   CVFIT: Regression Analysis

Minimization procedures can be adapted to nonlinear least-squares fitting by defining the function to be minimized as the least-squares error between the

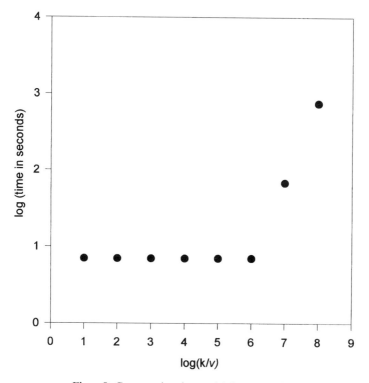

Figure 8. Computation time vs. k/v for a second-order reaction.

simulated data and experimental data. In the program CVFIT the Nelder–Mead Simplex procedure[10] has been adopted for nonlinear least-squares fitting of voltammograms. The simplex method is a relatively slow but highly robust manner in which to find the minimum of a function. The simplex procedure requires $n + 1$ initial sets of guesses for n parameters. The simplex method progresses by attempting to improve, step by step, on the "worst" of the $n + 1$ guesses. It moves along the error surface by making guesses in the direction of the best of the $n + 1$ guesses and replacing old guesses by better guesses. Unlike many other minimization procedures, it does not require a derivative of the error function.

An illustration of the error surface generated by a simplex fit is shown in Figure 9. The fit was a two parameter fit (reduction potential and heterogeneous rate constant) on a simulated data file.

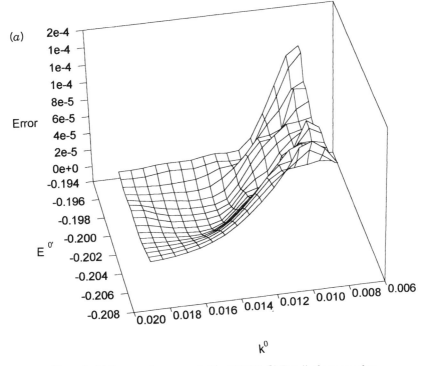

Figure 9. (a) Error surface generated by CVFIT. (b) Detail of error surface.

D. Examination of the EC Mechanism Fitting Through a Noise Analysis

In previous work, the EC mechanism (quasireversible case) was examined in some detail with a simulation/fitting approach. The system under examination was the reduction of vitamin B_{12}, which undergoes a fast cleavage reaction upon reduction. In this study, the rate of the following chemical reaction was determined by double potential step experiments. Subsequent fits of the voltammograms collected at various scan rates were analyzed by CVFIT for E, k, α, and area of electrode (to allow for small errors in the diffusion coefficients or concentrations). Table 2 shows the fitting results for six separate experiments at three different scan rates. The main result is that a voltammogram that appeared as "completely irreversible" actually can be analyzed to determine the reduction potential.

The feasibility of employing nonlinear curve fitting to extract information from cyclic voltammograms in any particular case can be investigated by

(b)

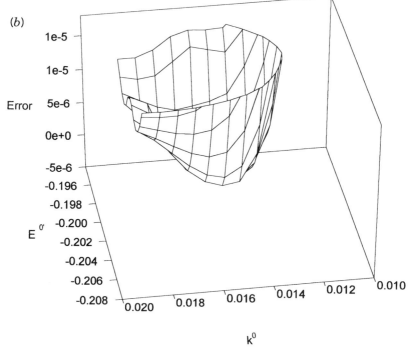

Figure 9. (*Continued*)

Table 2. Results of Fitting of Experimental Data for the EC Mechanism

E^0/V	$k^0/cm/s$	α
-1.529 ± 0.004	0.012 ± 002	0.78 ± 0.02

a noise study. Random data of a range comparable to the expected error of the experiment can be added to a simulated data set, and the desired parameters can be obtained through applying CVFIT to this artificial "experimental" data.

Since the results of the fit of the B_{12} data depend on subtle features of the voltammogram in this approach a "noise" study was performed. Simulated data (from the final fitted B_{12} parameters) to which uniform random deviation (noise) in a range that was 0%, 4%, 8%, and 20% of the peak current were fitted for E, k, and α. A typical simulated voltammogram for the reduction of

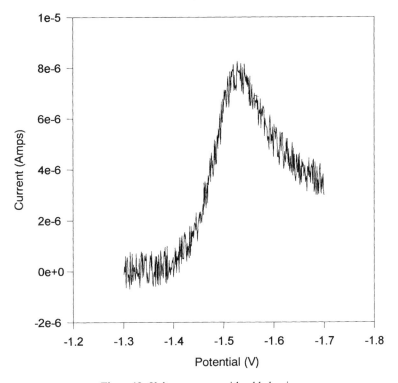

Figure 10. Voltammogram with added noise.

Table 3. Results of the Noise Analysis for the EC Mechanism

Noise Range	E^0/V	k^0/cms^{-1}	α
0%	-1.529	0.012	0.78
$\pm 2\%$	-1.533	0.014	0.77
$\pm 4\%$	-1.524	0.0096	0.80
$\pm 10\%$	-1.519	0.0081	0.83

B_{12} with added noise is shown in Figure 10. The results of the fit for "perfect" data were to obtain the exact values of the parameters used to generate the data, whereas when the noise was added, deviations in the parameters can be noted (Table 3). In the fitting of the experimental B_{12} data a small average deviation per data point is noted for fits of six separate experiments, as well as a small deviation in the fitted parameters. This is in agreement with the average deviation per data point obtained and is also in agreement with the variation in parameter values obtained for the "noise study" using a $\pm 2\%$ noise range.

A study of the fitting of simulations with added random data (noise) can inform one on the robustness of the fitting for any particular mechanism and is likely to point out the problems and give an indication of the quality of the data needed for any particular mechanism.

It may be wondered if the CVFIT approach can be useful beyond the relatively simple EC mechanism. More recently, Wandlowski and coworkers[20] have utilized CVFIT to obtain a quantitative mechanistic analysis of a scheme that included 12 parameters. However, this was accomplished in a stepwise manner, fitting voltammograms for smaller "pieces" of the puzzle and establishing certain parameters as constants, then moving on to larger pieces.

REFERENCES

1. D. K. Gosser, Jr., and P. H. Rieger, *Anal. Chem.*, **60**, 1159–1167 (1988).

2. D. K. Gosser, Jr., and F. Zhang, *Talanta*, **38**, 715–722 (1991).

3. D. K. Gosser, Jr. *Cyclic Voltammetry: Simulation and Analysis of Reaction Mechanisms*, VCH, New York (1993).

4. L. K. Bieniasz, *Comput. Chem.*, **16**, 11 (1992).

5. M. Rudolph, *J. Electroanal. Chem.*, **338**, 85 (1992).

6. J. E. Leffler and E. Grunwalk, *Rates and Equilibria of Organic Reactions*, Dover, New York (1963).

7. R. Marcus, *Angew. Chem. Ed. Engl.*, **32**(5), 691 (1993).

8. J. M. Saveant, *Single Electron Transfer and Nucleophilic Substitution, Advances in Organic Chemistry*, Vol. 26, Academic Press, London (1990).

9. F. Zhang, D. K. Gosser, Jr., and S. R. Meshnick, *Bio Pharm.*, **43**, 1805 (1992).

10. W. H. Press, B. P. Flannery, S. A. Teukolsky, and W. T. Vetterling, *Numerical Recipes: The Art of Scientific Computing*, Cambridge University Press, Cambridge (1985).

11. A. Ralston and A. Rabinowiz, *A First Course in Numerical Analysis*, McGraw–Hill, New York (1978).

12. C. G. Zoski, D. A. Sweigart, N. J. Stone, P. H. Rieger, E. Mocelin, T. F. Mann, and D. K. Gosser, Jr., *J. Am. Chem. Soc.*, **110**, 2109–2116 (1988).

13. J. F. Rusling, C.-N. Shi, D. K. Gosser, Jr., and S. Shukla, *J. Electroanal. Chem.*, **240**, 201–216 (1988).

14. D. Gosser, Jr., and Q. Huang, *Talanta*, **39**, 1155–1161 (1991).

15. K. Lee, D. Kuchynka, and J. Kochi, *Inorg. Chem.*, **29**, 4196 (1990).

16. K. Lee, C. Amatore, and J. Kochi, *J. Phys. Chem.*, **95**, 1285 (1991).

17. Y. Zhang, D. K. Gosser, Jr., P. H. Rieger, and D. Sweigart, *J. Am. Chem. Soc.*, **113**, 4062–4068 (1991).

18. J. Tommasino, D. Montauzon, H. He, A. Maisonnat, and R. Poilblanc, *Organometallics*, **11**, 4150 (1992).

19. C. Shi, W. Zhang, R. Birke, D. K. Gosser, Jr., and J. Lombardi, *J. Phys. Chem.*, **95**, 6276–6285 (1991).

20. Th. Wandlowski, D. Gosser, Jr., E. Akinele, R. De Levie, and V. Horák, *Talanta*, **40**(12), 1789–1798 (1993).

21. M. Schmittel, G. Gescheidt, and M. Rock, *Angew. Chem. Int. Ed. Engl.*, **33**(19) 1961 (1994).

CHAPTER

8

LIQUID–LIQUID ELECTROCHEMISTRY

PETR VANÝSEK

Northern Illinois University
Department of Chemistry
DeKalb, Illinois 60115

I. INTRODUCTION

If there are two phases in contact, each able to share and equilibrate a common species, then the site of the contact, the interface, will be a site of an interfacial potential. For example, the potential–building concept of a case in which a metal phase is in contact with a solution is quite familiar to an electrochemist. A silver wire in contact with a solution containing silver ions will have its potential constant, pinned down due to equilibrium between silver atoms in the wire and the silver ions in the solution. More obscure is the concept of

Modern Techniques in Electroanalysis, Edited by Petr Vanýsek, Chemical Analysis Series, Vol. 139.
ISBN 0-471-55514-2 © 1996 John Wiley & Sons, Inc.

building up a potential of a platinum wire immersed in aqueous solution of an indifferent electrolyte. The potential is not usually stable, certainly not as much as that of the above case of silver metal in a solution of silver salts. But it is still measurable and within certain range even predictable, if one can think of the surface oxides and dissolved oxygen as the potential pinning species.

The fact that two phases in contact, none of which is a metal, have also an interfacial potential, is not lost on biologists. The biological membrane, an interface between aqueous solution and oily lipid material, has an inherent membrane potential difference and its significance is well understood. It is the classically trained electrochemist who will more likely ignore the possibility of studying the potential difference residing on the interface between two immiscible liquids. In this chapter we will not attempt to justify the existence of liquid–liquid (L–L) electrochemistry, which has its solid niche in the physical electrochemistry field. Rather, we will attempt to show where the knowledge of the processes taking place on the interface can be of use to the analytical chemist.

One can envision an example of an analytically useful L–L interface in the ion-selective electrode (ISE) with a liquid ion exchange interface. For example, the potassium-selective device, an ISE, can be constructed from a certain organic solvent, suspended on a porous membrane. The purpose of the membrane is merely to provide some physical foundation for this organic solvent to stay in place. The organic solvent contains dissolved compound called *valinomycin*. Valinomycin has a ring structure that can engulf potassium ion. The size of the valinomycin molecule cavity is such that potassium ion is bound in it in preference over similar ions such as sodium, making the membrane with the organic solvent and valinomycin selective to potassium, with high sodium ion rejection. As the uncharged valinomycin binds with potassium ion, the resulting valinomycin–potassium ion complex is also charged, thereby fulfilling the prerequisite for an interfacial potential buildup, which says that the two phases in contact have to have a charged species in common. In this case the organic phase contains potassium ion bound to valinomycin, whereas the adjacent aqueous phase contains potassium ion without valinomycin, which is not soluble in water. Thus, potassium cation from the aqueous phase will be driven, at least to a certain extent, to the organic phase, because the potassium–valinomycin complex is more soluble in the organic phase than is the uncomplexed potassium cation in water. This can be viewed as extraction and in fact, extraction science is closely related to that of the L–L interfaces. However, in the above described process, which is actually an extraction of only a cation from one phase to another, without transport of a compensating charge, global (i.e., bulk) extraction would lead to violation of electroneutrality. Thus, only a local imbalance in charge can occur that will build up the potential difference on the interface. With a given amount of valinomycin dissolved in the oil phase, the potential difference on the interface, once the equilibrium is attained, will depend on the amount of potassium in the aqueous phase and hence the interfacial potential will become the desirable analytical signal, a measure of potassium concentration.

The practical ISE with a liquid membrane consists also of an inside compartment, filled with an aqueous solution containing potassium ions, bathing the membrane from the inside (Figure 1). The inner solution potassium ions also partake in the equilibrium with the oil–valinomycin phase. The established interfacial potential will, in general, be of opposite sign than that of the other, outside L–L interface; however, its value will remain constant as the potassium concentration inside the electrode assembly is constant. The overall potential E of the ISE selective to ion i is given, if there is not present an interfering ion complicating the relationship by the following expression, which is a form of Nernst equation[1]:

$$E = E^{0'} + \frac{RT}{nF} \ln a_i(\text{outside}). \qquad (1)$$

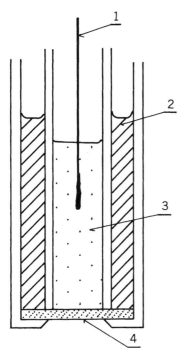

Figure 1. Schematic diagram of an ion-selective electrode with a liquid ion exchanger membrane. The outside compartment supplies the organic ion exchanger, and the inner compartment contains the solution of the "reference" side of the membrane. 1, Ag/AgCl electrode; 2, outside compartment; 3, inner compartment; 4, porous membrane soaked with the organic ion exchanger.

where $R = 8.314 \, J/(K \cdot mol)$ is the universal gas constant, T is the absolute temperature expressed in kelvins, $F = 96484 \, C/mol$ is the Faraday constant, equal to the (positive) charge of 1 mol of electrons, and n is the charge of the potential pinning species. Its value is 1 for the discussed case of potassium, it would be $+2$ for Ca^{2+}, and it takes sign, so it would be -1 for an F^- ISE. The argument of the logarithm contains activity of the ion i in the outside solution, rather than its concentration. Activity and concentration are, however, almost equal for low concentrations (< 1 mmol/liter). The standard potential of the ISE, $E^{0'}$, is a constant that includes a number of constant potentials, such as the constant potential of the inner L–L interface and the solution reference electrodes. However, the ISE potential change, E, ultimately reflects the changes in the analyte concentration surrounding the electrode-sensing element.

The interfacial potential is just one aspect of the immiscible liquid electrochemistry; and since it is in its widest electroanalytical applications described in a number of comprehensive texts,[2] we shall not deal with this subject matter.

The notion, further supported by solid evidence, that partitions of biological compartments (i.e., biological membrances) are the site of electrical potential is old. Du Bois-Reymond[3] performed the first experiment describing "animal electricity" in 1848. However, it was Koryta et al.[4] who recognized that the oil–liquid interface can be used as a simplified model for one side of a biological membrane. He also started using for the immiscible L–L interface the acronym ITIES, which stands for the interface between two immiscible electrolyte solutions. His initial goal was to establish experimental methods on the L–L interface that could produce new information about biological membranes. However, soon after publication of the first papers on the subject,[4–6] a number of methods used until then only in electrochemistry of metal–solution interface were applied to the study of L–L interfaces, ultimately moving in two directions: (i) study of the nanoscopic physical nature of the interface and (ii) application of such interfaces to problems in analytical chemistry. In this chapter we shall discuss the electroanalytical aspects of the interface.

One of the L–L properties that can be used for certain electroanalytical application is that the interface behaves, to the outside world of potentiostats and recorders, in the same way as the often studied and measured interfaces of metal electrodes bathed in solutions of salts. In the metal–solution interface case the electrical response observed as a result of applied electrical stimulus is mostly due to the oxidation or reduction of some species on the interface; in the L–L case, the change in the observed electrical response (mostly current or potential) due to perturbing electrical stimulus (usually potential or current) is the result of ion crossing (transport) across the interface rather than a redox process. Because either the charge transport or the redox process on the interface will give rise to a current in the connected electrical circuit and because the cause determining and limiting the process can in either case be, for example, diffusion, then an identical response of either type of interface, demonstrating diffusion limitation, will be observed. With the similarities in mind, it is of more value to point out the differences rather than the similarities, to capture the essence of the observed current–potential characteristics of the L–L interface.

II. THE VOLTAMMETRIC EXPERIMENT

Voltammetry is a method in which a potential difference is applied at the studied interface and its value is continuously and usually linearly changed with time (Figure 2). The resulting current response is recorded and analyzed. In cyclic voltammetry (the experiment that is most often performed), once the potential during the linear scan reaches a certain value, the direction of the potential sweep is automatically inverted and the effect of the reversal of the polarization is further studied. In a multisweep voltammetry, probably the most usual experimental procedure, a number of identical polarization cycles are repeated. This often leads to the discovery of changes over period of time or, when identical response is obtained on repeat cycles, it confirms that a dynamic steady state has been achieved.

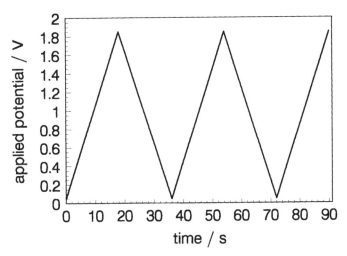

Figure 2. Voltage program used in performing the cyclic voltammetry experiment. This example corresponds to two and a half cycles between limits 0.05 and 1.85 V at scan rate of 100 mV/sec.

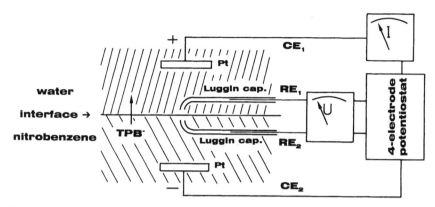

Figure 3. Schematic diagram of the experimental cell and the four-electrode apparatus used in voltammetric studies of ITIES. Reference electrodes RE_1 and RE_2 sense the potential across the interface, white CE_1 and CE_2 supply via a potentiostat a current needed to maintain the program potential. Transport of tetraphenylborate anion (TPB^-, indicated by an arrow) of the supporting electrolyte from nitrobenzene to water results from a sufficiently high positive potential applied to the aqueous phase (see Fig. 4). (Adapted from reference 73, with permission.)

The ITIES in most investigated experimental situations is formed as a free-standing horizontal boundary between two liquids, usually formed in a vertical glass tube of an electrochemical cell. The free-standing boundary arises because of the different densities of the two immiscible phases used.

The schematic diagram of an electrochemical cell that is often used for voltammetry shows a device (Figure 3) with two reference electrodes (RE_1 and RE_2) and two counterelectrodes made from platinum mesh (CE_1 and CE_2), connected to a special four-electrode potentiostat. The purpose of the potentiostat is to maintain the desired regulated potential between the two tips of the Haber–Luggin capillaries by virtue of passing current (or applying potential) between the two counterelectrodes. The potentiostat is necessary, because the solutions usually have considerable resistance and without compensation a large portion of the applied potential would be lost in the solution resistance.

The solvents suitable for ITIES experiments must be immiscible and chemically compatible (nonreactive), their densities should differ so that a free-standing interface can be established, and they both need to have high relative permittivity (ε) in order to keep dissolved salt in a dissociated form. This limits the choice of solvents because high permittivity of both usually means mutual solubility. Water ($\varepsilon = 78.58$) has been always one of the solvents. The other, very often used, is nitrobenzene ($\varepsilon = 34.82$), 1,2-dichloroethane ($\varepsilon = 10.36$), nitroethane ($\varepsilon = 30.3$, all at 25°C), and several other organic solvents of which o-nitrophenyl octyl ether ($\varepsilon = 24.2$) deserves special mention.

This compound has high viscosity and is used as a very useful plasticizer for liquid membranes of ion-selective electrodes. Samec et al.[7] performed a thorough study of this compound using the ITIES methodology.

To maintain conductivity of the phases and to suppress migration of the analyte species, supporting electrolytes are used in both phases. They are typically 0.1 mol/liter LiCl in the aqueous phase and a similar concentration of tetrabutylammonium tetraphenylborate (TBATPB) in the nitrobenzene phase. LiCl is a hydrophilic salt; hence the Li^+ and Cl^- ions remain confined mainly in the aqueous phase. Similarly, TBATPB is quite lipophilic. It dissociates in nitrobenzene, and the TBA^+ and TPB^- ions will remain mostly in that phase. As a consequence, an interface between two immiscible ionically conductive phases is established. These two phases do not share, except for some minute amounts, any common ions and its equilibrium interfacial potential is somewhat ambiguous, although it can be calculated.[8] But very importantly, as long as the interface does not pass any charge carriers, it can be polarized to a desired potential value and it behaves as an ideally polarizable interface, similar to that of an ideally polarizable metal electrode.

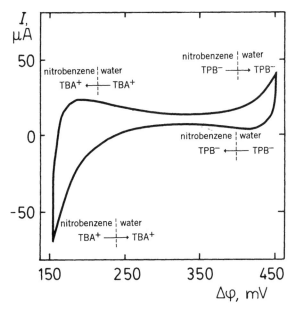

Figure 4. Supporting electrolyte voltammogram.[9] The aqueous phase is 0.1 mol/liter LiCl, and the nitrobenzene phase is 0.1 mol/liter tetrabutylammonium tetraphenylborate. The scan rate is 20 mV/sec. The processes responsible for individual current features are indicated close to the feature. (Adapted from reference 73, with permission.)

A voltammogram obtained for a water–nitrobenzene interface with the usual supporting electrolytes[9] is shown in Figure 4. The potential window in which the interface can be truly considered ideally polarizable is quite limited, a drawback in comparison to metal electrodes that usually provide much broader range. Let us explore what limits the potential range of the polarization window.

Consider first a scan from the middle of the potential window (about +250 mV for this particular choice of solvents and electrolytes) to the right in Figure 4, where the aqueous phase is increasingly positive with respect to the nonaqueous phase. [The convention of polarity is given by the difference $U_{RE_1} - U_{RE_2}$ (Figure 3), where RE_1 is always the reference electrode associated with the aqueous phase]. The containment of hydrophilic ions in water and hydrophobic ions in nitrobenzene is true only at the open circuit potential. In response to imposed potential, when the aqueous phase becomes sufficiently positive, anions in nitrobenzene (TPB^-) will cross the interface to the water phase. Li^+ ions could also cross from the water phase to the nitrobenzene phase, but this process requires a potential 20 mV higher than that for TPB^- transport (see Table 1). Higher potential will increase the rate of the ion transport, which will manifest itself as an increase in electric current (the rise at the right in Figure 4).

Table 1. Standard Potentials of Transfer $\Delta_\beta^z \varphi^0$ for Selected Individual Ions[72] from Water (Phase α) to Nitrobenzene (β)

Ion	$\Delta_\beta^z \varphi^0$ (mV)
Li^+	395
Mg^{2+}	361
Na^+	354
K^+	242
Rb^+	201
Cs^+	159
Choline	117
Acetylcholine	49
$TMeA^+$	35
TBA^+	-248
TPB^-	372
Picrate	47
Dodecylsulfate	-43
ClO_4^-	-83
Cl^-	-324

Upon reversal of the scan, TPB$^-$ and Li$^+$ will return to their original phases, causing a negative current peak near the extreme right. The height of the peak depends on the switching potential, which determines the amount of ions that have to be transported back to their original phase.

The structure of the liquid–liquid interface is such that there is a site of separated charge that behaves as a capacitor. As the potential of the aqueous phase is continuously decreased, the current, due to capacitive charging of the interface, is observed. This current is a common phenomenon observed whenever a constantly changing potential is applied on a capacitor. In this case the capacitor is the capacitance of the interface between the two liquids. As the aqueous phase becomes even less positive (and the nitrobenzene phase now becomes positive relative to the aqueous phase), organic phase cations (TBA$^+$) will flow in the aqueous phase. Cl$^-$ could also cross from water to nitrobenzene, but that process would require 52 mV more negative potential than the TBA$^+$ transfer. Again, current will be observed, and because of the direction of ion flow, it will be negative.

Upon reversal at the negative limit, positive current is observed due to the return of TBA$^+$ and possibly some amount of Cl$^-$ to their original phases. Subsequently, only charging current is observed and the overall cycle is complete. The previously described processes may be invoked repeatedly in subsequent cycles of voltammetric sweep.

The behavior is similar to what is observed in cyclic voltammetry on a metal electrode in a supporting electrolyte without electroactive species present. For example, for a mercury electrode in neutral KNO$_3$ solution, the negative boundary is due to the reduction of the electrolyte or solvent; the positive limit is defined by the oxidation of the mercury electrode itself. Within the confines of these limits, the redox behavior of certain species can be studied. Similarly, in ITIES, transport of ions that are less soluble in nitrobenzene than is TBATPB, but are also less soluble in water than is LiCl, can be observed. Regretfully, the range within which this transport can be studied is less than that afforded, for example, in the case of the mercury electrode.

Figure 5 shows the same system of supporting electrolytes as in Figure 4, but with the addition of about 0.5 mmol/liter of tetramethylammonium (TMeA$^+$) to the water phase. In this experiment, TMeA$^+$ is a semihydrophilic ion. If the potential of the aqueous phase is increased by scanning to the right, transport of TMeA$^+$ into nitrobenzene will occur. The transport across the interface is apparently rapid, not limited by the kinetic rate of ion transfer, but simply by the amount of ions that are able to reach the interface while diffusing from the solution bulk. As the aqueous phase is made more positive and the rate of TMeA$^+$ transport increases, depletion of TMeA$^+$ will occur on the aqueous side, resulting in a decrease of the current magnitude. Therefore a typical diffusion controlled peak is observed.

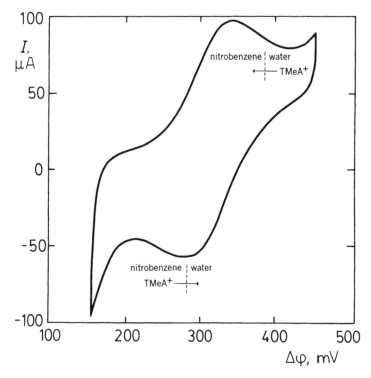

Figure 5. Transfer of tetramethylammonium between water and nitrobenzene.[9] $c(TMeA^+) =$ 4.7×10^{-4} mol/liter; the supporting electrolyte is the same as in Figure 4. Scan rate 20 mV/sec. Processes responsible for the rise of the peaks are indicated. (Adapted from reference 73, with permission.)

Upon reversing the scan at positive potentials (Figure 5), transport of $TMeA^+$ from nitrobenzene back to water will be observed. Again, because the transfer itself is quite a fast process relative to the scan rate used in the experiment, a diffusion-controlled peak will be observed.

The relationships describing ion transport across the interface are similar to those for transport of oxidizable species to an electrode and transport of the oxidized product away from the electrode. Therefore, the voltammograms for both processes have similar features—for example, a separation of the potentials at the extremes of the positive and negative peaks by $58/n$ mV for a kinetically reversible (rapid, relative to the used scan rate) process. The n denotes the charge of the oxidizable or the transported species.

III. THE POTENTIOMETRIC EXPERIMENT

The dynamic description of the ITIES is only one approach to understand the system. Another approach is to contemplate the system in equilibrium and the resulting interfacial potentials and ion repartitioning between the phases in contact. ITIES can be described in easy terms pertaining to equilibrium conditions; when current passes through such an interface, the processes at the interface and the mass and charge transport equations become more complex.

Consider an ITIES system containing a salt AX in water and a salt BX in the oil phase. Assume further that the cations designated as A^+ and B^+ are confined to their original phases, whereas an anion designated X^- can partition between the two phases. If ion X^- is shared by two immiscible phases α (water) and β in equilibrium, then the electrochemical potential of the ion X^- in both phases must be equal, because it is the definition of thermodynamic equilibrium. This leads to a relationship that involves the ratio of the ion activities in the two phases to the interfacial potential, $\Delta_\beta^\alpha \varphi$ ($\Delta_\beta^\alpha \varphi = \varphi(\alpha) - \varphi(\beta)$) that can be measured at the interface:

$$\Delta_\beta^\alpha \varphi = \Delta_\beta^\alpha \varphi_{Y^-}^0 + \frac{RT}{zF} \ln \frac{a_{Y^-}^\beta}{a_{Y^-}^\alpha} \tag{2}$$

where $\Delta_\beta^\alpha \varphi_{Y^-}^0$ is the standard potential of transfer of ion Y^- from phase α to β, a_{Y^-} is the activity of ion Y^- in the designated solvent, and z is the signed charge of the ion (for Y^{-1} it is -1). Values of $\Delta_\beta^\alpha \varphi_{Y^-}^0$ for individual ions can be determined from extraction data or from ITIES work, and selected potentials are listed in Table 1.

The potential of this interface is therefore a function of the concentration of B^-, similar to an electrode with a selective response to a particular ion. One may wish to compare this relationship with Eq. (1). In some analytical applications of ITIES, potentiometry at equilibrium is sufficient to determine the analyte concentration. However, the selectivity can be improved, and sometimes also an increase of sensitivity can be achieved by amperometric detection during which the interface is polarized from an external electrical source.

IV. INSTRUMENTATION

A. The Electrochemical Cell

One cell that is often being used in voltammetric studies is the one in Figure 3. The cell is typically made from glass. The size is such that the interface is about

1 cm^2. The most critical is the placement of the interface. Often a Teflon tube insert is used to form a sharp boundary between water–oil–glass–Teflon to keep the water–oil meniscus in place. For adjustment of the interface height, a piston, that can change to volume of the lower phase, is sometimes used.

B.   Potentiostat

A four-electrode potentiostat—that is, a potentiostat with a differential voltage input—is required in order to compensate for the resistance of both phases. Some workers also utilize positive feedback feature for the IR drop compensation. A commercial instrument that suits these needs is the Solartron 1286 or 1287 electrochemical interface. Because this instrument is costly and its purchase may not be justified if preliminary or only occasional experiments are desired, construction of a suitable circuit may be carried out in the laboratory.[10–15]

Perhaps the most practical alternative is to build a differential input adapter and use it together with any commercial three-electrode potentiostat that is likely to exist in a laboratory. A simple construction designed in our laboratory requires only three active components, a bipolar ± 15-V power source, and some hardware (see Figure 6). Two ultralow offset operational amplifiers (OP-07, Precision Monolithics, Inc.) are used as voltage followers to increase the input impedance of the reference electrode inputs. The outputs,

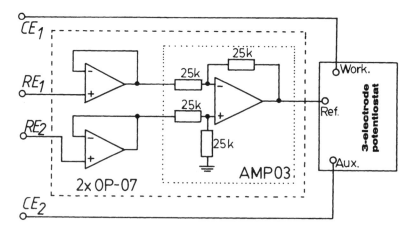

Figure 6. Differential adapter for a four-electrode potentiostat designed in our laboratory. The device outlined by the dashed rectangle can be used to convert many three-electrode commercial potentiostats into a four-electrode potentiostat. The outputs of OP-07 are also used to drive the shields of their inputs. The shown resistors are part of the commercial AMP-03 package in a dotted outline. The device is powered from a ± 15 V source.

also used to drive the shields of the input BNC connectors to lower the capacitive pickup of the shields, are fed into an AMP-03 precision unity gain differential amplifier (Precision Monolithics, Inc.). Because of the choice of active components, no adjustments or other components are required.

It should be noted that this circuit, if used with a commercial potentiostat that can supply high compliance voltage, could be destroyed if the high voltage were to appear on either reference electrode. Therefore, some precautionary measures should be taken to avoid this possibility. It is possible to construct an adapter able to operate at high compliance voltages, but its circuit would be more complicated.

The circuit in Figure 6 is connected to the electrochemical cell and a commerical potentiostat as shown. It is important not to invert RE_1 and RE_2; otherwise, the circuit will not operate.

Those who own the Princeton Applied Research EG&G 273 potentiostat may undertake a simple modification that will turn the potentiostat into a four-electrode instrument. The electrometer, with the cable connections for working electrodes, reference electrodes, and counterelectrodes, is already a differential electrometer with a symmetrical input. The unit has a built-in 1-$M\Omega$ resistor that prevents the two inputs from floating totally independently. The resistor is easy to locate, since it is soldered inside the electrometer box directly between the "sense" and the "working" miniature jacks. The modification consists only of removing the resistor and making a note of this modification on the box to inform of this nonstandard alteration. With an instrument modified in such a way, the working electrode connector will become the aqueous electrolyte counterelectrode CE_1, the "sense" jack will be the aqueous reference electrode RE_1, the reference electrode cable will become the nonaqueous reference connection RE_2, and the counterelectrode will be the nonaqueous counterelectrode CE_2. In this configuration the polarity set on the potentiostat will correspond to the polarity of the aqueous phase, relative to the nonaqueous phase. Conversion of the potentiostat to its regular three-electrode function is easy, by insertion of the shorting bar between the "sense" and the "working" miniature jacks.

C. Reference Electrodes

Because all electrified interfaces are studied in terms of potential differences and a "true" absolute potential, obtained without comparison to some other potential, cannot be measured, existence of reliable reference electrodes is paramount to any electrochemical study. Because there are two ionic phases in every ITIES experiment, two reference electrodes must be used.

The reference electrodes are placed in salt bridges inside Luggin–Haber capillaries, positioned in the proximity of the interface (see Figure 3). The reference electrode for the aqueous phase can be one of the many available and regularly described. The most convenient seems to be a chloridized silver wire immersed in a lithium chloride supporting electrolyte. The potential of the nonaqueous phase is sensed via a "reference" liquid interface. An aqueous phase with dissolved tetrabutylammonium chloride with Ag/AgCl electrode is in contact with nonaqueous supporting electrolyte of tetrabutylammonium tetraphenylborate. Thus the two phases share as their common ion tetrabutylammonium cation; and as long as its concentration in both phases remains constant, the potential of the "reference" interface remains constant as well.

V. ANALYTICAL APPLICATIONS

A. Potentiometry

Potentiometric measurements on ITIES are related to the principle of ISEs. The contribution of ITIES to the study of ISEs is in its ability to isolate and study a single interface at a time. The interfacial potential produced by a single shared ion is described by Eq. (2). If more ions can take part in the equilibrium, the equation becomes considerably more complex[16]:

$$\sum_{i=1}^{j} z_i c_i^0 \bigg/ \left\{ 1 + \frac{\gamma_i^\alpha}{\gamma_i^\beta} \exp\left[\frac{z_i F}{RT} (\Delta_\beta^\alpha \varphi - \Delta_\beta^\alpha \varphi_i^0) \right] \right\} = 0 \qquad (3)$$

This expression includes all ions, i (from 1 to j), involved in the equilibrium, where c_i^0 is the ion total concentration in both phases, z_i is the signed charge of the ion, γ_i is the ion activity coefficient in phase α or β, and $\Delta_\beta^\alpha \varphi_i^0$ is the standard potential of transfer of the ion involved (see Table 1). $\Delta_\beta^\alpha \varphi$, the interfacial potential resulting from the mixed equilibrium between all the ions, cannot, in general, be separated in a closed form. Relationship (3) can still be simplified to Eq. (2) if only one ion is able to cross the interface or if it dominates the equilibrium. If this ion is the analyte, the interface can be used for potentiometric determination with a straightforward potential dependence on concentration logarithm. If the simplification is not possible, a numerical method to solve for $\Delta_\beta^\alpha \varphi$ can be used.[8]

Inorganic cations that can be determined directly by potentiometry in the typical water–nitrobenzene configuration are K^+, Rb^+, and Cs^+. It is possible to determine more inorganic ions indirectly if they can be made more hydrophobic by complexation with a suitable ligand. For example, Pb^{2+} can be determined after complexation with polyethylene glycol.[17] Among inorganic anions that can be determined are SCN^-, ClO_4^-, BF_4^-, I^-, and NO_3^-.

A large number of organic cations and anions, which are more hydrophobic, can be determined. Examples are quaternary ammonium salts, choline, acetylcholine, picrate, and laurylsulfate.[18] Many organic dyes that exist as cations can also be detected.[19,20]

B. Cyclic Voltammetry

Figure 5, used in the introduction of polarization phenomena on ITIES, is also a representative example of the analytical applications of cyclic voltammetry. The position of the peaks is a function of the Gibbs energy of transfer (or the standard potential $\Delta_\beta^\alpha \varphi_i^0$, which is related to the Gibbs energy) of the particular ion, and it can be used for qualitative analysis. The quantity can be determined from the height of the peaks. The current measured at the peak (i_p) is given by

$$i_p = 269c^0 n^{3/2} A D^{1/2} v^{1/2} \tag{4}$$

The scan rate, v, is expressed in V/sec, A is the interface area in cm^2, D is the diffusion coefficient in cm^2 sec, n is the charge of the ion, and c^0 is the analyte concentration in the bulk solution in mol/liter. The prerequisite for determination of a given ion by cyclic voltammetry is that the ion can be transported before the supporting electrolytes from either phase will begin to transfer in the opposite phase. The same ions as those listed in potentiometry can be studied. Voltammetry allows us to see several ions simultaneously, as long as their standard potentials are different.

C. Dropping Electrolyte Interface

This technique is based on polarography. The interface is formed at the surface of an electrolyte drop expelled into the second immiscible electrolyte. Voltage scan was the polarization method used in early work.[4] Current scans, as performed in Freiser's group, are now more popular.[21,22] Figure 7 shows a device used for the voltammetric studies with the electrolyte dropping interface. In the illustrated case the dropping phase, water, is lighter than the nitrobenzene solution, and therefore the forming drop is ascending. Figure 8 shows a current scan polarogram on the water–dichloroethane interface. Curve A is for the supporting electrolyte, while curve B is obtained in the presence of an analyte. The wave location is useful in qualitative analysis, whereas its magnitude is a function of the concentration.

Recently, Suzuki[23] investigated for analytical purposes the transfer of ions from water to mixtures of water and formamide, using the dropping electrolyte technique.

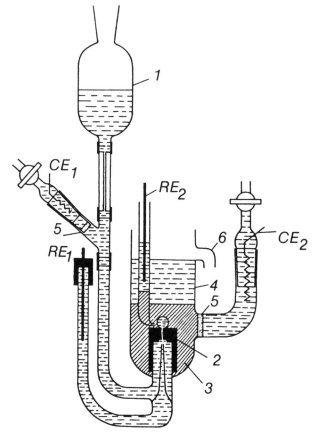

Figure 7. Device used to perform voltammetric experiments with the electrolyte dropping interface. **1**, The aqueous phase reservoir; **2**, Teflon attachment with calibrated opening; **3**, nonaqueous phase; **4**, used aqueous phase; **5**, glass frit; **6**, overflow. CE_1 and CE_2 Pt counterelectrodes; RE_1 and RE_2, Ag reference electrodes. (Adapted from reference 75, with permission.)

As the following section describes, small interfaces became quite useful in recent years. Hence, even a modification of the dropping interface, a micro-droplet, had been designed.[24]

D. Microinterfaces Between Two Immiscible Electrolytes

The purpose of using a small orifice separating two immiscible electrolytes is the same as that of using ultramicroelectrodes—that is, to lower or eliminate the effect of IR drop in the solutions. As the interface becomes progressively

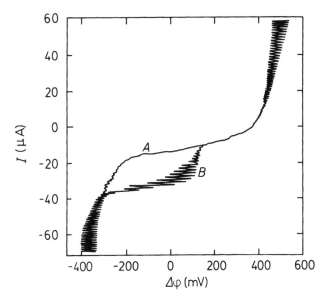

Figure 8. Current scan voltammogram on an ascending electrolyte drop. Curve A: Aqueous supporting electrolyte 1 mol/liter Na_2SO_4, dichloroethane phase 0.01 tetrahexylammonium tetraphenylborate. Curve B: Addition of 0.3 mmol/liter of 1-phenyl-3-methyl-4-trifluoroacetyl-5-pyrazolone anion in the aqueous phase. (Adapted from reference 22, with permission.)

smaller, the amount of analyte that can access the interface from the sides increases and thus there is more analyte available per unit area than with larger interfaces. Moreover, the diffusion is hemispherical rather than semilinear, and the limiting diffusion current is larger. Small interfaces were used in early studies by Girault and coworkers[25-28] and by Senda and coworkers.[29]

Recently, Osborne et al.[30] used the ultramicrointerface for analytical work, in which they amperometrically determined ionic species in aqueous solutions. They further developed a device, based on a microopening with 1,2-dichloroethane phase, that served as a sensor for creatinine assay.[31] The use of a microdroplet in combination of electrochemistry and spectroscopy has been described by Kitamura et al.[24]

One complication that makes the experiment challenging is restricting the area of the interface. Some of the work has been carried out on a small blunt capillary tip, where the hemispherical diffusion occurs on the outside of the capillary, whereas semi-infinite linear diffusion governs the transport inside the capillary. To take full advantage of hemispherical diffusion in both solutions, the microinterface has to be made in a thin flat separator. On one occasion, Girault used a plastic film with a specially etched hole, while we have used a hole in a thin glass wall.[32] In recent years the usefulness of the microinterface has brought sophistication in the restrictor preparation. Seddon et al.[33] describe the use of an excimer laser to fabricate microinterface restrictors.

Changes in interfacial tension that arise from polarization of the interface, uneven hydrostatic pressure, thermal expansion, and vibrations in the laboratory cause movement of the interface that complicates the properties of the device. An alternative is to use one phase that is solidified by a gel, such as polyvinyl chloride (PVC), in the case of nonaqueous nitrobenzene solution.[34,35] This gel-solidified microinterface was used successfully as an amperometric sensor. Figure 9 shows a voltammogram of picrate in water. The arrow indicates the potential at which the current is caused primarily by picrate transport. The lower part of the figure shows a calibration curve. Since the diffusion towards a microinterface is hemispherical, the steady-state current has a finite value and stirring of the solution is not required.

Samec et al.[27] constructed a microcell that allows experimental studies of ITIES in restricted volumes. Here they used quite viscous nonaqueous solvent, o-nitrophenyl octyl ether, that contributes to the stability of the interface. The solidified interfaces are often used in sensors and detectors, although the connection with ITIES studies is not always immediately obvious. For a recent paper in which the principle of solidified microinterface is used in a device that can be implanted into a beating heart, see the article by Cosofret et al.[36] The described device allows simultaneous monitoring of H^+ and K^+ ions.

The microinterfaces are an important tool in the study of kinetics (rates) of interfacial processes. Because they allow us to apply faster changing potentials than ordinary interfaces would, it is possible to detect higher rate constants. Beattie et al.[37] used borosilicate glass capillaries to study assisted potassium ion transfer by a crown ether dibenzo-18-crown-6. Similar to what occurs with valinomycin, this crown ether, selectively binds to potassium ion, which makes it possible to build on this principle a K^+ ion-selective membrane device. From their work it became apparent that the rate of transport was not, within the constraints of the experiment, limited by charge transfer, and the only limiting step was diffusion control. Therefore, these microinterface-based devices have the possibility to be used as fast (on solution diffusion time scale) sensing devices.

Various complexing agents add in one more important aspect to the utilization of the ITIES electrochemistry complex. Not only do they often bind to certain ions selectively (which makes it possible to design a unique sensor), they also facilitate the ion transport from one phase to another by making the ion, such as K^+ (surrounded, for example, by the valinomycin structure), quite

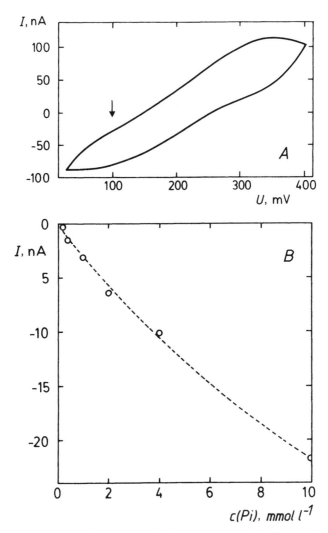

Figure 9. Amperometric detection of picrate ion in aqueous phase with 15% PVC–nitrobenzene gel microinterface. (**A**) Cyclic voltammogram of the sensor with 0.2 mmol/liter tetrabutylammonium picrate in gel and 0.4 mmol/liter picric acid in water. Scan rate 20 mV/sec. The arrow indicates the potential used in amperometry. (**B**) Calibration curve of the microinterface for amperometric determination of picrate. (Adapted from reference 34, with permission.)

hydrophobic—that is, more willing to enter the nonaqueous phase. For example, for Table 1 it can be seen that K^+ transport would be possible in the presence of Li^+ supporting electrolyte, because the standard potential difference between the two is 153 mV. However, in the presence of valinomycin, the K^+ value further decreases from the value of 242 mV, making the gap more apparent.

This trick has been used in determination of ions that normally would not be visible in the ITIES system, because their transfer would be hidden in the transfer of the supporting electrolytes. Thus, for example, Pb^{2+} was analytically determined on ITIES, using polyethylene glycol as the hydrophobicizing ligand.[17]

E.   Impedance Studies on Immiscible Interfaces

Impedance work in modern electrochemistry is a method of choice to probe surfaces and interfaces. Although it is seldom a method that would give direct analytical information of the surface, it is a powerful technique of contemporary electrochemistry and in this paragraph we shall discuss impedance of ITIES.

In 1939 Verwey and Niessen[38] described, in what was the first theoretical approach to ITIES, the interface as a diffuse layer. In the model, one phase contains an excess of negative charge, whereas the other phase contains an excess of positive charge; this excess, highest at the interfaces, drops off in a matter of gradually decreasing concentration function, similar to that of a diffuse double layer on the metal–solution interface, such as that described by Gouy [39,40] and Chapman.[41] Refinements of such models have appeared throughout the literature. The details of this are given by Samec and Mareček in reference 42. Recently, a modified Poisson–Boltzmann theory was used to model and study differential capacitance of the water–nitrobenzene[43] and water–1,2-dichloroethane interface.[44]

The methods that can be used to probe the structure of the interface rely on some kind of measurement of the charge parameters of the interface. Specifically, these are based on alternating current (ac) impedance methods, galvanostatic pulse method and interfacial tension methods. Here we shall discuss the rudiments of ac impedance.

Impedance is a kind of resistance of a system to a flow of electric current that is being imposed on the system by applied outside voltage. The simplest case of impedance is resistance R, simply calculated as the ratio of applied potential E and the resulting current $I (R = E/I)$, known as *Ohm's law*. In this concept the resistance is independent of time. However, in a capacitive or an inductive system the current reaches steady state only after a transient state. In capacitive situations the currents are generally high immediately after applying the outside voltage and then decrease to a steady-state value. In inductive circuits the initial current is typically lower than that in the steady-state case. In either case, the time-dependent behavior bears in itself information about the capacitive or inductive component of the system. It is possible to use a step perturbation (e.g., galvanostatic pulse[42]) or repetitively changing signal (ac current) to probe these properties. The latter is the basis for the ac impedance techniques.

For ac impedance, a specifically designed instrumentation is used, in which a signal of a small-amplitude (usually no more than 10 mV rms) ac voltage is superimposed (through the four-electrode potentiostat, described earlier) on the studied interface, and the resulting ac current is analyzed. The resulting current will, in general, have the same frequency as the input voltage. However, the peaks and valleys will not occur at the same time; there will be a phase shift between the two signals. Impedance response at a given frequency is therefore a two-dimensional result. The first number is the ratio of the voltage–current amplitude, which gives the $|Z|$ information, absolute value of the impedance. The phase shift (current–voltage), usually expressed in degrees, is the second component. Both can be expressed as a vector in polar coordinates (Figure 10).

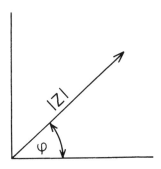

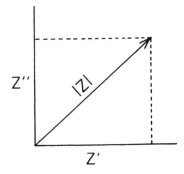

Figure 10. Representation of an impedance response in polar ($|Z|$ and φ) and Cartesian (Z', Z'', or Z_{real}, $Z_{imaginary}$) coordinates.

The other usual way of expressing this vector is to use rectangular (Cartesian) coordinates which can be obtained from the polar form by a simple geometric transformation. Since this graphical depiction of two-dimensional numbers is often used for complex numbers, the in-phase and out-of-phase components of the impedance are called the *real* and the *imaginary* impedance. In contemporary impedance spectroscopy the impedance response of a given system is collected in an automated fashion over a wide range of frequencies. Through the analysis of the impedance response as a function of frequency, it is possible to discern resistive, capacitive, and inductive components of the system. In electrochemical systems, it is mainly resistance and capacitance that are encountered.

The most critical step in impedance spectroscopy experiment, aside from correctly obtaining the data, is the actual data interpretation. The impedance response itself is only a signature of the interfacial structure. In fact it is possible to find the same signature originating from different systems. To find the adequate model requires finding an appropriate equivalent circuit. An equivalent circuit consists of usual discrete electronic components (e.g., resistors and capacitors) that have the same (equivalent) response as the system under study. In finding these equivalent circuits, one should use a set of common sense rules[45]: (1) The circuit should be the best analog of the geometrical/physical properties of the system (i.e., it should be possible to explain the meaning of each circuit component in terms of a physical model of the interface). (2) The individual circuit elements should depend on experimental variables in reasonable and interrelated ways. (3) The same equivalent circuit should apply over as wide a range of conditions as possible, and changes in the values of the element should correlate with changes in the electrochemical response.

One element that does not have a discrete circuit analog is the so-called Warburg impedance. It is a kind of resistance to ion transport limited by their diffusion. Samec and Mareček[42] found that the Warburg impedance can be used also in ITIES studies.

The model that describes a number of ITIES situations in the case of double-layer behavior, without an analyte ion transfer, is shown in Figure 11. The component R_s is the solution resistance, which is usually trivial to evaluate. C is the capacitance of the double layer, and it is typically this parameter that is most often evaluated to learn about the double-layer structure. The component in parallel, Z_f, is the faradaic impedance.[46,47] In general, Z_f includes a number of processes limiting charge transport. In the case of fast, simply diffusion-controlled process, the faradaic impedance becomes the Warburg impedance, Z_w. The Warburg impedance increases with decreasing frequency (at low frequencies, almost a dc condition, the interfacial layer will become depleted on charge transferring ions), and both the real and

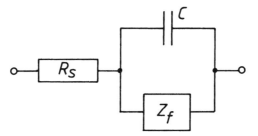

Figure 11. Equivalent circuit used in modeling impedance response of the liquid–liquid interface. R_s, solution resistance; C, interfacial capacitance; Z_f, faradaic impedance.

imaginary components change in the same manner:

$$Z_w = (1 - j)\sigma\omega^{-1/2} \tag{5}$$

where $\omega = 2\pi f$ is the circular frequency, $-j = \sqrt{-1}$. The function σ is the parameter that can be evaluated, using some kind of nonlinear least-squares data-fitting routine. It can be calculated as

$$\sigma = (4RT/n^2F^2Ac^0\sqrt{2D})\cosh^2(nF(E - E^0_{1/2})/2RT) \tag{6}$$

where A is the interfacial area, F is the Faraday constant, R is the gas constant, E is the interfacial potential, $E^0_{1/2}$ is the standard reversible potential, T is absolute temprature, c^0 is the bulk concentration of the ion that gives rise to the Warburg impedance, and D is the diffusion coefficient of the ion. Note that the units have to be in a self-consistent system. Note especially thar D is typically given in cm^2/sec, whereas molar concentration s is given in mol/liter, which is mol/dm^3.

When the impedance is obtained at the equilibrium potentials, Eq. (6) becomes

$$\sigma = (4RT/n^2F^2Ac^0\sqrt{2D}) \tag{7}$$

Equation (7) can be used for determining the diffusion coefficient of the ion being transported in the system, or to check for adequacy of the data interpretation, since a typical diffusion coefficient has a value of 10^{-6} to $10^{-5}\,cm^2/sec$. For example, Beattie et al.[37] reported D for dibenzo-18-crown-6 $[(4.55 \pm 0.25) \times 10^{-6}\,cm^2/sec]$ for the water–1,2-dichloroethane system.

Figure 12 shows experimental data obtained for the water–nitrobenzene interface is the presence of 0.1 mol/liter LiCl in water and 0.1 mol/liter

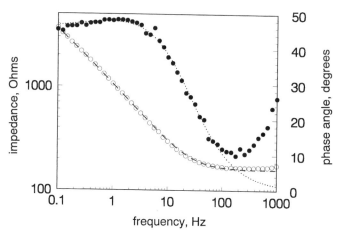

Figure 12. Bode plot (real versus imaginary impedance) of a typical liquid–liquid interface impedance experiment. 0.1 mol/liter LiCl in water and 0.1 mol/liter TBATPB in nitrobenzene, at 270 mV, versus AgCl/TBACl reference electrodes. ○ Impedance and ● phase angle are experimental results. Drawn lines result from an overlay of nonlinear least-squares fit to a circuit of Figure 11. $R_s = 158\,\Omega$, $C = 40.7\,\mu F$, and $Z_f = 2970\,\Omega$ (here Z_f is equal to Warburg impedance).

TBATPB in nitrobenzene, with an overlay of values obtained when equivalent circuit of Figure 11 was used. A set of data obtained at various potentials can be used to measure the double-layer capacitance as a function of applied potential. This data response is very important in studies of the interfacial structure.

VI. FUTURE DIRECTIONS

Besides applications primarily directed at the analytical applications, the ITIES has also caught the attention of theorists and workers in fields allied to electrochemistry. Benjamin[48] used computer methods to model the structure of the liquid–liquid interface. In a further study[49] he concentrated specifically on the theoretical investigation of structure, dynamics, and conformational equilibria at the water–1,2-dichloroethane interface, one of the most commonly studied solution pairs. A year later[50] he included in the dynamic studies of the interface transfer of ions. Later he also undertook molecular modeling of electron transfer across the 1,2-dichloroethane interface.[51] The transfer of electrons is not a commonly observed phenomenon, because it is ion transfer that is most often studied. However, this process has importance in the theoretical approach as well as in practical applications. Marcus[52–54] used experimental data of Geblewicz and Schiffrin[55] to derive relationships describing electron transport on ITIES. Girault[56] published work dealing with reorganization energy for heterogeneous electron-transfer reactions at liquid–liquid interfaces.

Higgins and Corn[57] used the method of second-harmonic generation on surfaces to learn more about molecules adsorbed at a polarized interface between water and 1,2-dichloroethane. This work spurred later experiments, both their own[58,59] and those by Richmond and coworkers[60,61] and by Sassaman and Wirth.[62]

The ions that cross the interface upon applied electrical potential can be fluorescing species. Often their fluorescence is more pronounced in one solvent, whereas in the other it is quenched.[63] This led Kakiuchi and Takasu[64–66] to the development of a very sensitive analytical method called *voltfluorometry*. Fluorescence measurements are intrinsically very sensitive, because of low background noise. In this application, presence or absence of minute amounts of a fluorescent dye can be observed, as a result of applied interfacial potential.

Another way of probing the liquid interfaces is through the use of a method called the *molecular vise*.[67] In this device a drop of a liquid, suspended on a capillary immersed in a second liquid, is somewhat aspirated inside the capillary. This decreases the interfacial surface area, which, in turn, results in arranging and gripping any molecules that may be adsorbed at the interface.

An interesting twist in the interfacial studies of ITIES was reported by Solomon and Bard.[68] Throughout this chapter it became obvious that a lot of effort is spent to understand the behavior of the ITIES and its microscopic structure. These authors were able to demonstrate the existence of an electron transfer reaction between water and 1,2-dichloroethane in the redox system of ferrocyanide/7,7,8,8-tetracyanoquinodimethane. Satisfied with that, they in turn placed a sharp pipette (5-μm tip) filled with 1,2-dichloroethane on the piezo stage of a scanning electrochemical microscope. Through the spatially resolved current, they were able to map a surface of alternating conductive/isolating regions. In some applications this is believed to be more practical than using sharp isolated metal tips.

This chapter gave an overview of the electrochemistry of ITIES, in particular as it applies to the aspect of analytical chemistry. As one can see, equally great interest in ITIES is in the basic physical electrochemistry as well as in some industrial applications. For those interested in these subjects, there are several reviews available that can be recommended for further reading.[47,69–74]

ACKNOWLEDGMENT

The work on this project was supported in part by the Office of Naval Research.

REFERENCES

1. W. Nernst, *Z. Phys. Chem.* (*Frankfurt*), **2**, 613–637 (1888).

2. D. A. Skoog and J. J. Leary, *Principles of Instrumental Analysis*, 4th ed., Saunders, Fort Worth, (1992).

3. E. Du Bois-Reymond, *Untersuchungen über thierische Elektrizität*, Vol. I, Berlin (1848).

4. J. Koryta, P. Vanýsek, and M. Březina, *J. Electroanal. Chem.*, **67**, 263–266 (1976).

5. J. Koryta, P. Vanýsek, and M. Březina, *J. Electroanal. Chem.*, **75**, 211–228 (1977).

6. Z. Samec, V. Mareček, J. Koryta, and M. W. Khalil, *J. Electroanal. Chem.*, **83**, 393–397 (1977).

7. Z. Samec, A. Trojánek, and E. Samcová, *J. Electroanal. Chem.*, **386**, 225–228 (1995).

8. P. Vanýsek, *Electroanalysis*, **2**, 409–413 (1990).

9. L. Zhang and P. Vanýsek, unpublished results.

10. Z. Samec, V. Mareček, and J. Weber, *J. Electroanal. Chem.*, **96**, 245–247 (1979).

11. Z. Figaszewski, Z. Koczorowski, and G. Geblewicz, *J. Electroanal. Chem.*, **139**, 317–322 (1982).

12. T. J. VanderNoot, D. J. Schiffrin, and R. S. Whiteside, *J. Electroanal. Chem.*, **278**, 137–150 (1990).

13. A. M. Baruzzi and J. Uhlken, *J. Electroanal. Chem.*, **282**, 267–273 (1990).

14. S. Wilke, *J. Electroanal. Chem.*, **301**, 67–75 (1991).

15. N. K. Harris, S. Jin, G. J. Moody, and J. D. R. Thomas, *Anal. Sci.*, **8**, 545–551 (1992).

16. L. Q. Hung, *J. Electroanal. Chem.*, **149**, 1–14 (1983).

17. Z. Sun and P. Vanýsek, *Anal. Chim. Acta*, **228**, 241–249 (1990).

18. P. Vanýsek, *Anal. Chem.*, **62**, 827A–835A (1990).

19. E. Wang and Z. Sun, *Trends Anal. Chem.*, **7**, 99–106 (1988).

20. Z. Sun and E. Wang, *Electrochim. Acta*, **33**, 603–611 (1988).

21. L. Sinru and H. Freiser, *Anal. Chem.*, **59**, 2834–2838 (1987).

22. H. Doe and H. Freiser, *Anal. Sci.*, **7**, 303–311 and 313–319 (1991).

23. M. Suzuki, *J. Electroanal. Chem.*, **384**, 77–84 (1985).

24. N. Kitamura, K. Nakatani, and H.-B. Kim, *Pure Appl. Chem.*, **67**, 79–86 (1995).

25. J. A. Campbell and H. H. Girault, *J. Electroanal. Chem.*, **266**, 465–469 (1989).

26. A. A. Stewart, G. Taylor, H. H. Girault, and J. McAleer, *J. Electroanal. Chem.*, **296**, 491–515 (1990).

27. A. A. Stewart, Y. Shao, C. M. Pereira, and H. H. Girault, *J. Electroanal. Chem.*, **305**, 135–139 (1991).

28. Y. Shao and H. H. Girault, *J. Electroanal. Chem.*, **334**, 203–211 (1992).

29. T. Ohkouchi, T. Kakutani, T. Osakai, and M. Senda, *Anal. Sci.*, **7**, 371–376 (1991).

30. M. C. Osborne, Y. Shao, C. M. Pereira, and H. H. Girault, *J. Electroanal. Chem.*, **364**, 155–161 (1994). See also erratum in *J. Electroanal. Chem.*, **371**, 291 (1994).

31. M. D. Osborne and H. H. Girault, *Microchim. Acta*, **117**, 175–185 (1995).

32. P. Vanýsek and I. C. Hernandez, *J. Electrochem. Soc.*, **137**, 2763–2768 (1990).

33. B. J. Seddon, Y. Shao, J. Fost, and H. H. Girault, *Electrochim. Acta* **39**, 783–791 (1994).

34. P. Vanýsek, I. C. Hernandez, and J. Xu, *Microchem. J.*, **41**, 327–339 (1990).

35. J. Xu, *Electrochemical Studies on Gel/Liquid Microinterfaces*, Thesis, Northern Illinois University, Department of Chemistry, DeKalb, IL (1990).

36. V. V. Cosofret, M. Erdösy, T. A. Johnson, R. P. Buck, R. B. Ash, and M. Neuman, *Anal. Chem.*, **67**, 1647–1653 (1995).

37. P. D. Beattie, A. Delay and H. H. Girault, *J. Electroanal. Chem.*, **380**, 167–175 (1995). See also minor erratum in *J. Electroanal. Chem.*, **387**, 147 (1995).

38. E. J. W. Vervey and K. F. Niessen, *Philos. Mag.*, **28**, 435–446 (1939).

39. G. Gouy, *C. R. Acad. Sci. (Paris)*, **149**, 654–657 (1910).

40. G. Gouy, *J. Phys.*, **4**, 457–467 (1910).

41. D. L. Chapman, *Philos. Mag.*, **28**, 435–446 (1913).

42. Z. Samec and V. Mareček, in *The Interface Structure and Electrochemical Processes at the Boundary Between Two Immiscible Liquids*, V. E. Kazarinov, ed., Springer, Berlin (1987), pp. 123–142.

43. Q. Z. Cui, G. Y. Zhu, and E. Wang, *J. Electroanal. Chem.*, **372**, 15 (1994).

44. Q. Z. Cui, G. Y. Zhu, and E. Wang, *J. Electroanal. Chem.*, **383**, 7–12 (1995).

45. A. J. Zhang, V. I. Birss, and P. Vanýsek, *J. Electroanal. Chem.*, **378**, 63–76 (1994).

46. A. J. Bard and L. R. Faulkner, *Electrochemical Methods. Fundamentals and applications*, Wiley, New York (1980).

47. V. E. Kazarinov, ed., *The Interface Structure and Electrochemical Processes at the Boundary Between Two Immiscible Liquids*, Springer, Berlin (1987).

48. I. Benjamin, *J. Chem. Phys.*, **96**, 577–585 (1992).

49. I. Benjamin, *J. Chem. Phys.*, **97**, 1432–1445 (1992).

50. I. Benjamin, *Science*, **261**, 1558–1560 (1993).

51. I. Benjamin, A molecular model for an electron transfer reaction at the water/1,2-dichloroethane interface, in *Structure and Reactivity in Aqueous Solutions*, C. J. Kramer and D. G. Truhlar, eds., ACS, Washington, DC (1995).

52. R. Marcus, *J. Phys. Chem.*, **94**, 1050–1055 (1990).

53. R. Marcus, *J. Phys. Chem.*, **94**, 4152–4155 (1990). See also erratum in *J. Phys. Chem.*, **94**, 7742 (1990).

54. R. Marcus, *J. Phys. Chem.*, **95**, 2010–2013 (1991). See also erratum *J. Phys. Chem.*, **99**, 5742 (1995).

55. G. Geblewicz and D. J. Schiffrin, *J. Electroanal. Chem.*, **244**, 27–37 (1988).

56. H. H. Girault, *J. Electroanal. Chem.*, **388**, 93–100 (1995).

57. D. Higgins and R. M. Corn, *J. Phys. Chem.*, **97**, 489–493 (1993).

58. R. M. Corn and D. A. Higgins, *Chem. Rev.*, **94**, 107–125 (1994).

59. R. R. Naujok, D. A. Higgins, D. G. Hanken, and R. M. Corn, *J. Chem. Soc., Faraday Trans. I*, **91**, 1411–1420 (1995).

60. J. C. Conboy, J. L. Daschbach, and G. L. Richmond, *J. Phys. Chem.*, **98**, 9688–9692 (1994).

61. J. C. Conboy, J. L. Daschbach, and G. L. Richmond, *Appl. Phys., A Solids Surf.*, **59**, 623–629 (1994).

62. J. L. Sassaman and M. J. Wirth, *Colloids Surf. A. Physicochem. Eng. Aspects*, **93**, 49–58 (1994).

63. P. Vanýsek, Electrified immiscible liquid boundaries. Conventional and microscopic boundaries, in *Electrochemistry in Colloids and Dispersions*, R. A. Mackay and J. Texter, eds., VCH, New York (1992), pp. 71–84.

64. T. Kakiuchi and Y. Takasu, *J. Electroanal. Chem.*, **365**, 293–297 (1994).

65. T. Kakiuchi and Y. Takasu, *Anal. Chem.*, **66**, 1853–1859 (1994).

66. T. Kakiuchi and Y. Takasu, *J. Electroanal. Chem.*, **381**, 5–9 (1995).

67. C. Bruckner-Lea and J. Janata, *Proceedings of the Electrochemical Society*, 94-1, paper 897 (1994).

68. T. Solomon and A. J. Bard, *Anal. Chem.*, **67**, 2787–2790 (1995).

69. H. H. J. Girault and D. J. Schiffrin, Electrochemistry of liquid–liquid interfaces, in *Electroanalytical Chemistry*, Vol. 15, A. J. Bard, ed., Marcel Dekker, New York (1989), p. 1.

70. J. Koryta, *Sel. Electr. Rev.*, **13**, 133–158 (1991).

71. Z. Samec, *Chem. Rev.*, **88**, 617–632 (1988).

72. P. Vanýsek, *Electrochemistry on Liquid/Liquid Interfaces*, Springer, Berlin (1985).

73. P. Vanýsek, *Trends Anal. Chem.*, **12**, 357–363 (1993).

74. H. H. Girault, Charge transfer across liquid–liquid interfaces, in *Modern Aspects of Electrochemistry*, Vol. 25, J. O' M. Bockris, B. E. Conway, and R. E. White, eds., Plenum, New York (1993), pp. 1–62.

INDEX

RETURN
TO ➡

CHEMISTRY LIBRARY
100 Hildebrand Hall • 642-3753

LOAN PERIOD 1	2 **1 MONTH**	3
4	5	6

ALL BOOKS MAY BE RECALLED AFTER 7 DAYS
Renewable by telephone

DUE AS STAMPED BELOW

MAY 2 4 1997	DEC 17 '04	
DEC 1 8 1997		
APR 02 '02 APR 25 '02	DEC 09 2005	
JUN 25 '02	APR 10 2006	
08/02/02	MAY 1 1	
MAR 26 '03		
MAR 26 '03		
NOV 06 '03		
DEC 1 8 2003		
JUL 1 6 '04		
NOV 1 9 '04		